Statistical Paradigms
Recent Advances and
Reconciliations

Statistical Science and Interdisciplinary Research

ISSN: 1793-6195

Series Editor: Sankar K. Pal *(Indian Statistical Institute)*

Description:
In conjunction with the Platinum Jubilee celebrations of the Indian Statistical Institute, a series of books will be produced to cover various topics, such as Statistics and Mathematics, Computer Science, Machine Intelligence, Econometrics, other Physical Sciences, and Social and Natural Sciences. This series of edited volumes in the mentioned disciplines culminate mostly out of significant events — conferences, workshops and lectures — held at the ten branches and centers of ISI to commemorate the long history of the institute.

Published*

*To view the complete list of the published volumes in the series, please visit:
http://www.worldscientific.com/series/ssir

Platinum Jubilee Series

Statistical Science and
Interdisciplinary Research — Vol. 14

Statistical Paradigms
Recent Advances and Reconciliations

Editors

Ashis SenGupta
Tapas Samanta
Ayanendranath Basu

Indian Statistical Institute, India

Series Editor: **Sankar K. Pal**

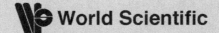 **World Scientific**

NEW JERSEY · LONDON · SINGAPORE · BEIJING · SHANGHAI · HONG KONG · TAIPEI · CHENNAI

Published by

World Scientific Publishing Co. Pte. Ltd.

5 Toh Tuck Link, Singapore 596224

USA office: 27 Warren Street, Suite 401-402, Hackensack, NJ 07601

UK office: 57 Shelton Street, Covent Garden, London WC2H 9HE

Library of Congress Cataloging-in-Publication Data

Statistical paradigms : recent advances and reconciliations / editors, Ashis SenGupta, Tapas Samanta, Ayanendranath Basu, Indian Statistical Institute, India.

   pages cm. -- (Statistical science and interdisciplinary research ; vol. 14)

  Includes bibliographical references.

  ISBN 978-9814343954 (hardcover : alk. paper) -- ISBN 9814343951 (hardcover : alk. paper)

  1. Parameter estimation. 2. Estimation theory. I. Sengupta, Ashis, editor. II. Samanta, Tapas, editor. III. Basu, Ayanendranath, editor.

  QA276.8.S736 2014

  519.5'44--dc23

                               2014024069

British Library Cataloguing-in-Publication Data

A catalogue record for this book is available from the British Library.

Printed in Singapore

Foreword

The Indian Statistical Institute (ISI) was established on 17th December, 1931 by a great visionary Prof. Prasanta Chandra Mahalanobis to promote research in the theory and applications of statistics as a new scientific discipline in India. In 1959, Pandit Jawaharlal Nehru, the then Prime Minister of India introduced the ISI Act in the parliament and designated it as an Institution of National Importance because of its remarkable achievements in statistical work as well as its contribution to economic planning.

Today, the Institute occupies a prestigious position in the academic firmament. It has been a haven for bright and talented academics working in a number of disciplines. Its research faculty has done India proud in the arenas of Statistics, Mathematics, Economics, Computer Science, among others. Over more than eighty years, it has grown into a massive banyan tree, like the institute emblem. The Institute now serves the nation as a unified and monolithic organization from different places, namely Kolkata, the Headquarters, Delhi, Bangalore, Chennai, and Tezpur, four centers, a network of five SQC-OR Units located at Mumbai, Pune, Baroda, Hyderabad and Coimbatore, and a branch (field station) at Giridih.

The platinum jubilee celebrations of ISI had been launched by Honorable Prime Minister Prof. Manmohan Singh on December 24, 2006, and the Govt. of India has declared 29th June as the "Statistics Day" to commemorate the birthday of Prof. Mahalanobis nationally.

Prof. Mahalanobis, was a great believer in interdisciplinary research, because he thought that this will promote the development of not only Statistics, but also the other natural and social sciences. To promote interdisciplinary research, major strides were made in the areas of computer science, statistical quality control, economics, biological and social sciences, physical and earth sciences.

The Institute's motto of 'unity in diversity' has been the guiding principle of all its activities since its inception. It highlights the unifying role of statistics in relation to various scientific activities.

In tune with this hallowed tradition, a comprehensive academic programme, involving Nobel Laureates, Fellows of the Royal Society, Abel prize winners and other dignitaries, has been implemented throughout the Platinum Jubilee year, highlighting the emerging areas of ongoing frontline research in its various scientific divisions, centers, and outlying units. It includes international and national-level seminars, symposia, conferences and workshops, as well as series of special lectures. As an outcome of these events, the Institute is bringing out a series of comprehensive volumes in different subjects under the title *Statistical Science and Interdisciplinary Research*, published by the World Scientific Press, Singapore.

The present volume titled *Statistical Paradigms: Recent Advances and Reconciliations* is the fourteenth one in the series. The volume consists of fourteen chapters, written by eminent scientists from different parts of the world. These chapters provide a current perspective of research and development, both from theoretical and application points of views, in different classical and emerging statistical paradigms, namely, parametric, non-parametric and semi-parametric, frequentist and Bayesian. Application areas considered include: Bioinformatics, Factorial Experiments and Linear Models, Hotspot Geoinformatics and Reliability. I believe the state-of-the art studies presented in this book reflecting advances in both theory and application of statistical science will be very useful to students, researchers as well as practitioners.

Thanks to the contributors for their excellent research contributions and to the volume editors Profs. Ashis SenGupta, Tapas Samanta and Ayanendranath Basu for their sincere effort in bringing out the volume nicely. Initial design of the cover by Mr. Indranil Dutta is acknowledged. Sincere efforts by Prof. Dilip Saha and Dr. Barun Mukhopadhyay for editorial assistance are appreciated. Thanks are also due to World Scientific for their initiative in publishing the series and being a part of the Platinum Jubilee endeavor of the Institute.

January 2013
Kolkata

Sankar K. Pal
Series Editor and
Former Director

Preface

The year 2006 marked the Platinum Jubilee of Indian Statistical Institute (ISI). ISI has a seminal role in promoting statistics in general worldwide. The International Conference on Statistical Paradigms: Recent Advances and Reconciliations (ICSPRAR-2008) was held during Jan 1–4, 2008, at ISI, Kolkata. This volume is number 14 in the series derived from the events celebrating the Platinum Jubilee of ISI, which is being published by World Scientific. It presents a collection of 14 papers in the said spirit, which emanated from the Conference and were peer-reviewed and edited. The papers are broadly arranged in four groups: Review, Parametric, Semi-parametric, and Nonparametric and Probability consisting of 2, 5, 4 and 3 entries, respectively. The chapters reflect advances in both theory and applications of several classical and emerging Statistical Paradigms — parametric, nonparametric and semi-parametric, frequentist and Bayesian — encompassing both theoretical advances and emerging applications in a variety of scientific disciplines. For advances in theory, the topics include: Bayesian Inference, Directional Data Analysis, Distribution Theory, Econometrics and Multiple Testing Procedures. The areas in emerging applications include: Bioinformatics, Factorial Experiments and Linear Models, Hotspot Geoinformatics and Reliability.

We acknowledge with thanks the support received from Ministry of Statistics and Program Implementation, Department of Science and Technology and Reserve Bank of India — all these being organizations of Government of India. We are also thankful to the Indian Statistical Institute for partial financial support. It is a pleasure to thank Prof. Sankar K. Pal, Ex-Director, ISI, for his support and encouragements prior to and during the conference and to Prof. Bimal K. Roy, Director, ISI for continuation of support thereafter. We are indeed grateful to the Referees for their cooperation and scholarly reports on the papers they have reviewed. The list of Referees includes: Drs. Sudipto Banerjee, Gopal Basak, Subir Bhandari, Sourabh Bhattacharya, Arijit Chakrabarti, Subhashis Ghosal, D.V. Gokhale, Sungsu

Kim, Arnab K. Laha, B. Purkait and Nityananda Sarkar. We also thank Prof. Pradipta Bandyopadhyay, Dr. Abhijit Mandal, Mr. Biswajit Basak, Mr. Imon Choudhury, Mr. Bhaskar Dutta, Ms. Debolina Ghatak. Ms. Debarati Ghosh, Mr. Taranga Mukherjee, Ms. Jitadeepa Nayak, Ms. Moumita Roy and Mr. Indranil Dutta for their help with the processing of the text. Last but not the least, we are deeply thankful to all the authors for contributing scholarly papers which have greatly enriched this volume.

We hope that this volume will serve as a useful and timely resource monograph to the researchers and practitioners of advances in the areas of Statistical Theory and Methods emanating from the rich spectrum of traditional and emerging statistical paradigms.

30 December 2013   Ashis SenGupta
Kolkata   Tapas Samanta
Ayanendranath Basu
Editors

List of Contributors

1. Manju Agarwal (Chapter 13), *Professor Research III MIT, Shib Nadar University, Greater Noida 203 207, India.*
   Email: agarwal_manjulata@yahoo.com.
2. Barry C. Arnold (Chapter 3), *Department of Statistics, University of California, Riverside, 1337 Olmsted Hall, CA 92521, USA.*
   Email: barry.arnold@ucr.edu.
3. S. K. Bhattacharjee (Chapter 6), *Department of Statistics, University of Rajshahi, Rajshahi 6205, Bangladesh.*
   Email: skbhattacharjee01@yahoo.com.
4. Ratan Dasgupta (Chapter 14), *Theoretical Statistics and Mathematics Unit, Indian Statistical Institute, 203 B. T. Road, Kolkata 700 108, India.*
   Email: rdgupta@isical.ac.in.
5. A. V. Dattatreya Rao (Chapter 7), *Department of Statistics, Acharya Nagarjuna University, Nagarjuna Nagar 522 510, India.*
   Email: avdrao@gmail.com.
6. R. L. Eubank (Chapter 11), *School of Mathematical and Statistical Sciences, Arizona State University, Tempe, AZ 85287-1804, USA.*
   Email: randystat@cox.net.
7. Jayanta K. Ghosh (Chapter 1), *Department of Statistics, Purdue University, 250 N. University Street, West Lafayette, IN 47907, USA.*
   Email: ghosh@stat.purdue.edu.
8. S. V. S. Girija (Chapter 7), *Department of Mathematics, Hindu College, Guntur 522 002, India.*
   Email: svsgirija@yahoo.com.
9. Dominique Guegan (Chapter 8), *University Paris 1 Panthéon-Sorbonne, 106 boulevard de l'hopital, 75013, Paris, France.*
   Email: dguegan@univ-paris1.fr.
10. Wenge Guo (Chapter 4), *Department of Mathematical Sciences, New Jersey Institute of Technology, Newark, NJ 07012, USA.*
    Email: wenge.guo@njit.edu.

11. D. J. Henderson (Chapter 12), *Department of Economics, State University of New York, Binghamton, NY 13902, USA.*
E-mail: djhender@binghamton.edu.

12. S. W. Joshi (Chapter 10), *Department of Computer Science, Slippery Rock University of Pennsylvania, Slippery Rock, PA 16057, USA.*
E-mail: sharadchandra.joshi@sru.edu.

13. R. E. Koli (Chapter 10), *Watershed Surveillance and Research Institute, JalaSRI, M. J. College, Jalgaon, Maharashtra 450 002, India.*
E-mail: rek.jalasri@gmail.com.

14. H. V. Kulkarni (Chapter 5), *Department of Statistics, Shivaji University, Kolhapur 416 004, India.*
Email: kulkarni_h_v@yahoo.co.in.

15. Antti Liski (Chapter 9), *Department of Signal Processing, Tampere University of Technology, P.O. Box 553, 33101 Tampere, Finland.*
Email: antti.liski@tut.fi.

16. Erkki Liski (Chapter 9), *Mathematics and Statistics, School of Information Sciences, FI-33014 University of Tampere, Finland.*
Email: epl@uta.fi.

17. Anandamayee Majumdar (Chapter 11), *School of Mathematical and Statistical Sciences, Arizona State University, Tempe, AZ 85287-1804, USA.*
Email: anandamayee.majumdar@gmail.com.

18. Pooja Mohan (Chapter 13), *RMS INDIA, A-7 Sector 16, Noida 201 301, India.*
Email: poojalovelyn@yahoo.com.

19. Md. Nurul Haque Mollah (Chapter 6), *Department of Statistics, University of Rajshahi, Rajshahi 6205, Bangladesh.*
Email: mnhmollah@yahoo.co.in.

20. G. P. Patil (Chapter 10), *Department of Statistics, Center for Statistical Ecology and Environmental Statistics, The Pennsylvania State University, University Park, PA 16802, USA.*
Email: gpp@stat.psu.edu.

21. S. C. Patil (Chapter 5), *Department of Statistics, Padmasree Dr. D. Y. Patil A. C. S. College, Pimpri, Pune 411 018, India.*
Email: birajdar-sangita@yahoo.com.

22. Sanat K. Sarkar (Chapter 4), *Department of Statistics, Temple University, Philadelpha, PA 19122, USA.*
Email: sanat@temple.edu.

23. I. Ramabhadra Sarma (Chapter 7), *Department of Mathematics, K. L. University, Vaddeswaram, Guntur, India.*
    Email: irbsarma44@yahoo.com.
24. Kanwar Sen (Chapter 13), *Department of Statistics, University of Delhi, Delhi 110 007, India.*
    Email: kanwarsen2005@yahoo.com.
25. Pranab K. Sen (Chapter 2), *Department of Biostatistics, University of North Carolina at Chapel Hill, 3101 McGavran-Greenberg Hall, CB #7420, Chapel Hill, NC 27599, USA.*
    Email: pksen@bios.unc.edu.
26. Aman Ullah (Chapter 12), *Department of Economics, University of California, Riverside, CA 92521, USA.*
    Email: aman.ullah@mail.ucr.edu.
27. Z. Zhong (Chapter 11), Microsoft, Seattle, WA, USA.
    Email: zhong@mathpost.la.asu.edu.

Contents

PART 1
REVIEWS

Chapter 1

Weak Paradoxes and Paradigms

Jayanta K. Ghosh

Purdue University, USA and Indian Statistical Institute, India

We consider two important counterexamples, one due to Basu (1988) and the other due to Wasserman (2004). We argue that while they provide insights about different weaknesses in Frequentist and Bayesian paradigms, too much stress on unbiasedness in the former and the full likelihood in the latter, the counterexamples are not as negative as appeared at first sight. It is shown how the paradigms can deal with the counterexamples without any fundamental change.

Keywords: Bayesian and Frequentist paradigms, Counterexamples, Pivotal variables, Unequal probability sampling.

Contents

1.1. Introduction

Many years ago I argued, tongue in cheek, that since Dr. D. Basu has found no paradox residing in the Bayesian paradigm, may be the paradigm has no paradox. I do not mean a paradox in the strict mathematical sense of leading to a logical inconsistency in the paradigm but rather an example which doesn't, strictly speaking, violate logic but nonetheless makes us uneasy and embarrassed. I call an example like this a weak paradox. I begin with such an example, which I learnt from Basu in a lecture he gave at the University of Illinois at Urbana-Champaign in 1965. For more references on this example, see Ghosh et al. (2006, p. 37).

Welch's example. Suppose X_1, X_2 are i.i.d. $U(\theta - \frac{1}{2}, \theta + \frac{1}{2})$, $-\infty < \theta < \infty$. Let $\bar{X} = (X_1 + X_2)/2$. Since θ is a location parameter, given any $0 < \alpha < 1$, we can find $c > 0$ such that

$$P_\theta\{\bar{X} - c \leq \theta \leq \bar{X} + c\} = 1 - \alpha \ \forall \theta \tag{1.1}$$

i.e., $[\bar{X} - c, \bar{X} + c]$ is a $100(1 - \alpha)\%$ confidence interval for θ.

Suppose now that $|X_1 - X_2| = 1$. This implies one of X_1, X_2 must be $\theta - \frac{1}{2}$ and the other must be $\theta + \frac{1}{2}$, which in turn implies $\bar{X}$ must be equal to θ. Hence for $|X_1 - X_2| = 1$, $[\bar{X} - c, \bar{X} + c]$ must cover θ. For such data, how can we be only $100(1 - \alpha)\%$ confident, and not 100% confident, that our interval contains θ?

Of course the probability in (1.1) is to be calculated over all samples and not just the given data. At least for $X_1, X_2 \sim N(\theta, 1)$, one may argue as in the next section that for any pair of observed X_1, X_2 it makes sense to interpret the probability in (1.1) in two ways. But not in Welch's example.

By a weak paradox I mean an example of this kind, where something that we normally do suddenly strikes us as rather odd to say the least, even if not logically inconsistent.

Some sticklers for propriety will say that the set of values of $\{X_1, X_2\}$ with $|X_1 - X_2| = 1$ has zero probability. One way of avoiding this is to make X_1, X_2 discrete with a location parameter θ. Berger (1985) has constructed such an example and it has been discussed in detail by Wasserman (2004). Rather than follow this route I will persevere with Welch's example further in Section 1.3 and come up with a confidence interval of this kind, which is optimal in a classical sense but also very, very bizarre. With probability α, it is an empty set and with probability $1 - \alpha$ it covers θ with probability one and you know from the data which case holds. Shouldn't this be embarrassing and qualify as a (weak) paradox in the classical paradigm? A whole lot of such examples are given by Fraser (2004).

What about a (weak) paradox in the Bayesian paradigm? One such is provided by Wasserman (2004) and is related to that in Robins and Ritov (1997). Ironically, it is closely related to Basu's example of circus elephants (Basu, 1988), where he tries to show how absurd the well-known Horvitz-Thompson estimate of classical statistics can be in some contexts. Wasserman presents a similar example which is embarrassing to Bayesians. I present both examples, Wasserman's and Basu's, briefly in the section on their examples.

In Sections 1.4 and 1.5, I show how the double interpretation of probability introduced in Section 1.2 can break down if we deviate from random sampling and use, say, unequal probability sampling. We also discuss Basu's and Wasserman's paradox, and suggest what a Bayesian might wish to do in Wasserman's example. Some concluding remarks sum up my attitude to these examples.

1.2. Pivotal Variables, Confidence and Credibility Interval

Given θ, let $X \sim N(\theta, 1)$ and $\theta \sim p(\theta) =$ uniform density on $\mathbb{R}$. Then it is easy to verify that
(1) given θ,

$$X - \theta \sim N(0, 1) \tag{1.2}$$

and that (2) given X,

$$X - \theta \sim N(0, 1). \tag{1.3}$$

Let z_α be the upper $100\alpha\%$ point of standard normal. By (1.2), $[X - z_\alpha, X + z_\alpha]$ covers θ with probability $(1 - \alpha)$ for fixed θ. Here θ is fixed and X is the random variable. By (1.3), $X - z_\alpha \leq \theta \leq X + z_\alpha$ with probability $(1 - \alpha)$ for fixed X. Here X is fixed and θ is the random variable.

The conclusion from (1.2) is that $[X - z_\alpha, X + z_\alpha]$ is a $100(1 - \alpha)\%$ confidence interval for θ and that from (1.3) is that it is also a $100(1 - \alpha)\%$ credibility interval for θ. Thus under the normal model we cannot have a Welch like example.

If the normal seems too restricted, we can use posterior normality (Ghosh et al., 2006, Ch. 4) to claim that $(\hat{\theta} \pm z_\alpha/\sqrt{na})$ is an approximate $100(1 - \alpha)\%$ confidence as well as credibility interval for θ. Here $a = -\frac{1}{n} \frac{\partial^2}{\partial \theta^2} \log f(\mathbf{X}|\theta)\big|_{\hat{\theta}}$ is observed Fisher information.

1.3. Welch's Example Revisited

Mukerjee and Ghosh (2004) have replaced Welch's confidence set by an equivariant confidence set which is optimal in the sense of being shortest, i.e., having minimum expected length for all θ, among all $100(1 - \alpha)\%$ equivariant confidence sets. Equivariance is defined below.

A confidence set $A(X_1, X_2) \subset R$ for θ is a $100(1 - \alpha)\%$ equivariant confidence set if

$$P_\theta\{\theta \in A(X_1, X_2)\} = 1 - \alpha \qquad (1.4)$$

and

$$A(X_1 + d, X_2 + d) = A(X_1, X_2) + d \qquad (1.5)$$

where $\theta \in A + d$ if and only if $\theta - d \in A$.

It is easy to show that $A(X_1, X_2)$ is equivariant if and only if it is of the form

$$A(X_1, X_2) = A(X_1 - X_2) + \bar{X} \qquad (1.6)$$

where $A(X_1 - X_2)$ is a set in $\mathbb{R}$ depending on X_1, X_2 through the (maximal) invariant function $(X_1 - X_2)$ under origin shifts. In particular, Welch's interval is of the form (1.6) and so is equivariant.

The expected length of an equivariant $A(X_1, X_2)$ is

$$E_\theta\left[\int_{A(X_1, X_2)} d\theta\right] = E_\theta\left[\int_{A(X_1 - X_2)} d\theta\right] = E_0\left[\int_{A(X_1, X_2)} dx\right]$$

which is a constant. Hence there is a shortest equivariant confidence set. It can be found by an application of the Neyman-Pearson lemma (Mukerjee and Ghosh, 2004). It has the following structure. With probability α, our confidence set is empty and with probability $(1 - \alpha)$ it includes all θ's compatible with the observed X_1, X_2.

1.4. Effect of Unequal Probability Sampling on Pivotal Quantities

Let (Y_i, λ_i), $i = 1, 2, \ldots, n$ be i.i.d. and for $i = 1, 2, \ldots, n$,

$$P\{\lambda_i = 0\} = p_0, \ P\{\lambda_i = 1\} = p_1, \ p_0 + p_1 = 1, \ p_0, p_1 > 0,$$

$$f_\theta(y_i | \lambda_i = 0) = \frac{2}{\sqrt{2\pi}} e^{-\frac{1}{2}(y_i - \theta)^2}, \ y_i > \theta$$

$$f_\theta(y_i | \lambda_i = 1) = \frac{2}{\sqrt{2\pi}} e^{-\frac{1}{2}(y_i - \theta)^2}, \ y_i < \theta.$$

The marginal density of Y_i is

$$f_\theta(y_i) = p_0 \frac{2}{\sqrt{2\pi}} e^{-\frac{1}{2}(y_i - \theta)^2}, \quad y_i > \theta$$

$$= p_1 \frac{2}{\sqrt{2\pi}} e^{-\frac{1}{2}(y_i - \theta)^2}, \quad y_i < \theta.$$

If we take the prior as in Section 1.2, i.e., $p(\theta) = 1 \ \forall \theta \in \mathbb{R}$, then it is easy to check the posterior is essentially as in Section 1.2 and hence, given $Y_1, \ldots, Y_n$, $\sqrt{n}(\bar{Y} - \theta) \sim N(0, 1)$. However, given θ, it is no longer true that $\sqrt{n}(\bar{Y} - \theta) \sim N(0, 1)$. The matching of probabilities is lost and $\sqrt{n}(\bar{Y} - \theta)$ is no longer a pivotal quantity.

However, the justification for the approximate pivotal quantity based on an MLE still holds. Also in this relatively simple case another way out would be to cluster the data into two groups, one with Y_i's equal to 0 and the other with Y_i's equal to 1. Conditionally for each group, one would get approximate pivotal quantities based on the corresponding MLE. The MLE based on all the data would be approximately a weighted average of the two group MLE's.

A third pivotal quantity would be based on the MLE calculated from the marginal distribution of the Y_i's. This would be inefficient relative to the first MLE based on conditional distributions or the asymptotically equivalent weighted average of the two group MLE's.

1.5. Examples of Basu and Wasserman Related to the Horvitz-Thompson Estimate

1.5.1. *Basu's example*

This subsection is borrowed from Basu (1988) but condensed as in Ghosh (2007).

Suppose that n units are drawn at random from a finite population containing N units and each drawn unit is replaced before the next draw. The ith unit has a probability p_i of being drawn.

The Horvitz-Thompson (HT) estimate, also called NHT (Rao, 1999), for the population total is

$$\frac{1}{n} \sum_{i=1}^{n} Y_i / p_i$$

where Y_i is the value of Y for the ith unit.

If each $p_i = 1/N$, this reduces to simple random sampling with replacement and the associated estimate for population total.

In Basu (1988), Basu presents a hilarious example of how absurd this can be in some cases. Basu imagines a circus owner who wants a rough estimate of the total weight of his fifty adult elephants. He picks out Sambo as a sort of average based on past measurements and then multiplies by fifty. This is his estimate, an estimate that would be approved by most Bayesians, even though it is not based on formal Bayesian calculations.

But at this point enters the circus statistician who is horrified by this ad hoc estimate. He tries to introduce an element of randomness by choosing a sampling plan that assigns a probability of 99/100 to Sambo and 1/4900 to each of the rest. The object is to choose Sambo with a high probability and yet have a valid estimate of variance. Not surprisingly, Sambo is selected and the statistician uses the HT estimate which in this case is

$$(\text{Sambo's weight}) \times \frac{100}{99}$$

which is a serious underestimate.

For a plausible data on these fifty elephants, we calculated the mean squared error (MSE) of both estimates – the subjective biased estimate of the owner and the unbiased estimate of the circus statistician. They are 755047.7 kg^2 and 4.064×10^{12} kg^2 respectively. As expected the biased subjective estimate does much better. Bayesians believe in real life this will happen much more often than expected by classical statisticians. On the other hand a classical statistician need not be embarrassed by the silly unbiased estimate, which would be nobody's choice. For that particular sampling plan, an unbiased estimate is very expensive, it drives up the variance towards infinity. For more details on this phenomenon see two beautiful papers, Doss and Sethuraman (1989) and Liu and Brown (1993).

1.5.2. *Wasserman's example*

We take this example from Wasserman (2004), who says this is a simplified version of the example in Robins and Ritov (1997), involving missing observations and unequal probabilities.

Consider n i.i.d. random vectors $(X_i, R_i, Y_i), i = 1, 2, \ldots, n$. N is a finite but very large number, any realistic sample size n is much smaller. The parameter vector is $(\theta_1, \theta_2, \ldots, \theta_N), 0 \leq \theta_j \leq 1$. Let

$$\mathbf{p} = (p_1, p_2, \ldots, p_N)$$

be a known vector of probabilities satisfying

$$0 < \delta \leq p_j \leq 1 - \delta < 1, \quad 1 \leq j \leq N, \tag{1.7}$$

where δ is a fixed positive number. Each (X_i, R_i, Y_i) is generated as follows. X_i is uniformly distributed over the so called "labels" $\{1, 2, \ldots, N\}$. R_i is a Bernoulli random variable with probability p_{X_i} that $R_i = 1$.

Given $R_i = 0$, Y_i is not observed (i.e., Y_i is missing). Given $R_i = 1$, Y_i is observed, Y_i is Bernoulli with probability θ_{X_i} of being one.

The object is to estimate $\psi = \frac{1}{N} \sum_{j=1}^{N} \theta_j$.

The likelihood is easy to write down but there is very little information on most θ's. So Bayes estimates of θ's would be poor, and taking their average will give the Bayes estimate for ψ, which is unlikely to be consistent.

On the other hand a Horvitz-Thompson estimate, namely,

$$\hat{\psi} = \frac{1}{n} \sum_{i=1}^{n} \frac{R_i Y_i}{p_{X_i}}$$

is unbiased and has variance $\leq \frac{1}{n\delta^2}$. This implies $\hat{\psi}$ is consistent. But "it cannot be derived from a Bayesian or likelihood point of view" ... since the p_{X_i}'s "drop out of the log-likelihood". Wasserman observes, "These ... examples illustrate an important point. Bayesians are slaves to the likelihood function. When the likelihood goes away, so will Bayesian inference."

This is like problems arising in survey sampling, where no likelihood function is available. However, alternative subjective methods have been proposed. Basu (1988), in his presentation of the circus example, refers to face validity of some of his estimates, and is quite critical of the Horvitz-Thompson estimate, when Assumption (1.7) does not hold. Ghosh and Meeden (1997) provide an approach which tries to connect the seen data with the unknown data about which some sort of prediction is needed. They introduce a posterior, without introducing a prior or a formal likelihood. They validate its use by its Frequentist behavior.

Below I use a rule of this kind which takes care of missingness of the data. Both the missingness and the many p_j's are also part of the problem in the RRW (Robins, Ritov, Wasserman) example. An entirely different approach that I have not explored at all is to break the average into averages based on somewhat smaller data, say, having n/k observations, and then model the likelihood of averages of parameters based on sets of independent observations.

Suppose then we simplify the problem by assuming the p_j's are same and the common value is p. Then the effective sample size is np, and the sample mean is $\frac{1}{np} \sum_{i=1}^{n} R_i Y_i$, which has what Basu calls face validity. But this is just the Horvitz-Thompson estimate for this special case. In the general case, using Assumption (1.7), we can divide the p_j's in k classes such that within a class they are nearly equal, k tends to ∞, k/n tends to 0 as n tends to ∞. Then the Horvitz-Thompson estimate can be approximated by a weighted mean of k quantities, each of which is like the above simple estimate with some face validity.

To conclude, Bayesians who are aware of the Bayesian literature on the Horvitz-Thompson estimate will accept it if (1.7) is valid.

Having said this let me try an alternative to $\hat{\psi}$. The p_i's are what are causing trouble as in Section 1.5.1. In the following we try to reduce the enormous number N of p_i's by grouping them together into k classes, k much smaller than the sample size n. The whole range of p_i's, namely $[\delta, 1 - \delta]$ is divided up into k classes $[a_1 = \delta, a_2), [a_2, a_3), \ldots, [a_k, a_{k+1} = 1 - \delta]$. We take p_i to be the mid-point of the ith class. Though n is much smaller than N, we assume it is also a large number, certainly much larger than k. Then the frequency n_j of p_{X_i}'s falling in say the jth class is also large. We may rethink of our data as follows. For each $j = 1, 2, \ldots, k$, we have n_j observations of the form (X, R, Y) with $R \sim$ the fixed Bernoulli (p_j) and $Y \sim$ Bernoulli (θ_X), the second set of Bernoulli's are not fixed. Hence for these n_j observations $R_i Y_i$ is Bernoulli $(p_j \theta_{X_i})$ and $\frac{1}{n_j} \sum_{i=1}^{n_j} R_i Y_i$ is a "good" estimator for

$$\frac{1}{N_j} \sum_{p_i \text{ in } j\text{th class}} R_i \theta_{X_i}.$$

It makes sense to Bayesians, see Ghosh and Meeden (1997). Also, n_j/n is a good estimator for N_j/N. Hence a "good" estimator for $\psi = \frac{1}{N} \sum_{i=1}^{N} \theta_{X_i}$ is

$$T = \sum_{j=1}^{k} \frac{n_j}{n} \frac{1}{n_j} \left(\sum_{i=1}^{n_j} R_i Y_i \right) / p_j.$$

For a Bayesian it is a natural approximation to the HT estimator since the R's are one of k Bernoulli random variables (approximately). We present below some numerical comparisons.

We take $N = 1000$, $n = 100$, $k = 10$ and $\delta = 0.2$ and generate $\theta_1, \ldots, \theta_N$ as a sample from a uniform distribution over a subset $[a, b]$ of $[0, 1]$ for some selected values of a and b. The vector $(p_1, \ldots, p_N)$ is taken as $p_1 = 0.2$, $p_N = 0.8$, $p_{i+1} - p_i = (0.8 - 0.2)/(N - 1)$ for all i. We calculate the estimates of the MSE of $\hat{\psi}$ and T using 1000 replications. The following table gives the MSE of $\hat{\psi}$ and T for different values of a and b. The table shows that the estimator T performs at least as well as the Horvitz-Thompson estimator $\hat{\psi}$. However, it is remarkable that the simple minded HT estimate defies substantial improvement.

Table 1.1. Estimates of the MSEs of $\hat{\psi}$ and T for $\theta_1, \ldots, \theta_N$ generated from $U[a, b]$.

$[a, b]$	MSE $(\hat{\psi})$	MSE (T)
$[0.1, 0.9]$	0.00885	0.00878
$[0.6, 0.9]$	0.01156	0.01149
$[0.1, 0.4]$	0.00519	0.00516
$[0.35, 0.65]$	0.00902	0.00896

1.6. Concluding Remarks

We present a few (weak) paradoxes for both classical and Bayesian paradigms. We also note how classical statisticians and Bayesians have reacted to them. Given examples like that of Welch (see also Cox, 1958), some like Tukey have suggested that in these examples we don't use these particular confidence intervals, some, following Fisher, have recommended conditioning and some others, like Kiefer, have come up with a novel Frequentist revision of the fixed confidence into a data dependent one. For reactions to Basu's example see Rao (1999).

My feeling about Wasserman's example is similar. I have tried to take a first step, which isn't likelihood based but Bayesian in the spirit of Ghosh and Meeden (1997). But unequal probability sampling remains a challenge to Bayesians.

Acknowledgments

I thank Wasserman (Wasserman, 2004) and Robins and Ritov (Robins and Ritov, 1997) for this provocative example. I would also like to acknowledge help on various matters from Arijit Chakrabarti and Tapas Samanta.

References

1. Basu, D. (1988). *Statistical Information and Likelihood*. A collection of critical essays by Dr. D. Basu (J. K. Ghosh, Ed.), Lecture Notes in Statistics, Springer-Verlag, New York.
2. Berger, J. O. (1985). The frequentist viewpoint and conditioning. In *Proceedings of the Berkeley Conference in Honor of Jerzy Neyman and Jack Kiefer, Vol. I* (L. M. Le Cam and R. A. Olshen, Eds.), 15–44, Wadsworth, Inc., Monterey, California.
3. Cox, D. R. (1958). Some problems connected with statistical inference. *Ann. Math. Statist.* **29**, 357–372.

4. Doss, H. and Sethuraman, J. (1989). The price of bias reduction when there is no unbiased estimate. *Ann. Statist.* **17**, 440–442.
5. Fraser, D. A. S. (2004). Ancillaries and conditional inference. *Statist. Sci.* **19**, 333–369.
6. Ghosh, J. K. (2007). Role of randomization in Bayesian analysis – An expository overview. In *Bayesian Ststistics and Its Applications* (S. K. Upadhyay et al., Eds.) 198–204, Anamaya Publishers, New Delhi.
7. Ghosh, J. K., Delampady, M. and Samanta, T. (2006). *An Introduction to Bayesian Analysis – Theory and Methods.* Springer-Verlag, New York.
8. Ghosh, M. and Meeden, G. (1997). *Bayesian Methods for Finite Population Sampling.* Chapman and Hall, London.
9. Liu, R. C. and Brown, L. D. (1993). Nonexistence of informative unbiased estimators in singular problems. *Ann. Statist.* **21**, 1–13.
10. Mukerjee, R. and Ghosh, J. K. (2004). Unpublished note.
11. Rao, J. N. K. (1999). Some current trends in sample survey theory and methods (with discussion). *Sankhyā* **B 61**, 1–57.
12. Robins, J. M. and Ritov, Y. (1997). Toward a curse of dimensionality appropriate (CODA) asymptotic theory for semi-parametric models. *Stat. Med.* **16**, 285–319.
13. Wasserman, L. (2004). *All of Statistics – A Concise Course in Statistical Inference.* Springer, New York.

Chapter 2

Nonparametrics in Modern Interdisciplinary Research: Some Perspectives and Prospectives

Pranab K. Sen

University of North Carolina at Chapel Hill, USA

Over the past five decades, nonparametric methods have emerged as a viable and often more adaptable way of statistical modeling and analysis of more complex and sometimes interdisciplinary studies. The contemporary evolution of information (and bio-)technology has posed some highly nonstandard interdisciplinary research problems, genuinely in need of statistical appraisal. Some nonparametric perspectives and prospectives in some challenging interdisciplinary research, along with a highlight of some developments in the past fifty years, are appraised here.

Keywords: Bayesian nonparametrics, Curse of dimensionality, Dimensional asymptotics, Rank-permutation principle, Union-intersection principle.

Contents

2.1. Introduction

Prior to 1960s, nonparametrics, usually referred to as *distribution-free* or *rank based* methods, were mostly confined to specific simple models where suitable *hypotheses of invariance* led to statistical inference (mostly, testing) procedures without explicitly assuming the form of underlying distributions (as is customarily the case in parametric statistical inference). Yet in such simple set ups, the need for *asymptotics* arose in diverse ways, thus providing the impetus for innovative theoretical as well as methodological research which remains unabated even today.

During the 1960s and 1970s, nonparametrics were annexed to *biological assays*, a variety of biomedical problems, multivariate analysis as well as sequential and *time-sequential analysis*. In the same vibrant mode, during the 1970s and 1980s, nonparametrics entered the arena of *clinical trials* and *survival analysis*, thus opening the doors for the foundation of *semiparametrics* as well. The 1990s have witnessed a much broader and more unified developments of nonparametrics, aiming towards a *beyond parametrics* conglomerate, annexing semiparametrics as well as *Bayesian nonparametrics* in a natural way. During the past ten years, the advent of information and biotechnology has posed some challenging statistical tasks where beyond parametrics approaches seem to be more appealing. It is in this vein, some perspectives and prospectives of nonparametrics are appraised here.

To motivate our appraisal, in Section 2.2, we start with glimpses of nonparametrics during 1960s and 1970s. Section 2.3 relates to the genesis of semiparametrics in a broader domain of nonparametrics along with their conjugate impact on clinical trials and survival analysis. Section 2.4 is devoted to nonparametric *smoothing* and related topics which later on led to the evolving area of *knowledge discovery and data mining* (KDDM) or *statistical learning*. The main focus on nonparametric perspectives and prospectives in modern interdisciplinary research is laid down in Sections 2.5 and 2.6; statistical undercurrents in data mining are appraised along with. *High-dimensional low sample size* (HDLSS) perspectives are overwhelming in this context, and nonparametric prospective are assessed along with. An overview of general nonparametric perspectives and prospective is presented in the concluding section.

2.2. Glimpses of Nonparametrics I

Looking at the current (rather puzzling) status of statistical appraisals in the evolving field of *genomics* and *bioinformatics*, I could not check my temptation in quoting the (Mosteller, 1948) phrase *quick and dirty methods* attributed to the development of nonparametric methods prior to 1945. However, those intuitively rich heuristics led to an evolution of most innovative theory and novel methodology in a far broader field than contemplated in the 1940s and partly in 1950s, Apart from a systematic development of the basic structure of the hypotheses of invariance (under appropriate group of transformations that maps the sample space onto itself) supporting *distribution-free tests* for some simple cases (e.g., one-, two-, several sample, goodness-of-fit, bivariate independence, and some simple blocked designs problems), there were some sparkling developments in this period: Wald and Wolfowitz (1944) treatise of *permutation tests,* Hoeffding's (1948) *U*-statistics, and Pitman's (1948) *local alternatives* based *asymptotic relative efficiency* (ARE) foundations are among these landmarks of developments. In the 1950s, a more sustained and systematic development evolved culminating with the Chernoff and Savage (1958) asymptotics, stealing the limelight of nonparametrics for a while.

In the 1960s, evolution of nonparametrics not only highlighted innovative theory and novel methodology but also encompassed hitherto new areas of applications. The asymptotic distribution theory of linear rank statistics culminated with the Hájek (1968) article; his earlier research (Hájek, 1961) on *permutational central limit theorems* unified diverse developments during the past twenty years. In another way, Hájek (1962) most elegantly laid down the foundation of *contiguity of probability measures* (due to LeCam, 1960) and incorporated that in nonparametics in a systematic way; as of today, this contiguity has played a focal role in statistical inference far beyond the traditional boundaries of parametric, semiparametric and nonparametric inference. In a second front, Huber (1964) came up with a formulation of *robustness,* albeit in a local sense, which set a new direction of research, and nonparametrics quickly annexed robustness in a more global perspective in its vagaries. *Weak convergence* of (weighted) *empirical processes* entered the arena of nonparametric in the 1960s, more vigorously in the next decade, and it reshaped the foundation of asymptotic theory not only in nonparametrics but also in almost every walk of statistical inference, probability theory and stochastic processes. It is still an active area of theoretical research.

There were some other notable developments in nonparametrics in the 1960s. Prior to 1963, nonparametric methods were mostly confined to hypothesis testing problems and tolerance limits setups (Savage 1962). A general theory of estimation of parameters based on appropriate rank statistics cropped up in 1963 (Hodges and Lehmann, 1963; Sen, 1963) and the next ten years witnessed a phenomenal growth of research literature on *R-estimation theory* encompassing conventional location, shift, scale and regression models (Adichie, 1967; Sen, 1968; Jurečková, 1969, 1971; Koul, 1969, among others, reported systematically in Jurečková and Sen, 1996). Together with the Huber (1964) robust *M-estimators* and the so called *L-estimators* based on linear combinations of order statistics (partly developed in the 1950s), such nonparametric estimators drew the attention of theoretical statisticians as well as researchers working with statistical methodology in various applied research setups. Among these novel applications, a notable case was *biological assays* or bioassays. A distribution-free method of (point as well as interval) estimation of *relative potency* of a new drug with respect to a standard drug in *dilution assays* (Sen, 1963) opened the doors for a flow of subsequent developments for over a decade. Bio-assays, *growth-curve models* and *repeated measurement designs* (or *longitudinal data models*) were all annexed to nonparametrics in the estimation domain as well.

The invariance structures, laying the foundation of suitable hypotheses of invariance, yield genuinely distribution-free tests in simple (mostly) univariate models; however, they stumbled into impasses in more general complex problems where nuisance parameters often mar the basic invariance structure, Also, in multivariate analogues of some of these simple univariate problems, interdependence of coordinate variables may invalidate such simple invariance structures. Chatterjee and Sen (1964) developed some *rank-permutation principles* which paved the way for *multivariate nonparametrics*; a sustained and vigorous development on these Multivariate nonparametric methods took place for quite sometimes, and the first monograph in this line by Puri and Sen (1971) set the tone for more nonparametrics in diverse nonstandard setups. An *alignment principle* underlying *R*-estimation theory along with multivariate nonparametrics led to more adaptable and efficient nonparametrics for *subhypothesis testing* problems in univariate as well as multivariate setups (albeit with some compromise on the genuine distribution-freeness). Nonparametric estimation theory in multivariate problems as well as *statistical functionals* evolved at about the same time.

Another significant development in the 1960s to 1970s is the conjugation of *martingale theory* in nonparametrics, leading to simpler and often more powerful asymptotics, Ranks are not independent even under hypotheses of invariance, so that rank statistics may not have typically *independent increments* as commonly perceived in sums of independent variables. Under hypotheses of invariance, many rank statistics have martingale property, though the increment may not be in general homogeneous. For independent and identically distributed random vectors, *U*-statistics, like the sample mean, may have reversed martingale property, while von Mises functional, by virtue of their close proximity to *U*-statistics, can be well approximated by suitable reversed martingales. In general statistical functional, under appropriate regularity conditions, can be approximated by such martingales or reverse martingales. Such martingale characterizations and approximations provide much simpler proofs of asymptotic distribution theory for nonparametric statistics, at least under hypothesis of invariance as well as under local alternatives. Contiguity properties referred to earlier provide automatic access to such asymptotics under contiguous alternatives. The wealth of *martingale central limit theorems* and *invariance principles* developed in the early 1970s were thus immediately adapted to nonparametrics, These results had a tremendous impact on the development of *sequential nonparametrics*. The bulk of development in this area along with nonparametric sequential estimation theory (Sen and Ghosh, 1971) were reported in the Sen (1981) monograph.

Another notable nonparametric development in the 1970s is the incorporation of nonparametrics in *clinical trials, survival analysis* and *reliability analysis*. Under hypothesis of invariance, *progressively censored* linear rank statistics have a natural martingale structure (Chatterjee and Sen, 1973) that has been fully exploited to formulate *time-sequential nonparametrics*. Just a few months prior to this development, the seminal work of Cox (1972) innovated the so called *proportional hazards model* (PHM) which led to the development of semiparametrics. In this variation, the PHM paved the way for incorporation of more general *multiplicative intensity processes* in survival analysis. Thus, a general class of *counting processes* commonly perceived in survival analysis and reliability theory has gone through an evolutionary development during the past forty years; martingale characterizations have been fully explored in this context as well. Various *censoring schemes* (Type I, Type II, random censoring; Kaplan and Meier, 1958) developed earlier were reformulated in this counting process format, and more unified methodological developments evolve at a good pace. This conjugation of nonparametrics and semiparametrics is working out well even at the present time. *Repeated significance testing* (RST), *group sequential procedures, interim analysis* and *longitudinal follow-up studies* have all been annexed in this spectrum of development.

2.3. Glimpses of Nonparametrics II

Major developments in asymptotic theory engulfed theoretical advances in statistical inference in general (and nonparametrics in particular) during the 1960s and 1970s. Yet such asymptotics did not fully explore their scope and impact of theoretical research in applied research and professional work. This previously perceived shyness of applied researchers and professionals in adapting nonparametrics in their work seemed to have faded to some extent in the 1980s. This phenomenon was partly due to the diversity of applied research with increasing demand for statistical appraisals where, generally, conventional parametric approaches often turned out to be inadequate to untenable. Thus, heuristic considerations started favoring nonparametrics to a greater extent. Model complexities and other extraneous considerations may often call for greater flexibility of the underlying distributions, instead of assuming specific parametric forms, thus leading to nonparametrics in a genuine sense. The very introduction of semiparametrics was to draw a middle line with partly parametric setups and partly nonparametric, though that way raising concern over the robustness of such a compromise. *Transformation of variables and statistics*, a common device in applied statistical research (to stabilize the variance and or to accelerate the asymptotic approximations in moderate samples) sometimes becomes somewhat out of the place and unusable due to the very nature of the response variables, model complexity or imprecise modeling of study data. These developments led to the evolution of *statistical smoothing* as presented in the next section.

Another important development in this arena is the Efron (1979) innovation of *resampling plans* which unified the earlier developed *jackknife* method (Quenouillie 1949) in the light of the related *bootstrap* methodology; the former stands on simple random sampling without replacement and the latter with replacement, thus both being *subsampling* from the original sample in a general sense. Whereas jackknifing aims to reduce the bias and estimate the sampling variance, bootstrapping, in addition, may also be used to estimate the sampling distribution itself. In a broader sense, both of these resampling plans are essentially nonparametric in flavor, albeit there are also some parametric versions of resampling plans. Bootstrap methods emerged at a time when statistical methodology was trying to cope with the immense annexation of computer technology, thus opening a new frontier: *computational statistics*. Nonparametrics seem to have a special berth in this resampling bandwagon. To appreciate the full impact of resampling methods in statistical practice, there has been a steady growth of research literature dealing with the theoretical foundation of the practice of resampling plans in almost all walks of life and science, and the quest is far from being over. There is ample room for further theoretical innovation to validate and match the need for implementation of resampling plans in more complex (such as data mining and HDLSS) setups.

Counting processes and associated (semi-)martingale theory, primarily arising in semiparametrics in the field of clinical trials, characterized the growth of nonparametric research during the last two decades of the past century, and this evolution is still unabated. The Cox (1972) PHM led to some martingale characterizations in the mid 1970s but it was not until the end of the decade, a systematic study of counting processes related to PHM and more general multiplicative intensity processes emerged. Once characterized that way, in the 1980s theoretical developments led to a sustained growth of research literature in survival and reliability analysis; an encyclopedic compilation of research on statistical inference based on counting processes culminated with the monograph of Andersen et al. (1993). More general multivariate counting processes in the same vein have been developed mostly in the past 15 years. The simple Cox PHM has revolutionized the use of counting processes in actual applications. Yet, it has its limitations in applications. For example, in survival analysis, if two survival functions are crossing then the PHM assumption may not be tenable. Therefore, in the next round of research, mostly in the late 1980s and 1990s, various modifications of the PHM have been advocated. In this vein, *time-dependent regressors* and *time-varying regression parameters* models have been advocated; they eliminate the stringent PHM clause to a greater extent, albeit statistical analysis becomes more complex and probably more asymptotic in nature (Murphy and Sen, 1991).

The annexation of longitudinal data models and repeated measurement designs to nonparametrics evolved in the 1970s. However, a more systematic effort to consolidate nonparametric (and semiparametric) methods in this area dawned in the 1980s. Initially, the bulk of multivariate nonparametric methods, developed in the 1960s and 1970s were incorporated to explore repeated measurement designs. At the same time, Huber's robust statistical methodology also found a proper place in the dissemination of longitudinal data models. The basic (parametric) structure of *generalized linear models* (GLM) and some specific *nonlinear models* were often found to be unrealistic if not too stringent, especially when an exponential family of densities (including (multi-)normal laws) were not even approximately tenable, thus raising the robustness issues in right perspective. In another front, *mixed-effects* and *random-effects* models were introduced initially in strictly parametric (and mostly, normal errors) setups and subsequently in more general nonparametric setups with much less restrictive distributional assumptions. In this way, *measurement error* and *error-in-variables* models have also undergone a systematic transition from parametrics to nonparametrics (and semiparametrics too); see Sen and Saleh (2010). Longitudinal data and time-series models in actual applications are often perceived to have distinct *heteroscedastic* errors in addition to possible dependence and nonnormality. In this respect too, nonparametric methods have greater scope and robustness perspectives. In a variety of applications, especially in biostatistics and (environmental) health sciences, *clustering* of units is quite common: Within cluster observations are usually dependent (often accountable by allowing an additive *random cluster-effect*) while inter-cluster observations may be taken to be stochastically independent. In this respect too, nonparametric approaches may be more convenient to smooth out some of these irregularities in a more interpretable manner. Some aspects of these developments are appraised in the next section.

One of the greatest advents of nonparametrics in the past thirty years has been the "reaching out" to biostatistics, epidemiology, environmental science as well as other public health, clinical and biomedical sciences in a vibrant way. In these fields, often interdisciplinary research crops up in rather complex setups. Some of these may relate to observational studies whose planning may be quite different from conventional design of experiments. There may be a persistent qualitative undercurrent so quantifying them and putting in a statistical framework may often be a challenging task. Further, collection of data may entail some irregularities violating to some extent the principles of objective data collection, thus posing roadblocks for statistical modeling and analysis. Investigator's bias (or errors), recording or instrumental errors, less than perfect questionnaire adaption, compliance with the study scheme, lost in follow-up studies, nonstandard study protocols (for example, case-control and cohort studies), misclassification and missing observations, and some vast number of extraneous factors (many of which are not identifiable) can create a genuine impasse for the implementation of standard statistical packages, albeit there is a genuine need for statistical appraisal. All these distractions make conventional parametric approaches less appealing, non-robust and often unsuitable in the context of specific applications. *Variable selection* or *identification* of most relevant subset of factors or responses (when there are many) may, by no means, be an easy task. More of these problems crop up in data mining problems. Nonparametrics fare better in this respect. Understanding the very objective(s) of such a non-statistical study, translating the setup into a statistically interpretable one and then implementing appropriate statistical reasoning for quantitative analysis based statistical conclusions are by no means routine tasks, and often they create genuine impasses for statistical resolutions.

Translating a non-statistical problem in health and environmental sciences into a statistical one is one of the most basic tasks which require considerable skills in data modeling and data interpretation. In clinical research, other than clinical trials, there is less emphasis on survival time but more on prevalence, propensity and other aspects. In that way, generally, one encounters a multidimensional binary or categorical data model, resulting in a much larger number of parameters, often far exceeding the sample size that is affordable from time and cost considerations. The use of biomarkers, sometimes in large numbers, adds to the model complexity whereby no simple parametric model seem to be adequate or adaptable. No wonder that semiparametrics have invaded this field with the main objective of reducing the dimensionality of the parameter space by slicing out a nonparametric component that may not be of prime interest. Yet in this development, often the very basic formulation of semiparametric modeling may be questionable in specific contexts. Likewise, *generalized linear models* such as based on *logit* transformations in binary data models may lose its natural appeal in a semiparametric setup, lacking interpretability of the associated parameters of interest. In the study of the progression of chronic diseases or disorders, while it may be quite tempting to incorporate some simple stochastic processes in statistical modeling, in real life cases, due to various extraneous constraints, such processes are often inappropriate.

In *health related quality of life* (HRQoL) studies, the greatest hurdle is to handle high-dimensional qualitative responses and translating them into (quantitatively) statistically analyzable ones. In environmental toxicity studies, some *spatio-temporal* processes are easy to conceive but far more difficult to incorporate. A large number of explanatory variables, paucity of data, measurement errors, and other basic factors can invalidate such simple models. On the same count, simple parametric models are often quite inadequate in such complex studies. Nonparametric and semiparametric methods are increasingly adapted in such complex studies, resulting in greater model flexibility, robustness and adaptability perspectives. These developments, dating back to the 1980s have vastly enriched the field of biostatistics far beyond the traditional frontiers of traditional biometry, and yet there are innumerous impasses in this trajectory. Simple (equal probability) random sampling (with or without replacement) setups underlying conventional statistical designs are hardly tenable, and as a result, stationarity (homogeneity) and independence of increments as commonly perceived in simple parametric formulations are to be dispensed with in nonparametric and semiparametric approaches in order to impart more model flexibility and greater scope for adaptability in diverse fields of applications.

In many experimental or observational studies, it is not unusual to think of some *prior information* and incorporate it for better modeling and statistical appraisal. Of course, this refinement is expected when the incorporated prior information is relevant. This provides the general motivation for *Bayes methods*. Prior information in conjugation with the model assumptions provide the *posterior* distribution of parameters of interest and thereby lead to improved statistical inference. On the other hand, if the chosen prior is not appropriate, the posterior distribution may not conform to the true situation and derived statistical conclusions may not be representative, may even be inconsistent. To eliminate this shortcoming, uncertain priors are sometimes used to derive *shrunken* estimators that are very good under validity of chosen priors and robust for incorrect priors. This idea of *shrinkage estimators* (Stein, 1956) has revolutionized the scope of adoption of Bayes methods and led to some variants such as *empirical Bayes* (EB) and *hierarchical Bayes* (HB) methodology, far beyond the quarters of parametrics (Sen and Saleh, 1985), incorporating greater model flexibility and robustness perspectives. The so called *Bayes factor* has become a household word in current biostatistics research in clinical trials and environmental health sciences. In a related vein, Ferguson (1973) initiated an innovative idea of using *Dirichlet processes* in empirical distributional processes and thereby opening the doors for *Bayesian nonparametrics* in a highly proactive way, encompassing both theoretical innovations and adaptable methodology (Ghosh and Ramamoorthi, 2003). We shall append some further discussion on Bayesian nonparametrics in a later section.

2.4. Smoothing and Statistical Functionals

A statistical function is a functional of the distribution function F, defined on an appropriate domain. Hoeffding's U-statistics and von Mises' V-statistics (Hoeffding, 1948) are classical examples of simple *statistical functionals* which are *estimable* and optimal in a broad nonparametric sense. These functionals involve some *kernels* of finite degree that simplify the study of their statistical properties. A comparatively more complex statistical functional is the sample quantile (or median) which may not pertain to a kernel of finite degree. Nevertheless, such statistical functionals are generally expressible as functions of the empirical distribution function or some weighted versions. Statistical functionals arise commonly in the context of R-, M- and L-estimation theory as well as hypothesis testing theory based on such statistics. Studies of statistical properties of statistical functionals generally depend on their differentiability perspectives in a suitable norm (and in appropriate function spaces). Kolmogorov-Smirnov and Cramer-von Mises goodness of fit test statistics are bona fide members of this class.

Corresponding to a distribution function F, defined on an appropriate domain $\mathcal{F}$, we denote a statistical functional by $T(F)$, The sample or empirical distribution function F_n is an unbiased and optimal estimator of F, and hence, it may be quite intuitive to estimate $T(F)$ by $T(F_n)$, the so called plugged in estimator. If $T(F)$ is a linear functional, then $T(F_n)$ is unbiased for $T(F)$, and hence $T(F)$ is estimable in the sense of Hoeffding (1948). However, in general, we need some other regularity conditions for characterizing $T(F_n)$ as a suitable estimator of $T(F)$. The salient point in this context is that F may be sufficiently smooth (i.e., continuous, differentiable etc.) but F_n, being a step-function, may not be adequately smooth to validate $T(F_n)$ as a suitable estimator. A notable example is the density function $f(x) = (d/dx)F(x) = f(F^{-1}(F(x)))$ which is thus a statistical functional defined for absolutely continuous F. On the other hand, F_n is a step function with increment n^{-1} at the realized values of the sample observations (and flat elsewhere). Thus, F_n is not differentiable and does not provide an estimator of f. This feature of (weighted) empirical distributions led to a continual evolution, a little over fifty years ago, of nonparametric estimation of density function and other allied functionals based on certain *smoothing* principles.

The basic idea came from the classical histogram method applied to F_n but any fixed bandwidth would result in a big bias relative to the mean square error, and as such, the *sieves* and *nearest neighborhood* methods evolved with the main objective of balancing the asymptotic bias and mean square error in an interpretable way (Rosenblatt, 1956; Nadaraya, 1964; Watson, 1964). In this way, the rate of convergence has to be compromised to the order $n^{-\lambda}$, for some $\lambda < 1/2$, typically being 2/5 for the univariate case and even slower for multivariate models. The *bandwidth selection* problem in concordance with the so called *kernel method* cropped up in a natural way, though largely being of asymptotic flavor. Lot of methodological developments during the past four decades has rendered this field as one of the most active areas of current statistical research. Such methodological advances have also been matched by extensive data analytic tools.

There are some especial considerations for density estimation when the observed random variables are nonnegative (where *survival function* is more commonly used in statistical modeling and analysis), or when there is a finite (though unknown) *end point* of the underlying distribution without much prior information on the *order of terminal contact*; the latter plays a basic role in determining the rate of convergence of the estimates. The bibliography is extensive and there are some excellent texts covering these developments. As such, we avoid citing the extensive list of references, but refer to Tsybakov (2008) which cites other references.

Nonparametric estimation of density functions led to many related developments. First, there are some functions of the density function which are thus functionals too. The empirical survival function, the complement of the empirical distribution function, is by definition a step (down) function and it is often desired to have a smooth estimator of the underlying survival function so that other functional related to that can be estimated properly. For example, the *hazard function*, the negative logarithmic derivative of the survival function, needs the estimation of the density function or a smooth and differentiable survival function. On the same count, *mean remaining life* (MRL) at various age, a functional of the survival function, would be more interpretative when based on smoothed estimates of the survival function. *Monotone hazard function*, *decreasing MRL* (DMRL) and some other functionals are better studied with smoothed survival function instead of the empirical ones. *Nonparametric maximum likelihood estimators* (NPMLE) of such functionals under constraints are often intractable and smoothing provides alternative estimators having good interpretability and robustness properties. We may refer to Chapter 7 of Silvapulle and Sen (2005) for a general account of related developments.

Smoothing procedures are also quite popular in studies of *nonparametric regression* problems. In parametric setups, the dependent variable Y is related to a set of independent or explanatory variables $\mathbf{x}$ in an assumed regression (linear, generalized linear, nonlinear) model of known functional form but involving unknown parameters (algebraic constants) and there is some superimposed chance or error variables that depicts the regression relationship subject to chance fluctuation. Thus, two sets of regularity assumptions are made in this context: (i) a specified functional form of the regression and (ii) some assumptions regarding the error distributions. The advantage of this formulation is a finite (and smaller) dimensional parameter space pertaining to the environment $n \gg p$ and often allowing exact statistical inference on the parameters. The prime disadvantage is the lack of robustness against possible departures from the assumed model, even to a smaller extent. In a nonparametric regression model, the nature of the regression function is allowed to be quite flexible, along with much less stringent regularity assumptions on the associated error distributions. For example, one may set

$$m(\mathbf{x}) = E\{Y|\mathbf{x}\} = \int ydF(y|\mathbf{x}), \ \mathbf{x} \in R^k, \qquad (2.1)$$

treating $m(.)$ as an arbitrary function subject to some local smoothness conditions. Sometimes monotonicity or other restraints are imposed. In this setup, the usual linear or parametric regression function is replaced by a regression curve or surface of arbitrary form, and as a result the parameter space may no longer be of finite dimension., thus inviting the *curse of dimensionality* problem.

The crux of the problem is to estimate the distribution function $F(y|\mathbf{x})$ nonparametrically. Viewed from this perspective, the methodology is analogous to the density estimation procedures, and a vast literature has cropped up during the past 3 decades. In this context, the regressors $\mathbf{x}$ may also be taken to be stochastic as well as subject to measurement errors. In this full generality, although the regression is quite flexible, it has slower rate of convergence, and generally, needs much larger sample size to have practical adaption in real life problems. As an intermediate and somewhat compromising formulation, semiparametric models and other variants have also been developed in the past two decades. For example, instead of the expectation in the definition of $m(.)$, one may take the median or some quantile of the distribution, resulting in the so called *median (quantile) regression*. Also, some semiparametric structures may be imposed on the distribution function $F(y|\mathbf{x})$ so as to obtain better statistical resolutions when such semiparametric modeling is justified. There is, however, some loss of robustness due to this semiparametric formulation.

2.5. Curse of Dimensionality and Nonparametrics

The functionals, treated in the preceding section, are the precursors of the so called high-dimensional low sample size (HDLSS) perspectives in statistical modeling and analysis. For statistical functional, the parameter space is an appropriate Euclidean one whereas in HDLSS, p, the dimension of response (observed) variables is generally exceedingly large and n, the number of observations is disproportionately small. This is therefore classified under the $p \gg n$ environment. In a sense, clinical trials, interim analysis and repeated significance tests are the precursors of HDLSS problems, albeit, there is a basic difference in the two setups. In clinical trials and other survival analysis setups, although independent and homogeneous increments of associated stochastic processes may not be taken for granted, there is a finite number of tests on accumulating datasets where martingale theory has been extensively adapted in methodological formulations. To the contrary, typically, in bioinformatics and genomics, such HDLSS problems are complex and require highly nonstandard statistical resolutions, some of which are yet not transparent.

As illustrations, we may consider several examples from clinical as well as genomic studies. First, a clinical study relating to the role of bacteria in lactational mastitis with emphasis on the use of antibiotic treatment (Kvist, L. J. et al., 2008). Breastfeeding is a significant factor for health outcomes of the child as well as mother. Breastfeeding associated pain, generally requiring a prolong use of anti-fungals (antibiotic), is a major cause for curtailed breastfeeding. The use of antibiotic may introduce side-effects and some bacterial change may create health hazards for both the mother and child. Therefore, it may be desirable to compare the human milk microbiome of healthy and pain-inflicted women. This clinical study (Stuebe et al., 2009) required extensive specimen of various types of bacteria in breastfeeding mothers milk along with possible stool specimen for both the mother and the infant. In this way, there are some (m) major bacteria conceived to have impactful interaction with mastitis, and microbiological studies of their prevalence in two groups of breastfeeding mothers (control and case) during the first three months after childbirth were planned, wherein a number of culture (k) are destined to detect the presence of each bacterium beyond a threshold level. Thus, the binary outcome of yes or no, entails some loss of information as otherwise could have been extracted if the bacterial count could be recorded in a quantitative scale. Also, other demographic informations are to be collected at the baseline. Typically, with m bacteria, k cultures and n_0 subjects in the control group and n_1 in the case group, we have a multivariable and multifactor binary response model (Roy, 1957). However, the study is expensive. Thus, typically, n_0, n_1 are small, while m and k are generally not. For each combination of bacteria and case-control groups, we have k cultures to assess the relative picture. For such an $m \times k \times 2$ classification, each of the observations has a binary response, i.e., 0 or 1 depicting the presence of the bacterium below or above a threshold level. There are other explanatory variables that might shed light on the bacteria-mastitis interaction. In a conventional modeling, there are some $p = 2^{2mk} - 1$ parameters that account for the main-effects as well as interactions of all orders. In this setup, typically, p is far larger than $n_0 + n_1$ so that we encounter an usual HDLSS model in a categorical data model setup. For example, out of a large pool of bacteria, one may choose about 10 of them which can be assessed with microbiological cultures, and for various perspectives, some 5 or more of them are to be incorporated. Thus, p could be as large as 10^{100} which completely outnumbers the number of subjects, usually in the 50s at best. This makes it inappropriate to use standard discrete multivariate analysis. It may be tempting to import fancy semiparametric models. However, there remains a grave concern about their general interpretability in such HDLSS setups. Some simple nonparametric summaritative measures seems to have a greater appeal from statistical motivation and interpretation, and they are better understood by the clinical researchers too. For each combination of culture (c) and bacterium (b) we have a binary response model relating to two independent groups ($g = 1, 2$). This leads to a set of $m \times k \times 2$ odd ratios and we are to compare the mk odd ratios of the control group with the corresponding ones in the case group. For each of the mk buckets, we have the common n_0 subjects in the control group and n_1 in the case, so that these mk statistics are not independent. As such, one needs to draw statistical conclusions bearing in mind the dependence of the multitude of test statistics. This is exactly where HDLSS models are coming into the picture in a natural way.

Consider a second example, a microarray data model where there are a few (n) arrays, and within each array there are p, some thousands of genes. Thus, here clearly $p \gg n$, and due to excessive cost of data collection, typically, the number of arrays is small. Also, the arrays may correspond to differential levels of biological or environmental stressors. For each gene, the *gene expression* level, measured by color intensity or luminosity, is a non-negative variable which has typically a highly skewed distribution. Even making the usual Box-Cox transformation, the distribution may not conform closely to a normal one. However, if there is gene-environment interaction, then the distributions of the gene-expressions over the arrays should have some change - not necessarily in the location or scale factor. Particularly, if the gene expressions are measured on the unit interval [0, 1], such changes could be in a right or left tiltedness instead of a shift in location or scale alone. This invalidates usual parametric procedures for statistical modeling and analysis. In the developments in microarray studies, a majority of researchers either take it for granted that the gene-expressions are independent and normally distributed, or they appeal to some Bayesian setups which justify normality in some way (Efron, 2004). Even if a multivariate normality assumption based multivariate analysis of variance (MANOVA) procedure is used, the basic environment of $p \gg n$ implies that usual sum of product matrices are highly singular. This, in turn, creates impasses for direct adaption of likelihood ratio tests or their variants (Sen, 2006). Basically, one needs to have couple of safeguards: The tests should be invariant under any strictly monotone transformation on the responses, and they are adaptable in the $p \gg n$ environment. This automatically puts nonparametrics in a more favorable stand. In order to assess the gene-environment interaction, it is therefore desirable to make use of suitable nonparametric statistics which may not require basic parametric models nor that the responses are precisely measurable. From this consideration, Kendall's tau statistic has a natural appeal, and this has been exploited in Sen (2008). Basically, here also, the tau statistics for different genes against environmental (biological) stressors may not be independent. Again a HDLSS model setup has been exploited fully through the use of the Chen-Stein Theorem (Chen, 1975) along with an extension to a Poisson process (Sen, 2008). A basic consideration in this setup is the characterization that at a very high level of crossing the marginal p-values are pairwise independent. This basic result (motivated by some results of Sibuya (1959) on bivariate extreme statistics) allows the adaptation of the Chen-Stein Poisson approximation to a slightly more general Poisson process one, and that in turn, seems to have a better control of *false discovery rate* (FDR).

Consider a third illustrative example of high-dimensional categorical data model. The 2002-2003 SARS epidemic model originated in Southern China and it had an identified single-stranded and positive-sense RNA coronavirus, possibly from different demographic strata (sectors/countries). The scientific focus was the statistical comparison of sets of genomic (SARSCoV) sequences over the strata. The acquired data pertained to 4 countries and there were 192 genes; the sample sizes being 6, 3, 4 and 12. All these are quite small compared to the number of genes. Conventional categorical MANOVA procedures were rule out inapplicable as the total number of parameters (4×4^{192}) were disproportionately large compared to the individual (and even the combined) sample size. Moreover, because of the origin of SARS epidemic in China, it was conceived that the impact might be more in Hong Kong and Beijing areas than in Taipei or Thailand. This naturally led to the need for constrained statistical inference in such a HDLSS setup. The study was motivated by the use of Gini-Simpson index based Hamming distances that summarize the variation in terms of suitable concentration measures and enables the use of some simple statistical procedures. The null hypothesis of homogeneity of the 4 groups generates an invariance structure which in turn leads to suitable permutation procedures. Then such permutation procedures have been adapted to constrained statistical inference setups (Sen et al., 2007).

These illustrations, though simple, provides us with good incentives with the statistical tasks needed to be accomplished before we can claim a berth on this HDLSS express. In the concluding section, guided by all these findings, it is intended to provide an overall assessment of the statistical challenges that are encountered in interdisciplinary research, and some suggestions for possible resolutions with due emphasis on nonparametric approaches.

2.6. Concluding Remarks and Observations

The conventional setup of i.i.d. observations, convenient from mathematical analysis point of view, as well as conjugate priors, commonly used in Bayesian approach, may not appear to be realistic in most of the interdisciplinary research setups. The design of a study with a scientific objective may have overwhelming undercurrents due to extraneous factors which may invariably lead to complex sampling schemes, and in some cases. even less planned observational studies. Lacking this simplicity of the so called equal probability sampling schemes, conventional parametric solutions are more likely to be vulnerable than other alternative approaches. Also, optimality considerations, restrained to simpler designs, may be out of reach in the majority of such interdisciplinary research setups. In genomics studies, for example, in microarray problems, the gene expressions are mostly assumed to be normally distributed. As has been noted in the previous section, typically, the gene expressions have distributions defined over the unit interval $[0, 1]$, and mostly with remarkable skewness. Although a Box-Cox type of transformation may render more affinity to symmetric patterns, assuming such transformed distributions to be normal may be too stringent from real applications perspective. Secondly, in such $K \gg n$ environments, if n is moderately large then marginally for each gene an asymptotic normality may be reasonable. However, as K is very large, simultaneously assuming joint-normality for all these gene expressions is not only unrealistic but also it suffers from the following drawback. Typically, the normal approximation may be tenable in the central part but the tails may not come under the validity of such approximations. Note that for large (positive) x, for a standard normal distribution function $\Phi(x)$, the Mills ratio states that

$$1 - \Phi(x) \sim x^{-1}\phi(x)$$

so that for small α, one has

$$-\log(1 - \alpha) \sim (1/2)\{\log(2\pi) + \{\Phi^{-1}(1 - \alpha)\}^2\} + \log \Phi^{-1}(1 - \alpha).$$

Or, in other words, for small α,

$$2\alpha \sim \log(2\pi) + \{\Phi^{-1}(1 - \alpha)\}^2.$$

For a two-sided test, for normal distribution, we have similarly,

$$\alpha \sim \log(2\pi) + \{\Phi^{-1}(1 - \alpha/2)\}^2$$

However, if the actual distribution has heavier tails (as may typically be the case), the above normal approximation will give a much smaller p-value compared to the actual p-value. Likewise, for lighter tail distributions, it could be in the opposite way.

If we look into the Benjamini-Hochberg (1995) type of FDR controlling procedures, we may notice that it places a lot of weights on the behavior of smaller ordered p-values for the K marginal tests, often assuming further that these marginal tests are independent. As such, if normal approximation is used for assessing marginal p-values, but they correspond to a more heavier tail distribution, these smaller p-values based on normal approximation could be quite smaller than the actual p-values, so that the conclusions may be quite inappropriate. For lighter tail distributions it could be the other way. This unpleasant feature of normal approximation based FDR conclusions deserves most critical appraisal. If the marginal test statistics are not independent, the joint distribution of these p-values based on marginal normal approximation may even be more vulnerable to tail behaviors of multivariate distributions.

It is no wonder that in bioinformatics there has been a lot of emphasis on the use of resampling plan based statistical learning or data mining tools. If the null hypotheses generate some invariance structure, permutation tools are indeed valuable in this respect. However, with large K and small n, the properties of such resampling plan based FDR procedures need to be assessed more thoroughly. The situation is comparatively better for nonparametric procedures whenever such invariance structure holds under the null hypotheses. In some of these cases, the permutation distributions for the marginal test statistics agree with their unconditional counterparts, and hence, the vulnerability of the normal approximation can be eliminated to a greater extent, albeit, there could be more demanding computational complexities. Moreover, in general, in a HDLSS setup, such permutation distributions may not agree with the null distributions, so that the criticism labeled against normal approximations may also reside in such resampling based inference procedures (Sen, 2008). Actually, in many of these problems, not only there is a huge number of tests in a multiple hypothesis testing (MHT) context crop up but also, in many cases, structural complexities lead to so called constrained alternatives. It may be therefore necessary to use these test statistics in such a way that these restricted alternatives are properly addressed in the formulation. Otherwise, there could be substantial loss of power. In this respect, the classical Roy (1953) union-intersection principle (UIP) can have significant advantages over other conventional tests. In a nonparametric setup, some of these restraints can be handled in a broader way, and the UIP has been extensively applied to formulate appropriate tests in MHT setups - we may refer to Sen (2007, 2009b).

Data manipulation prior to any quantitative analysis is very common in bioinformatics. Such manipulations not only distort simple structures underlying recorded data sets but also makes it more difficult to apply statistical validation tools. This is the main disputed territory bordering statistical methodology and statistical learning. Variable selection procedures, as commonly used, for convenience of analysis, often take shelter under the Bayesian setups and appeal to *Monte Carlo Markov Chain* (MCMC) through which various algorithms are incorporated for quantitative answers to complex qualitative phenomena. There has been a lot of development in this field of research. It is my hope that gradually and more clearly this stream of development will eliminate the basic shortcomings of Bayesian data mining in HDLSS models. Of course, there is a basic qualm on the role of statistical inference in bioinformatics. Researchers in molecular biology and genomics may have some preconceived hypotheses and may be more eager in using some quantitative analysis tools to justify their ideas. Statistical science is undoubtedly invaluable in this assessment task. Nevertheless, it needs a lot of annexation of data analysis in a statistically validated setup. In many situations, like image analysis or brain tomography, the patterns can be successfully guessed by efficient computer algorithms, and statistical science can add more confidence to such educated guess. However, clinical use of such images (on human subjects) needs much more care on an individual basis instead of a population basis. That aspect by itself shifts the emphasis on subject-wise decisions instead of conventional statistical inference. There is a definite need to have more annexation of statistical reasoning in many clinical decisions, and in this respect too, nonparametrics may fare better than conventional parametrics. The most imminent area is *computational biology* and statistics is indispensable in this quest, though we may wonder *"Whither biostochastics in computational biology and bioinformatics?"* (Sen, 2009a). The bottom line is that statistical researchers working in this interdisciplinary field must have a basic understanding of the biological undercurrents, and likewise, the (molecular) biological researchers should be well aware of the stochastic evolutionary forces underlying such complex biological setups so that they can through effective collaboration extract the stochastics from the mass of biological diversity and interpret their basic findings in a much more meaningful way. I can foresee a clearcut case for *beyond parametrics* in statistical reasoning which may be more adaptable in this complex setup.

References

1. Adichie, J. N. (1967). Estimates of regression parameters based on rank tests. *Ann. Math. Statist.* **38**, 894–904.
2. Andersen, P. K., Borgan, O., Gill, R. and Keiding, N. (1993). *Statistical Methods Based on Counting Processes*. Springer, New York.
3. Benjamini, Y. and Hochberg, Y. (1995). Controlling the false discovery rate: a practical and powerful approach to multiple testing. *J. Roy. Statist. Soc.* **B 57**, 289–300.
4. Chatterjee, S. K. and Sen, P. K. (1964). Nonparametric tests for the bivariate two sample location problem. *Calcutta Statist. Assoc. Bull.* **13**, 18–58.
5. Chatterjee, S. K. and Sen, P. K. (1973). Nonparametric testing under progressive censoring. *Calcutta Statist. Assoc. Bull.* **22**, 13–50.
6. Chen, L. H. Y. (1975). Poisson approximations for dependent trials. *Ann. Probab.* **2**, 534–545.
7. Chernoff, H. and Savage, I. R. (1958). Asymptotic normality and efficiency of certain nonparametric test statistics. *Ann. Math. Statist.* **29**, 972–994.
8. Cox, D. R. (1972). Regression models and life tables (with discussion). *J. Roy. Statist. Soc.* **B 34**, 187–220.
9. Efron, B. (1979). Bootstrap methods: another look at the jackknife. *Ann. Statist.* **7**, 1–26.
10. Efron, B. (2004). Large-scale simultaneous hypothesis testing: The choice of a null hypothesis. *J. Amer. Statist. Assoc.* **99**, 96–104.

11. Ferguson, T. (1973). A Bayesian analysis of some nonparametric problems. *Ann. Statist.* **1**, 209–230.
12. Ghosh, J. K. and Ramamoorthi, R. V. (2003). *Bayesian Nonparametrics.* Springer, New York.
13. Hájek, J. (1961). Some extensions of the Wald-Wolfowitz-Noether theorem. *Ann. Math. Statist.* **32**, 506–523.
14. Hájek, J. (1962). Asymptotically most powerful rank order tests. *Ann. Math. Statist.* **33**, 1124–1147.
15. Hájek, J. (1968). Asymptotic normality of simple linear rank statistics under alternatives. *Ann. Math. Statist.* **39**, 325–346.
16. Hájek, J., Sidak, Z. and Sen, P. K. (1999). *Theory of Rank Tests*, 2nd Ed. Academic Press, London.
17. Hodges, J. L. and Lehmann, E. L. (1963). Estimates of location based on rank tests. *Ann. Math. Statist.* **34**, 598–611.
18. Hoeffding, W. (1948). A class of statistics with asmptotically normal distribution. *Ann. Math. Statist.* **19**, 293–325.
19. Huber, P. J. (1964). Robust estimation of a location parameter. *Ann. Math. Statist.* **35**, 73–101.
20. Huber, P. J. and Ronchetti, E. M. (2009). *Robust Statistics*, 2nd Ed. John Wiley, New York.
21. Jurečková, J. (1969). Asymptotic linearity of a rank statistic in regression parameter. *Ann. Math. Statist.* **40**, 1889–1900.
22. Jurečková, J. and Sen, P. K. (1996). *Robust Statistical Procedures: Asymptotics and Interrelations.* John Wiley, New York.
23. Kaplan, E. L., Meier, P. (1958). Nonparametric estimation from incomplete observations. *J. Amer. Statist. Assoc.* **53**, 457–481, 562–563.
24. Koul, H. L. (1969). Asymptotic behavior of Wilcoxon type confidence regions in multiple linear regression. *Ann. Math. Statist.* **40**, 1950–1979.
25. Kvist, L. J., Larsson, B. W., Hall-Lord, M. L., Steen, A. and Schalén, C. (2008). The role of bacteria in lactational mastitis and some considerations of the use of antibiotic treatment. *Intern. Breastfeed. Jour.* **3**, 6.
26. LeCam, L. (1960). Locally asymptotically normal families of distributions. *Univ. of Calif. Publ. in Statist.* **3**, 37–98.
27. Mosteller, F. (1948). A k-sample slippage test for an extreme population. *Ann. Math. Statist.* **19**, 58–65.
28. Murphy, S. A. and Sen, P. K. (1991). Cox regression model with time-dependent covariates. *Stochastic Process. Appl.* **39**, 153–180.
29. Nadaraya, E. A. (1964). On estimating regression. *Theor. Probab. Applicat.* **9**, 141–142.
30. Pitman, E. J. G. (1948). Lecture Notes, Columbia Univ.
31. Puri, M. L. and Sen, P. K. (1971). *Nonparametric Methods in Multivariate Analysis.* John Wiley, New York.
32. Quenouillie, M. H. (1949). Approximate tests of correlation in time series. *J. Roy. Statist. Soc.* **B 11**, 68–84.
33. Rosenblatt, M. (1956). Remarks on some nonparametric estimates of a density function. *Ann. Math. Statist.* **27**, 832–837.

34. Roy, S. N. (1953). A heuristic method of test construction and its use in multivariate analysis. *Ann. Math. Statist.* **24**, 220–238.

35. Roy, S. N. (1957). *Some Aspects of Multivariate Analysis.* John Wiley, New York and Asia Publ., Bombay.

36. Savage, I. R. (1962). *Bibliography on Nonparametric Statistics.* Harvard University Press, Cambridge, MA, USA.

37. Sen, P. K. (1963). On the estimation of relative potency in dilution (direct) assays by distribution-free methods. *Biometrics* **19**, 532–552.

38. Sen, P. K. (1968). Estimates of regression coefficients based on Kendall's tau. *J. Amer. Statist. Assoc.* **63**, 1379–1389.

39. Sen, P. K. (1981). *Sequential Nonparametrics: Invariance Principles and Statistical Inference.* John Wiley, New York.

40. Sen, P. K. (2006). Robust statistical inference for high-dimensional data models with applications to genomics. *Austr. J. Statist.* **35**, 197–214.

41. Sen, P. K. (2007). Union-intersection principle and constrained statistical inference. *J. Statist. Plan. Infer.* **137**, 3941–3952.

42. Sen, P. K. (2008). Kendall's tau in high-dimensional genomic parsimony. *IMS Coll.* **3**, 251–266.

43. Sen, P. K. (2009a). Whither biostochastics in computational biology and bioinformatics. *Fron. Appl. & Comp. Math.* pp.15–31.

44. Sen, P. K. (2009b). High-dimensional discrete statistical models: UIP, MCP and CSI perspectives. *Adv. in Mult. Stat. Meth.* 1–24.

45. Sen, P. K. and Ghosh, M. (1971). On bounded length sequential confidence intervals based on one sample rank order statistics. *Ann. Math. Statist.* **42**, 189–202.

46. Sen, P. K. and Saleh, E. (1985). On some shrinkage estimators of multivariate location. *Ann. Statist.* **13**, 272–281.

47. Sen, P. K. and Saleh, E. (2010). The Theil-Sen estimator in a measurement error perspective. In *Nonparametrics and Robustness in Modern Statistical Inference and Time Series Analysis: A Festschrift in honor of Professor Jana Jurecková* (J. Antoch, M. Hušková and P. K. Sen, Eds.), *Inst. Math. Statist. Collection Ser.* **7**, 224–234.

48. Sen, P. K., Tsai, M. and Jou, Y.-S. (2007). High-dimensional low sample size perspectives in constrained statistical inference: The SARSCORV genome in illustration. *J. Amer. Statist. Assoc.* **102**, 685–694.

49. Sibuya, M. (1959). Bivariate extreme statistics. *Ann. Inst. Stat. Math.* **11**, 195–210.

50. Silvapulle, M. J. and Sen, P. K. (2005). *Constrained Statistical Inference: Inequality, Order and Shape Restrictions.* John Wiley, New York.

51. Stein, C. (1955). Inadmissibility of the usual estimator for the mean of a multivariate normal distribution. *Proc. 3rd Berkeley Symp. Math. Statist. Probab.* **1**, 197–206.

52. Stuebe, A. M., Michels, K. B., Willett, W. C., Manson, J. E., Rexrode, K. and Rich-Edwards, J. W. (2009). Duration of lactation and incedence of myocardial infraction in middle-to-late adulthood. *Amer. J. Obstet. Gynecol.* **200**, 138.

53. Tsybakov, A. B. (2008). *Introduction to Nonparametric Estimation.* Springer, New York.
54. Wald, A. and Wolfowitz, J. (1944). Statistical tests based on the permutation of the observations. *Ann. Math. Statist.* **15**, 358–372.
55. Watson, G. (1964). Smooth regression analysis. *Sankhyā* **A 26**, 359–372.

PART 2
PARAMETRIC

Chapter 3

Bounds on Distributions Involving Partial Marginal and Conditional Information: The Consequences of Incomplete Prior Specification

Barry C. Arnold

University of California, Riverside, USA

Elicitation of prior distributions will necessarily be imprecise and frequently will yield incompatible probability assessments. Identification of the class of distributions compatible with the given information can be used to generate a spectrum of possible Bayes estimates from which to choose. In the absence of compatibility, a Bayes estimate based on a minimally incompatible prior is recommended.

Keywords: Bayesian inference, Compatible probabilities, Cross-product ratios, Kullback-Leibler distance, Multiple assessments.

Contents

3.1. Introduction

Suppose that we are in a classical Bayesian inference setting. That is, we have available a realized value $\underline{x}$ of a random variable $\underline{X}$ whose distribution is described by the model

$$f_{\underline{X}}(\underline{x}) \in \{f(\underline{x};\underline{\theta} : \underline{\theta} \in \Theta\}.$$

On the basis of the observation $\underline{X} = \underline{x}$, we wish to estimate $\eta = g(\underline{\theta})$ where $g : \Theta \to \mathbb{R}$ is a known function.

Assuming squared error loss, the standard Bayesian approach to this problem is to elicit or postulate a prior density for $\underline{\theta}$, say $\pi(\underline{\theta})$, and to use as an estimate of η, the quantity

$$E(g(\underline{\theta})|\underline{X} = \underline{x}).$$

Turning the Bayesian crank in this fashion may be complicated by the fact that the conditional density of $\underline{\theta}$ given $\underline{X} = \underline{x}$ might be analytically intractable. However, since one-dimensional random variables are typically easier to simulate than multivariate variables, Gibbs sampler approaches will often allow us to approximate simulations from the density of $\underline{\theta}$ given $\underline{X} = \underline{x}$. But in fact, focusing on difficulties associated with posterior simulation overlooks a potentially more troublesome aspect of this inference paradigm. The question "Where does the prior come from?" is often raised by those encountering Bayesian inference for the first time. And it is a fair question. Recently, there has been a strong focus on the use of what are called objective prior distributions. These are not infrequently designed to lead to inferences that are not greatly at variance with corresponding frequentist results. The classical Bayesian view of prior distributions (as espoused by de Finetti and Savage, for example) is that they represent approximations to distributions determined by the subjective beliefs of the experimenter (or informed expert) based on all available information and experience. There are two commonly advocated approaches to the problem of selecting a prior density to reflect subjective beliefs. The first approach consists of identifying a flexible parameterized family of possible prior densities for $\underline{\theta}$ and then choosing one density from this family to match elicited features of the experimenter's inchoate subjective distribution. For a toy example, we can return to Bayes' original problem involving tossing a coin with unknown probability p of coming up Heads on a particular toss. A frequently proposed flexible family of priors for p in this setting is the family of Beta densities (though you could quite reasonably question whether such a family displays adequate flexibility to approximate all possible plausible prior belief structures). We could then ask the experimenter to provide us with two features of his subjective distribution, chosen from a list perhaps including the mean, the mode, the median, a quantile, a standard deviation, etc. We would then select a Beta density chosen to match these two distributional characteristics of the subjective distribution.

The second possible approach involves use of an inquisition related to a series of hypothetical bets about the true value of the parameter. Thus we might ask, is it more likely that p is less than $1/2$ than that p is greater than $1/2$. And then continue with a series of questions refining our knowledge of the underlying subjective beliefs. There are several obvious difficulties with this approach. One problem is that in order to get a good grasp of the nature of the prior distribution, an inordinately large number of questions must be asked. But a potentially more important difficulty is associated with the fact that human beings are not necessarily (perhaps almost never) completely coherent in their probability assessments. Thus there may well exist no distribution which is consistent with the list of bet selections provided by the experimenter (one is tempted to say, the victim of this inquisition). The present discussion then will admit that this kind of inconsistency is likely and will seek minimally incompatible prior distributions, that is, distributions which differ, in some precise manner of measurement, the least from matching the available prior information provided by the experimenter.

The information provided is thus expected to be in the form of a finite list of equalities and inequalities regarding the probabilities that $\underline{\theta}$ belongs to various subsets of Θ. Thus for subsets $A_1, A_2, \ldots, A_m, B_1, B_2, \ldots, B_m, C_1, C_2, \ldots, C_p$ and numbers $\delta_1, \delta_2, \ldots, \delta_p$, we are given that

$$P(\underline{\theta} \in A_i) \leq P(\underline{\theta} \in B_i), \quad i = 1, 2, \ldots, m, \tag{3.1}$$

and

$$P(\underline{\theta} \in C_i) = \delta_i, \quad i = 1, 2, \ldots, p. \tag{3.2}$$

Our goal then will be to select a distribution for $\underline{\theta}$ that is minimally inconsistent with these probability statements or, in the case of compatibility, to identify all possible compatible distributions.

If there is a well defined minimally incompatible distribution, say $\pi^*(\underline{\theta})$, we will then report (using squared error as a loss function) our Bayes estimate of $g(\underline{\theta})$ to be

$$E_{\pi^*}(g(\underline{\theta})|\underline{X} = \underline{x}), \tag{3.3}$$

where the subscript on E indicates the particular prior, π^*, involved in the computation.

If there is one or, usually, many prior densities that are compatible with the given prior information (3.1) and (3.2), we denote the class of all such priors by $\mathcal{F}^*$ and in such a case we can only reasonably report that our estimate of $g(\underline{\theta})$, say $\tilde{g}(\underline{\theta})$ satisfies

$$\inf\{E_\pi(g(\underline{\theta})|\underline{X} = \underline{x}) : \pi \in \mathcal{F}^*\} \le \tilde{g}(\underline{\theta})$$

$$\le \sup\{E_\pi(g(\underline{\theta})|\underline{X} = \underline{x}) : \pi \in \mathcal{F}^*\}. \tag{3.4}$$

Hopefully the given interval will be relatively narrow, but if it is not, it merely indicates that the available prior information in the form (3.1)-(3.2) is inadequate to the task of estimating $g(\underline{\theta})$ and further elicitation (augmentation of information in the form (3.1)-(3.2)) will be called for.

It is often plausible to also seek conditional prior information as well as expectation and conditional expectation information. Thus we might elicit values or bounds for prior quantities such as:

$$P(\underline{\theta} \in C_i|\underline{\theta} \in D_i), \tag{3.5}$$

$$E(h_i(\underline{\theta})) \tag{3.6}$$

$$E(k_i(\underline{\theta})|\underline{\theta} \in G_i) \tag{3.7}$$

etc. We would then seek compatible or nearly compatible distributions given (3.1)-(3.2) together with values or bounds for quantities of the form (3.5)-(3.7).

With this motivation at hand, we begin a discussion of bounds on distributions involving partial marginal and/or conditional information.

3.2. The Finite Discrete Case in k-Dimensions

We begin by discussing results appropriate for the situation in which we have a k-dimensional random quantity which has a finite number of possible values. Rather than use $\underline{\theta}$ to denote the random vector, we will use $\underline{X}$, a more customary notation. Of course the results will apply with only a notational change to the case in which $\underline{\theta}$ is a k-dimensional unknown parameter with a subjective distribution that is to be determined by partial distributional information. It is also relevant to mention that the discrete finite case is not just a toy example. More general distributions can often be adequately approximated by a finite discrete distribution and in many cases, where analytic results are elusive, such an approximation will be the avenue of choice. With that apology, or lack thereof, we turn to the finite discrete case.

Suppose that a k-dimensional random vector $\underline{X} = (X_1, X_2, \ldots, X_k)$ has support $\underline{\chi} = \chi_1 \times \chi_2 \times \cdots \times \chi_k$ where each χ_i has finite cardinality, say n_i. The set of possible values for $\underline{X}$ will then have cardinality $n = \prod_1^k n_i$.

First we will consider the case in which the information is precise, later we will consider imprecise information, recognizing that in practice imprecise information is more likely to be encountered than precise information.

Suppose that the information provided can be summarized in the form:

(i) $P(\underline{X} \in A_i) = \delta_i, \quad i = 1, 2, \ldots, m_1,$ for specific sets $A_1, A_2, \ldots,$ $A_{m_1} \subset \underline{\chi},$

(ii) $P(\underline{X} \in B_i | \underline{X} \in C_i) = \eta_i, \quad i = 1, 2, \ldots, m_2,$ for specific sets $B_1, B_2, \ldots, B_{m_2}$ and $C_1, C_2, \ldots, C_{m_2},$

(iii) $E(\psi_i(\underline{X})) = \xi_i, \quad i = 1, 2, \ldots, m_3$ for specific functions $\psi_1, \psi_2, \ldots, \psi_{m_3},$

(iv) $E(\phi_i(\underline{X}) | \zeta_i(\underline{X}) = \lambda_i) = \omega_i,$ for specific functions ϕ_i, ζ_i and constants $\lambda_i, \omega_i, \quad i = 1, 2, \ldots, m_4.$

These conditions can all be written as linear equality constraints on the quantities

$$p(\underline{x}) = P(\underline{X} = \underline{x}), \quad \underline{x} \in \underline{\chi}.$$

Thus the conditions (i)-(iv) can be written as:

(i′) $\sum_{\underline{x} \in A_i} p(\underline{x}) = \delta_i, \quad i = 1, 2, \ldots, m_1,$

(ii′) $\sum_{\underline{x} \in B_i \cap C_i} p(\underline{x}) - \eta_i \sum_{\underline{x} \in C_i} p(\underline{x}) = 0, \quad i = 1, 2, \ldots, m_2,$

(iii′) $\sum_{\underline{x}} \psi_i(\underline{x}) p(\underline{x}) = \xi_i, \quad i = 1, 2, \ldots, m_3,$

(iv′) $\sum_{\zeta_i(\underline{x}) = \lambda_i} \phi_i(\underline{x}) p(\underline{x}) - \omega_i \sum_{\zeta_i(\underline{x}) = \lambda_i} p(\underline{x}) = 0, \quad i = 1, 2, \ldots, m_4.$

If we then stack the $p(\underline{x})$'s to form a column vector $\underline{p}$ of dimension $n = \prod_1^k n_i$, then all the information in (i)-(iv) or equivalently (i′)-(iv′) can be summarized succinctly in the form

$$A\underline{p} = \underline{b}, \tag{3.8}$$

where A is a known matrix of dimension $m \times n$ where $m = m_1 + m_2 + \cdots + m_5$ and $\underline{b}$ is a known vector of dimension m. We will then seek to identify a vector $\underline{p}$ with non-negative coordinates that sum to 1 and which satisfies (3.8).

If m is large, it is likely that there will be no solution to the system of linear equations (3.8), in which case we would search for a minimally incompatible solution. Thus we seek a vector $\underline{p}$ satisfying:

$$A\underline{p} \simeq \underline{b},$$

$$\underline{p} \geq \underline{0},$$

and

$$\underline{p} \cdot \underline{1} = 1.$$

The last equality could have been incorporated in the equation $A\underline{p} = \underline{b}$, but it is probably reasonable to insist on it being satisfied exactly and not approximately. A reasonable approach to solving this problem is to select a distance measure between vectors of dimension m, say $d(\underline{a}, \underline{b})$ and seek to find $\underline{p}$ to minimize

$$d(A\underline{p}, \underline{b})$$

subject to the constraints $\underline{p} \geq \underline{0}$ and $\underline{p} \cdot \underline{1} = 1$. A simple and attractive choice for the distance function is one of the form

$$d(A\underline{p}, \underline{b}) = \sum_{j=1}^{m} (A_{(j)}\underline{p} - b_j)^2, \tag{3.9}$$

but some other distance function might be more appropriate in some circumstances.

However it is actually more likely that imprecise rather than precise information will be available. Thus one could expect to have information supplied in the form

(a) $P(\underline{X} \in A_i) \leq P(\underline{X} \in B_i), \quad i = 1, 2, \ldots, m_1,$

(b) $\eta_{1i} \leq P(\underline{X} \in C_i | \underline{X} \in D_i) \leq \eta_{2i},$

etc.

Again, all these inequality constraints can be rewritten as linear inequality constraints on $\underline{p}$. Taking a discrepancy function of the form (3.9), we will seek to minimize it subject to a large number of linear inequality constraints, a standard quadratic programming problem.

In principle, both problems (precise and imprecise) are thus solvable, though in practice the size of m and especially the size of n may make the problem unwieldy. Further details on this general approach may be found in Arnold, Castillo and Sarabia (1999, 2002).

We now turn to consider some special cases where alternative approaches are feasible.

3.3. Conditional Specification in the Case $k = 2$

In the case in which $k = 2$, considerable attention has been paid to the situation in which only conditional information is available. Thus it is assumed that we are given information of the form

$$a_{ij} = P(X = x_i | Y = y_j), \quad i = 1, 2, \ldots, n_1, \; \jmath = 1, 2, \ldots, n_2,$$

$$b_{ij} = P(Y = y_j | X = x_i), \quad i = 1, 2, \ldots, n_1, \; \jmath = 1, 2, \ldots, n_2,$$

which can be arranged to form two $n_1 \times n_2$ matrices A and B.

If all the a_{ij}'s and b_{ij}'s are positive, then there are several ways to determine whether A and B are compatible in the sense that there will exist a bivariate distribution for (X, Y) with conditional probabilities given by A and B. A good survey may be found in Arnold, Castillo and Sarabia (1999). We mention here only two:

(i) A and B are compatible iff they have identical cross-product ratios, i.e., iff for every $i_1 < i_2$ and $j_1 < j_2$,

$$\frac{a_{i_1 j_1} a_{i_2 j_2}}{a_{i_1 j_2} a_{i_2 j_1}} = \frac{b_{i_1 j_1} b_{i_2 j_2}}{b_{i_1 j_2} b_{i_2 j_1}}.$$

(ii) A and B are compatible iff there exist vectors $\underline{c}$ and $\underline{d}$ of dimensions n_1 and n_2 respectively such that for every i, j

$$\frac{a_{ij}}{b_{ij}} = \frac{c_i}{d_j}. \tag{3.10}$$

If some of the a_{ij}'s or b_{ij}'s are zero then, for compatibility, the incidence matrices of A and B must coincide and there must exist $\underline{c}$ and $\underline{d}$ such that (3.10) holds for every i, j for which $a_{ij} b_{ij} > 0$. This can also be described in terms of what Arnold, Castillo and Sarabia (2004) call rank one completion. Thus A and B are compatible if they have identical incidence matrices and there exists a positive matrix C of rank 1 such that

$$c_{ij} = \frac{a_{ij}}{b_{ij}} \quad \text{for every } (i, j) \text{ for which } a_{ij} b_{ij} > 0.$$

3.4. Conditional Specification when $k > 2$

Consider $\underline{X}$ of dimension k with possible values $\underline{\chi} = \chi_1 \times \chi_2 \times \cdots \times \chi_k$. We introduce the notation $\underline{X}_{(i)}$ to denote the vector $\underline{X}$ with the i-th coordinate X_i deleted. An analogous definition relates $\underline{x}_{(i)}$ to $\underline{x}$ in $\underline{\chi}$. An analog to the situation described in Section 3.3 involves specification of matrices

$A^{(1)}, A^{(2)}, \ldots, A^{(k)}$ where, for each ℓ, $A^{(\ell)}$ is of dimension $n_\ell \times (n/n_\ell)$ with elements

$$a^{(\ell)}_{i, \underline{j}_{(\ell)}} = P(X_\ell = x_i | \underline{X}_{(\ell)} = \underline{x}_{(\ell)}).$$

It may be observed that in Section 3.3, in the case of compatibility, the vectors $\underline{c}$ and $\underline{d}$ were in fact proportional to the marginal densities of X and Y respectively. In the present situation, compatibility will be encountered if there exist non-negative arrays which are proportional to the marginal densities of $\underline{X}_{(1)}, \underline{X}_{(2)}, \ldots, \underline{X}_{(k)}$.

3.5. Restriction to Parametric Families

Especially in Bayesian contexts, it is often deemed desirable to seek a multivariate density that is compatible or nearly compatible with given marginal and conditional probabilistic information, but with the restriction that the joint density selected must be a member of some specific family of densities, usually chosen for analytic tractability or for analytic tractability of the associated posterior density.

The only difference caused by the restriction to parametric families is that, instead of searching for $\underline{p}$ which will be most nearly compatible with the given information, we search for $\underline{p}(\underline{\lambda})$ where $\underline{\lambda}$ denotes the parameter indexing the family of acceptable distributions. Some examples may be found in Arnold and Gokhale (1998).

3.6. The Absolutely Continuous Case, $k = 2$

There is a rich literature dealing with the characterization of bivariate densities based on partial or complete conditional information. The emphasis is usually on determining whether the information is compatible and particular interest focuses on the case in which a unique compatible distribution exists. Throughout this section, though the focus is on absolutely continuous examples, it is not difficult to envision discrete versions of the problems discussed.

3.6.1. *Complete conditional information*

Consider two candidate families of conditional densities $a(x, y)$ and $b(x, y)$. When is it true that there exists a joint density for (X, Y) such that

$$f_{X|Y}(x|y) = a(x, y), \quad x \in S(X), \ y \in S(Y)$$

and

$$f_{Y|X}(y|x) = b(x,y), \quad x \in S(X), \ y \in S(Y)$$

where $S(X)$ and $S(Y)$ denote the supports of X and Y respectively. Parallel to (3.10) we have from Arnold and Press (1989):

Theorem 3.1. *(Compatible conditionals) A joint density $f(x,y)$ with $a(x,y)$ and $b(x,y)$ as its conditional densities will exist iff*

  (i) $\{(x,y) : a(x,y) > 0\} = \{(x,y) : b(x,y) > 0\} = N$, say,
 (ii) There exist functions $u(x)$ and $v(y)$ such that for every $(x,y) \in N$ we have

$$\frac{a(x,y)}{b(x,y)} = u(x)v(y) \tag{3.11}$$

  in which $u(x)$ is integrable over $S(X)$.

In the absence of compatibility, we would seek a bivariate density $f(x,y)$ which is minimally incompatible with the given $a(x,y)$ and $b(x,y)$. In particular we might seek a member of some given parametric family of bivariate densities, say $\{g(x,y;\underline{\lambda}) : \underline{\lambda} \in \Lambda\}$ that will be minimally incompatible. Thus, for example, one might choose $\underline{\lambda}^* \in \Lambda$ to minimize $D(\underline{\lambda})$ where

$$D(\underline{\lambda}) = \int_{S(X)} \int_{S(Y)} [(g(x|y;\underline{\lambda}) - a(x,y))^2$$

$$+ (g(y|x;\underline{\lambda}) - b(x,y))^2]\psi(x,y) \ dxdy, \tag{3.12}$$

for some suitable choice of weight function $\psi(x,y)$.

Alternatively, one might compute conditional expectations of the form $E_a(X|Y = y)$, $E_b(Y|X = x)$, $E_g(X|Y = y;\underline{\lambda})$ and $E_g(Y|X = x;\underline{\lambda})$ and choose a value of $\underline{\lambda}$ to minimize the discrepancy between these regression functions, i.e., minimize

$$\widetilde{D}(\underline{\lambda}) = \int_{S(Y)} [E_a(X|Y = y) - E_g(X|Y = y;\underline{\lambda})]^2 \psi_1(y)dy$$

$$+ \int_{S(X)} [E_b(Y|X = x) - E_g(Y|X = x;\underline{\lambda})]^2 \psi_2(x)dx, \tag{3.13}$$

for suitable weight functions ψ_1 and ψ_2.

3.6.2. *Partial conditional information*

Suppose that we are only given some of the values for $a(x,y)$ and $b(x,y)$. Thus we have

$$a(x,y), \quad (x,y) \in A_1 \subset S(X) \times S(Y)$$

and

$$b(x,y), \quad (x,y) \in A_2 \subset S(X) \times S(Y).$$

Here only obvious modifications are required if we are to seek a minimally incompatible density chosen from a parametric family $\{g(x,y;\underline{\lambda}) : \underline{\lambda} \in \Lambda\}$. Thus we would in this case choose $\underline{\lambda}$ to minimize

$$D_{A_1,A_2}(\underline{\lambda}) = \int \int_{A_1} (g(x|y;\underline{\lambda}) - a(x,y))^2 \psi(x,y) \, dxdy$$

$$+ \int \int_{A_2} (g(y|x;\underline{\lambda}) - b(x,y))^2 \psi(x,y) \, dxdy. \tag{3.14}$$

If A_1 and/or A_2 are relatively sparse subsets of $S(X) \times S(Y)$, it will be easy to find a completely compatible density. For example, if $A_2 = \{(x,y) : x = x_0, y \in S(Y)\}$ then compatibility is assured.

3.6.3. *Given one conditional density and a regression function*

Suppose that we are given $a(x,y)$, a family of conditional densities for X given $Y = y$, indexed by $y \in S(Y)$ and the regression function of Y given X, say $\psi(x)$. Can we find a bivariate density for (X,Y) such that

$$f_{X|Y}(x|y) = a(x,y), \quad x \in S(X), \ y \in S(Y) \tag{3.15}$$

and

$$E(Y|X = x) = \psi(x), \quad x \in S(X). \tag{3.16}$$

Or failing this, can we find a minimally incompatible density, one which comes as close as possible to satisfying (3.15)-(3.16). Or perhaps there will be more than one compatible density. In such a case we would like to identify all possible compatible densities. Arnold, Castillo and Sarabia (1999) provide certain examples where these questions can be resolved, but in general the questions are still open.

In fact the problem is closely related to the problem of identifying mixing distributions. If (3.15)-(3.16) are to hold then there must exist a density function $g(x)$ for X such that

$$\psi(x) = \int_{S(Y)} y\, a(x,y) g(y) dy$$

where ψ and a are known and g is to be determined.

Alternatively, we might wish to identify a density in given parametric family $\{g(x,y;\underline{\lambda}) : \underline{\lambda} \in \Lambda\}$ that is minimally incompatible with (3.15)-(3.16). To achieve this, a possible objective function to be minimized by a judicious choice of $\underline{\lambda}$ might be

$$\widetilde{\widetilde{D}}(\underline{\lambda}) = \int_{S(X)} \int_{S(Y)} (g(x|y;\underline{\lambda}) - a(x,y))^2 \phi(x,y) dx dy$$

$$+ \int_{S(X)} (E_g(Y|X = x;\underline{\lambda}) - \psi(x))^2 \tau(x) dx$$

for suitable weight functions $\phi(x,y)$ and $\tau(x)$.

3.6.4. *Given a conditional density and a conditional median or mode function*

Suppose that we seek a joint density for (X,Y) with conditional densities of X given Y of the form

$$f_{X|Y}(x|y) = a(x,y) \qquad (3.17)$$

and with with conditional median function $\zeta(x)$, i.e., where

$$P(Y \leq \zeta(x)|X = x) = 1/2. \qquad (3.18)$$

As of now, this problem is unresolved. It is of course possible to seek a minimally incompatible density to be chosen from a given parametric family of densities by setting up a suitable objective function, but the precise compatibility problem is still open.

Arnold, Castillo and Sarabia (2008) consider a parallel problem, involving somewhat different conditional distributions, which can be solved. They suppose that we are given

$$\varphi(x,y) = P(X \leq x|Y \leq y)$$

for $x \in S(X)$ and $y \in S(Y)$, and for some $p \in (0,1)$, the conditional percentile function $\psi_p(x)$ defined implicitly by:

$$P(Y \leq \psi_p(x)|X \leq x) = p, \quad x \in S(X).$$

With this set-up, provided that ψ_p is invertible, they are able to identify all possible compatible bivariate distributions. An analogous version would deal with survival functions instead of distribution functions, which might be more appropriate in reliability settings.

If instead of a conditional median, we consider a conditional mode function, some progress can be made. Thus, if we are given

$$f_{X|Y}(x|y) = a(x,y), \quad x \in S(X), \ y \in S(Y)$$

and we are given the conditional mode function of Y given X:

$$\text{Mode}\{f_{Y|X}(y|x)\} = \xi(x), \quad x \in S(X).$$

Assume that the mode is the solution to the equation $\frac{\partial}{\partial y} f_{Y|X}(y|x) = 0$ or equivalently to $\frac{\partial}{\partial y} \log f_{Y|X}(y|x) = 0$. Assuming that $\xi(x)$ is invertible, Arnold, Castillo and Sarabia (2008) verify that the corresponding density of Y must be given by

$$f_Y(y) = \exp\left\{-\int_0^y g(\xi^{-1}(u), u) du\right\}.$$

Some specific examples are displayed in Arnold,Castillo and Sarabia (2008).

Remark (Another open question) What if we are given $P(X \leq x | Y \leq y)$ and the conditional mode function of Y given $X \leq x$?

3.6.5. *An alternative approach to the partial knowledge case*

Suppose that instead of being given $a(x,y)$ and $b(x,y)$ completely, we are only given

$$a(x,y), \quad x \in S(X), \ y \in I(Y)$$

and

$$b(x,y), \quad x \in I(X), \ y \in S(Y)$$

where $I(X) \subsetneq S(X)$ and $I(Y) \subsetneq S(Y)$.

Assume that $I(X) \times I(Y)$ has positive measure to avoid trivial solutions. In this case we are given only some of the conditional densities, but those that are given are given completely. Arnold and Gokhale (1998) suggest the use of a Kullback-Leibler distance function as a measure of incompatibility, i.e.,

$$\int_{S(Y)} [\int_{I(X)} b(x,y) \log[\frac{b(x,y)f_X(x)}{f_{X,Y}(x,y)}] d\lambda_1(x)] \varphi_1(y) dy$$

$$+ \int_{S(X)} [\int_{I(Y)} a(x,y) \log[\frac{a(x,y)f_Y(y)}{f_{X,Y}(x,y)}] d\lambda_2(y)] \varphi_2(x) dx$$

for suitable weight functions $\lambda_1, \lambda_2, \varphi_1$ and φ_2, and propose choosing $f_{X,Y}(x,y)$ to minimize this. Alternatively, one could use the same discrepancy function but restrict the choice to members of some prescribed parametric family of bivariate densities. Note that in the parametric setting $I(X)$ and $I(Y)$ can be of measure zero, indeed they can be finite sets. See Arnold and Gokhale (1998) for examples.

3.6.6. *Multiple assessments*

We can extend our development to consider cases in which more than one set of candidate conditionals are provided, by multiple experts. Thus we might have information from k experts in the following form. Expert j supplies

$$\{a_j(x,y) : x \in S(X), \ y \in I_j(Y)\}$$

and

$$\{b_j(x,y) : x \in I_j(X), \ y \in S(Y)\},$$

$j = 1, 2, \ldots, k$. We would like to find a joint density (in general, or from a specified parametric family) most nearly compatible with all of the experts. If our experts are judged to be of equal credibility, it is reasonable to use as an objective function, the sum of the discrepancy measures corresponding to each of the k experts. An alternative approach would consist of computing a most nearly compatible density for each expert and taking the average of these k nearly compatible densities. The two approaches will typically yield different results. Further investigation would be desirable to determine which of the two approaches should be recommended in specific cases.

3.7. Back to the Bayes-ic Problem

In this paper, a major motivation for the study of incomplete conditional specification settings was that, in prior elicitation scenarios for multidimensional parameters, conditional information is often elicited and it inevitably will be of a partial nature. We do not ever have time to completely pin down the expert's prior beliefs. In most practical situations we can however expect that sufficiently detailed elicitation will be employed to guarantee that the information will be incompatible. Unless the experimenter has in mind a completely specified analytic prior that he/she uses to analytically determine answers to the probing questions used in the process. It follows that there will be less interest in the problem of how to choose from a collection of compatible priors, and more interest in selecting a minimallly incompatible prior. Such a prior may well be selected from some convenient flexible family of priors. The objective functions used in the present paper to identify minimally incompatible priors are ad hoc. In practice, attention should be paid to how the prior is to be used and the definition of minimal incompatibility should take this into account.

Acknowledgement

The author is indebted to Enrique Castillo and Jose Maria Sarabia for numerous discussions and suggestions on material related to topics discussed in this paper.

References

1. Arnold, B. C., Castillo, E and Sarabia, J. M. (1999). *Conditional specification of statistical models.* Springer, New York.
2. Arnold, B. C., Castillo, E and Sarabia, J. M. (2002). Exact and near compatibility of discrete conditional distributions. *Comput. Statist. Data Anal.* **40**, 231–252.
3. Arnold, B. C., Castillo, E and Sarabia, J. M. (2004). Compatibility of partial or complete conditional probability specifications. *J. Statist. Plann. Inf.* **123**, 133–159.
4. Arnold, B. C., Castillo, E and Sarabia, J. M. (2008). Bivariate distributions characterized by one family of conditionals and conditional percentile or mode functions. *J. Mult. Anal.* **99**, 1383–1392.
5. Arnold, B. C. and Gokhale, D. V. (1998). Distributions most nearly compatible with given families of conditional distributions. *Test* **7**, 377–390.
6. Arnold, B. C. and Press, S. J. (1989). Compatible conditional distributions. *J. Amer. Statist. Assoc.* **84**, 152–156.

Chapter 4

Stepdown Procedures Controlling a Generalized False Discovery Rate

Wenge Guo[1] and Sanat K. Sarkar[2]

[1] *New Jersey Institute of Technology, USA*
[2] *Temple University, USA*

Often in practice when a large number of hypotheses are simultaneously tested, one is willing to allow a few false rejections, say at most $k-1$, for some fixed $k > 1$. In such a case, the ability of a procedure controlling an error rate measuring at least one false rejection can potentially be improved in terms of its ability to detect false null hypotheses by generalizing this error rate to one that measures at least k false rejections and using procedures that control it. The k-FDR which is the expected proportion of k or more false rejections and a natural generalization of the false discovery rate (FDR) is such a generalized notion of error rate that has recently been introduced and procedures controlling it have been proposed. Many of these procedures are stepup procedures. Some stepdown procedures controlling the k-FDR are presented in this article.

Keywords: Error rates, False rejections, Simultaneous tests of hypothesis.

Contents

4.1. Introduction

For simultaneous testing of null hypotheses using tests that are available for each of them, procedures have traditionally been developed exercising a control over the familywise error rate (FWER), which is the probability of rejecting at least one true null hypothesis (Hochberg and Tamhane, 1987), until it is realized that this notion of error rate is too stringent while testing a large number of hypotheses, as it happens in many modern scientific investigations, allowing little chance to detect many false null hypotheses. Therefore, researchers have focused in the last decade or so on defining alternative, less stringent error rates and developing multiple testing methods that control them.

The false discovery rate (FDR), which is the expected proportion of falsely rejected among all rejected null hypotheses, introduced by Benjamini and Hochberg (1995), is the first of these alternative error rates that has received the most attention (Benjamini, Krieger and Yekutieli, 2006; Benjamini and Yekutieli, 2001, 2005; Blanchard and Roquain, 2008; Finner, Dickhaus and Roters, 2007, 2009; Gavrilov, Benjamini and Sarkar, 2009;, Genovese, Roeder and Wasserman, 2008; Genovese and Wasserman, 2002, 2004; Sarkar, 2002, 2004, 2006, 2008a, 2008c;, Storey, 2002, 2003; and Storey, Taylor and Siegmund, 2004). Recently, the ideas of controlling the probabilities of falsely rejecting at least k null hypotheses, which is the k-FWER, and the false discovery proportion (FDP) exceeding a certain threshold $\gamma \in [0, 1)$ have been introduced as alternatives to the FWER and methods controlling these new error rates have been suggested (Dudoit, van der Laan and Pollard, 2004; Guo and Rao, 2010; Guo and Romano, 2007; Hommel and Hoffmann, 1987; Lehmann and Romano, 2005; Korn, Troendle, McShane and Simon, 2004; Romano and Shaikh, 2006a, b; Romano and Wolf, 2005, 2007; Sarkar 2007, 2008b; and van der Laan, Dudoit and Pollard, 2004).

Sarkar (2007) has advocated using the expected ratio of k or more false rejections to the total number of rejections, the k-FDR, which is a less conservative notion of error rate than the k-FWER. Several procedures controlling the k-FDR have been developed under different dependence assumptions on the p-values. Sarkar (2007) has utilized the kth order joint distributions of the null p-values and developed procedures under the MTP_2 positive dependence (due to Karlin and Rinott, 1980) and arbitrary dependence conditions on the p-values. Sarkar and Guo (2009) considered a mixture model involving independent p-values and provided a simple and intuitive upper bound to the k-FDR through which they developed newer procedures controlling the k-FDR. Sarkar and Guo (2010) relaxed the requirement of using the kth order joint distributions of the null p-values as well as the MTP_2 condition used in Sarkar (2007). They utilized only the bivariate distributions of the null p-values and developed different k-FDR procedures, assuming a positive dependence condition, weaker than the MTP_2 and same as the one under which the procedure of Benjamini and Hochberg (1995) controls the FDR (Benjamini and Yekutieli, 2001; Sarkar, 2002), and also an arbitrary dependence condition on the p-values.

Suppose that $H_1, \ldots, H_n$ are the null hypotheses that we want to simultaneously test using their respective p-values $p_1, \ldots, p_n$. Let $p_{(1)} \leq \cdots \leq p_{(n)}$ be the ordered p-values and $H_{(1)}, \ldots, H_{(n)}$ the associated null hypotheses. Then, given a non-decreasing set of critical constants $0 \leq \alpha_1 \leq \cdots \leq \alpha_n \leq 1$, a stepup multiple testing procedure rejects the set of null hypotheses $\{H_{(i)}, i \leq i^*_{SU}\}$ and accepts the rest, where $i^*_{SU} = \max\{i : p_{(i)} \leq \alpha_i\}$, if the maximum exists, otherwise accepts all the null hypotheses. A stepdown procedure, on the other hand, rejects the set of null hypotheses $\{H_{(i)}, i \leq i^*_{SD}\}$ and accepts the rest, where $i^*_{SD} = \max\{i : p_{(j)} \leq \alpha_j \ \forall \ j \leq i\}$, if the maximum exists, otherwise accepts all the null hypotheses. The critical constants are determined subject to the control at a pre-specified level α of a suitable error rate which, in this case, is the k-FDR defined as follows. Let R be the total number of rejected null hypotheses, among which V are falsely rejected and S are correctly rejected. Then, the k-FDR is defined as

$$k\text{-FDR} = \mathrm{E}(k\text{-FDP}), \quad \text{where } k\text{-FDP} = \frac{VI(V \geq k)}{R \vee 1}, \qquad (4.1)$$

with $R \vee 1 = \max\{R, 1\}$, which reduces to the original FDR when $k = 1$.

Most of the procedures developed so far for controlling the k-FDR are stepup procedures, except a few developed in Sarkar and Guo (2010) that are stepdown procedures developed for independent as well as dependent p-values. In this article, we will focus mainly on developing some more stepdown procedures controlling the k-FDR under the independence as well as some forms of dependence conditions on the p-values.

4.2. Preliminaries

In this section, we will present two lemmas related to a general stepdown procedure which will be useful in developing stepdown procedures controlling the k-FDR in the next section.

Let n_0 and $n_1(= n - n_0)$ be respectively the numbers of true and false null hypotheses. Define $\hat{q}_1, \cdots, \hat{q}_{n_0}$ and $\hat{r}_1, \cdots, \hat{r}_{n_1}$ to be the p-values corresponding to the true and false null hypotheses respectively and let $\hat{q}_{(1)} \leq \cdots \leq \hat{q}_{(n_0)}$ and $\hat{r}_{(1)} \leq \cdots \leq \hat{r}_{(n_1)}$ be their ordered values.

First, we have the following lemma.

Lemma 4.1. *Consider a stepdown procedure with critical constants $0 \leq \alpha_1 \leq \cdots \leq \alpha_n \leq 1$. Let R be the total number of rejections, of which V are false and S are correct. For a fixed $k > 0$, let J be the (random) largest index such that $\hat{r}_{(1)} \leq \alpha_{k+1}, \cdots, \hat{r}_{(J)} \leq \alpha_{k+J}$ (with $J = 0$ if $\hat{r}_{(1)} > \alpha_k$). Then, given $J = j$, $V \geq k$ implies that (i) $S \geq j$ and (ii) $\hat{q}_{(k)} \leq \alpha_{k+j}$.*

Proof. If $j = 0$, part (i) obviously holds. If $R = n$, then also part (i) holds. Now suppose $j > 0$ and $R < n$. Then, $p_{(R+1)} > \alpha_{R+1}$. Let us assume that $R < k + j$, then $R - k + 1 \leq j$. Thus $\hat{r}_{(R-k+1)} \leq \alpha_{R+1}$. Noting that $V \geq k$, then $\hat{q}_{(k)} \leq \alpha_R$. Therefore, $p_{(R+1)} \leq \max\{\hat{q}_{(k)}, \hat{r}_{(R-k+1)}\} \leq \alpha_{R+1}$. It leads to a contradiction. Thus $R \geq k+j$. Observe that $\hat{r}_{(j)} \leq \alpha_{k+j} \leq \alpha_R$, then $S \geq j$. Thus, part (i) of the lemma follows.

To prove part (ii), we use reverse proof. Assume $\hat{q}_{(k)} > \alpha_{k+j}$. Noting that $\hat{r}_{(j)} \leq \alpha_{k+j}$ and $\hat{r}_{(j+1)} > \alpha_{k+j+1}$ when $j < n_1$, thus $R < k + j$ and then $V < k$. Therefore, if $V \geq k$, then $\hat{q}_{(k)} \leq \alpha_{k+j}$. Thus part (ii) of the lemma is also proved. $\qquad\square$

The following second lemma is taken from Sarkar and Guo (2010).

Lemma 4.2. *Given a stepdown procedure with critical constants $0 \leq \alpha_1 \leq \cdots \leq \alpha_n \leq 1$, consider the corresponding stepdown procedure in terms of the null p-values $\hat{q}_1, \ldots, \hat{q}_{n_0}$ and the critical values $\alpha_{n_1+1} \leq \cdots \leq \alpha_n$. Let V_n denote the number of false rejections in the original stepdown procedure and $\hat{R}_{n_0}$ denote the number of rejections in the stepdown procedure involving the null p-values. Then we have, for any fixed $k \leq n_0 \leq n$,*

$$\{V_n \geq k\} \subseteq \left\{\hat{R}_{n_0} \geq k\right\}. \tag{4.2}$$

4.3. Stepdown k-FDR Procedures

We will develop some new stepdown procedures in this section that control the k-FDR. Before we do that, we want to emphasize a few important points.

First, while seeking to control k or more false rejections, we are tolerating at most $k - 1$ of the null hypotheses to be falsely rejected. In other words, we can allow the first $k - 1$ critical values to be arbitrarily chosen to be as high as possible, even all equal to one. However, it is not only counterintuitive to have a stepwise procedure with critical values that are not monotonically non-decreasing but also it is unrealistic to allow the first $k - 1$ most significant null hypotheses to be rejected without having any control over them. So, the best option is to keep these critical values at the same level as the kth one; see also Lehmann and Romano (2005) and Sarkar (2007, 2008b). The stepdown procedures that we are going to develop next will have their first k critical values same. Second, the k-FDR procedures developed here are all generalized versions of some stepdown FDR procedures available in the literature. So, by developing these procedures we are actually providing some general results related to FDR methodologies. Third, although an FDR procedure also controls the k-FDR, the k-FDR procedures that we develop here are all more powerful than the corresponding FDR procedures.

Theorem 4.1. *Assume that the p-values satisfy the following condition:*

$$P\left\{\hat{q}_i \leq u \big| \hat{r}_1, \cdots, \hat{r}_{n_1}\right\} \leq P\left\{\hat{q}_i \leq u\right\} \leq u, \quad u \in (0,1), \qquad (4.3)$$

for any $i = 1, \cdots, n_0$. Then, the stepdown procedure with the critical constants

$$\alpha_i = \begin{cases} \frac{k\alpha}{n} & \text{if } i = 1, \ldots, k \\ \min\left\{\frac{kn\alpha}{(n-i+k)^2}, 1\right\} & \text{if } i = k+1, \ldots, n \end{cases} \qquad (4.4)$$

controls the k-FDR at α.

Proof. When $n_0 < k$, there is nothing to prove, as in this case the $k\text{-}FDR = 0$ and hence trivially controlled. So, we will assume $k \leq n_0 \leq n$ while proving this theorem.

Using Lemma 4.1 and noting $V \leq n_0$, we have

$$\begin{aligned} E\left(k\text{-}FDP \big| J = j\right) &= E\left(\frac{V}{V+S} \cdot I(V \geq k) \big| J = j\right) \\ &\leq \frac{n_0}{n_0 + j} Pr\left(V \geq k \big| J = j\right) \\ &\leq \frac{n_0}{n_0 + j} Pr\left(\hat{q}_{(k)} \leq \alpha_{j+k} \big| J = j\right), \qquad (4.5) \end{aligned}$$

for any fixed $j = 0, 1, \ldots, n_1$. Let N be the number of p-values corresponding to true null hypotheses that are less than or equal to constant α_{j+k}. Then, using Markov's inequality and condition (4.3), we have

$$\begin{aligned} P\left(\hat{q}_{(k)} \leq \alpha_{j+k} \big| J = j\right) &= P\left\{N \geq k \big| J = j\right\} \leq \frac{E(N | J = j)}{k} \\ &= \frac{1}{k} \sum_{i=1}^{n_0} P\left(\hat{q}_i \leq \alpha_{j+k} \big| J = j\right) \leq \frac{n_0 \alpha_{j+k}}{k}. \qquad (4.6) \end{aligned}$$

Thus, from (4.4), (4.5) and (4.6), we have

$$E\left(k\text{-}FDP|J=j\right) \le \frac{n_0^2 n\alpha}{(n_0+j)(n-j)^2}, \tag{4.7}$$

which is less than or equal to α, since $n_0 \le n - j$ and $(n_0 + j)(n - j) = n_0 n + j(n - n_0 - j) \ge n_0 n$. This proves the theorem. $\square$

Remark 4.1. Note that the critical constants in (4.4) satisfy the following inequality:

$$\alpha_i \ge \alpha_i^* = \begin{cases} \frac{k\alpha}{n} & \text{if} \quad i = 1, \dots, k \\ \frac{k\alpha}{n-i+k} & \text{if} \quad i = k+1, \dots, n, \end{cases} \tag{4.8}$$

where α_i^*'s are the critical constants of the stepdown k-FWER procedure in Lehmann and Romano (2005). In other words, the k-FDR procedure in Theorem 4.1 is more powerful than the k-FWER procedure in Lehmann and Romano (2005), as one would expect, although the latter does not require any particular assumption on the dependence structure of the p-values.

Romano and Sheikh (2006b) gave a stepdown FDR procedure under the same condition as in (4.3). This procedure is generalized in Theorem 4.1 to a k-FDR procedure. Condition (4.3) is slightly weaker than the independence assumption between the sets of true and false p-values. No other assumptions are made here regarding the dependence structure within each of these sets. If, however, the null p-values are independent among themselves with each being distributed as $U(0, 1)$, this procedure can be improved to the one given in the following theorem.

Theorem 4.2. *Let*

$$G_{k,s}(u) = P\left\{U_{(k)} \le u\right\} = \sum_{j=k}^{s} \binom{s}{j} u^j (1-u)^{s-j}, \tag{4.9}$$

the cdf of the kth order statistic based on s i.i.d. $U(0,1)$. The stepdown procedure with the following critical constants

$$\alpha_i = \begin{cases} G_{k,n}^{-1}(\alpha) & \text{if} \quad i = 1, \dots, k \\ G_{k,n-i+k}^{-1}\left(\frac{n\alpha}{n-i+k}\right) & \text{if} \quad i = k+1, \dots, n \end{cases} \tag{4.10}$$

controls the k-FDR at α if the $\hat{q}_i$'s are i.i.d. $U(0,1)$ and independent of $(\hat{r}_1, \ldots, \hat{r}_{n_1})$.

Proof. For any fixed $j = 0, 1, \ldots, n_1$, we have from (4.5) and (4.10)

$$E\left(k\text{-}FDP \middle| J = j\right) \leq \frac{n_0}{n_0 + j} G_{k,n_0}(\alpha_{j+k})$$

$$= \frac{n_0}{n_0 + j} G_{k,n_0}\left(G_{k,n-j}^{-1}\left(\frac{n\alpha}{n-j}\right)\right)$$

$$\leq \frac{n_0}{n_0 + j} G_{k,n-j}\left(G_{k,n-j}^{-1}\left(\frac{n\alpha}{n-j}\right)\right)$$

$$\leq \frac{n_0 n\alpha}{(n_0 + j)(n - j)} \leq \alpha. \tag{4.11}$$

The second inequality follows from the fact that $n_0 \leq n - j$ and the cdf $G_{k,s}$ is increasing in s. This proves the theorem. $\qquad\square$

Remark 4.2. Benjamini and Liu (1999) obtained a stepdown procedure assuming complete independence of all the p-values. We generalize this procedure in Theorem 4.2 from an FDR to a k-FDR procedure, but under a slightly weaker assumption allowing the false p-values to have an arbitrary dependence structure.

We now go back to Sarkar and Guo (2010) and generalize a stepdown k-FDR procedure given there assuming independence of the p-values. More specifically, we have the following theorem.

Theorem 4.3. *The stepdown procedure with critical constants $\alpha_1 = \cdots = \alpha_k \leq \cdots \leq \alpha_n$, where α_i/i is decreasing in i and*

$$\frac{i\alpha_k}{k} G_{k-1,i-1}(\alpha_{n-i+k}) \leq \alpha \text{ for all } k \leq i \leq n, \tag{4.12}$$

controls the k-FDR at α when the p-values are positively dependent in the sense that $E\{\phi(p_1, \ldots, p_n) \mid \hat{q}_i \leq u\}$ is nondecreasing in u for every $\hat{q}_i$ and any nondecreasing (coordinatewise) function ϕ, and the $\hat{q}_i$'s are i.i.d. $U(0,1)$.

Proof.

$$k\text{-}FDR = E\left\{\frac{V}{R} \cdot I\left(V \geq k\right)\right\} = E\left\{\sum_{r=k}^{n} \frac{1}{r} \sum_{i=1}^{n_0} I\left(\hat{q}_i \leq \alpha_r, V \geq k, R = r\right)\right\}$$

$$= \sum_{i=1}^{n_0} \sum_{r=k}^{n} \frac{1}{r} P\left\{p_{(k)} \leq \alpha_k, \cdots, p_{(r)} \leq \alpha_r, p_{(r+1)} > \alpha_{r+1}, \hat{q}_i \leq \alpha_r, V \geq k\right\}$$

$$= \sum_{i=1}^{n_0} \frac{1}{k} P\left\{p_{(k)} \leq \alpha_k, \hat{q}_i \leq \alpha_k, V \geq k\right\} + \sum_{i=1}^{n_0} \sum_{r=k+1}^{n} E\left[P\left\{p_{(k)} \leq \alpha_k, \cdots,\right.\right.$$

$$p_{(r)} \leq \alpha_r, V \geq k | \hat{q}_i\right\} \left\{\frac{I(\hat{q}_i \leq \alpha_r)}{r} - \frac{I(\hat{q}_i \leq \alpha_{r-1})}{r-1}\right\}\right] \qquad (4.13)$$

$$= \sum_{i=1}^{n_0} \frac{1}{k} P\left\{\hat{q}_i \leq \alpha_k, V \geq k\right\} + \sum_{i=1}^{n_0} \sum_{r=k+1}^{n} \frac{1}{r} P\left\{p_{(k)} \leq \alpha_k, \cdots, p_{(r)} \leq \alpha_r,\right.$$

$$\alpha_{r-1} < \hat{q}_i \leq \alpha_r, V \geq k\right\} - \sum_{i=1}^{n_0} \sum_{r=k+1}^{n} \frac{1}{r(r-1)} P\left\{p_{(k)} \leq \alpha_k, \cdots,\right.$$

$$p_{(r)} \leq \alpha_r, \hat{q}_i \leq \alpha_{r-1}, V \geq k\right\}$$

(with $p_{(n+1)} = 1$ and $\alpha_{n+1} = 1$).

Let $R_{n-1}^{(-i)}$ and $V_{n-1}^{(-i)}$ denote respectively the total numbers of rejections and false rejections in the stepdown procedure based on the ordered values $p_{(1)}^{(-i)} \leq \cdots \leq p_{(n-1)}^{(-i)}$ of $\{p_1, \ldots, p_n\} \setminus \{p_i\}$ and the $n-1$ critical values $\alpha_2 \leq \cdots \leq \alpha_n$. Note that

$$P\left\{\hat{q}_i \leq \alpha_k, V \geq k\right\} = P\left\{\hat{q}_i \leq \alpha_k, V_{n-1}^{(-i)} \geq k - 1\right\}, \qquad (4.14)$$

$$P\left\{P_{(k)} \leq \alpha_k, \cdots, P_{(r)} \leq \alpha_r, \alpha_{r-1} < \hat{q}_i \leq \alpha_r, V \geq k\right\}$$

$$= P\left\{P_{(k)}^{(-i)} \leq \alpha_k, \cdots, P_{(r-1)}^{(-i)} \leq \alpha_{r-1}, \alpha_{r-1} < \hat{q}_i \leq \alpha_r, V_{n-1}^{(-i)} \geq k - 1\right\}$$

$$= P\left\{R_{n-1}^{(-i)} \geq r - 1, \alpha_{r-1} < \hat{q}_i \leq \alpha_r, V_{n-1}^{(-i)} \geq k - 1\right\}$$

$$= P\left\{R_{n-1}^{(-i)} \geq r - 1, V_{n-1}^{(-i)} \geq k - 1 \mid \hat{q}_i \leq \alpha_r\right\} \alpha_r$$

$$- P\left\{R_{n-1}^{(-i)} \geq r - 1, V_{n-1}^{(-i)} \geq k - 1 \mid \hat{q}_i \leq \alpha_{r-1}\right\} \alpha_{r-1}$$

$$\leq P\left\{R_{n-1}^{(-i)} \geq r - 1, V_{n-1}^{(-i)} \geq k - 1 \mid \hat{q}_i \leq \alpha_r\right\} \left(\alpha_r - \alpha_{r-1}\right), \qquad (4.15)$$

with the last inequality following from the positive dependence assumption made in the theorem and the fact that the set

$$\left\{R_{n-1}^{(-i)} \geq r - 1, V_{n-1}^{(-i)} \geq k - 1\right\}$$

is a decreasing set in the p-values, and

$$P\left\{P_{(k)} \le \alpha_k, \cdots, P_{(r)} \le \alpha_r, \hat{q}_i \le \alpha_{r-1}, V \ge k\right\}$$

$$\ge P\left\{P_{(k)}^{(-i)} \le \alpha_k, \cdots, P_{(r-1)}^{(-i)} \le \alpha_{r-1}, \hat{q}_i \le \alpha_{r-1}, V^{(-i)} \ge k-1\right\}$$

$$= P\left\{R_{n-1}^{(-i)} \ge r-1, \hat{q}_i \le \alpha_{r-1}, V_{n-1}^{(-i)} \ge k-1\right\}$$

$$= P\left\{R_{n-1}^{(-i)} \ge r-1, V_{n-1}^{(-i)} \ge k-1 \mid \hat{q}_i \le \alpha_{r-1}\right\} \alpha_{r-1}, \qquad (4.16)$$

for $k+1 \le r \le m$. Therefore, using (4.14)-(4.16) in (4.13), we have

$$k\text{-}FDR \le \sum_{i=1}^{n_0} \frac{1}{k} P\left\{\hat{q}_i \le \alpha_k, V_{n-1}^{(-i)} \ge k-1\right\}$$

$$+ \sum_{i=1}^{n_0} \sum_{r=k+1}^{n} P\left\{R_{n-1}^{(-i)} \ge r-1, V_{n-1}^{(-i)} \ge k-1 \mid \hat{q}_i \le \alpha_r\right\} \left\{\frac{\alpha_r}{r} - \frac{\alpha_{r-1}}{r-1}\right\}$$

$$\le \sum_{i=1}^{n_0} \frac{1}{k} P\{\hat{q}_i \le \alpha_k, V_{n-1}^{(-i)} \ge k-1\} \le \sum_{i=1}^{n_0} \frac{\alpha_k}{k} P\{\hat{R}_{n_0-1}^{(-i)} \ge k-1\}$$

$$\le \sum_{i=1}^{n_0} \frac{\alpha_k}{k} P\{\hat{q}_{(k-1)}^{(-i)} \le \alpha_{n_1+k}\} = \frac{n_0 \alpha_k}{k} G_{k-1,n_0-1}(\alpha_{n_1+k}), \qquad (4.17)$$

which is controlled at level α if the α_i's are chosen subject to (4.12). The third inequality in (4.17) follows from Lemma 4.2. Thus the theorem is proved. $\qquad \square$

Remark 4.3. A variety of stepdown procedures can be obtained using critical values satisfying the conditions in Theorem 4.3 once the distributional assumptions in the theorem hold. For instance, one may choose the critical constants $\alpha_i = (i \vee k)\beta/n$ with the β determined subject to

$$\frac{\beta}{n} \max_{k \le n_0 \le n} \left\{n_0 G_{k-1,n_0-1}\left(\frac{(n-n_0+k)\beta}{n}\right)\right\} = \alpha. \qquad (4.18)$$

This is what Sarkar and Guo (2010) proposed under the independence of the p-values. Similarly, we can consider the critical constants $\alpha_i = \left(\frac{i}{n}\right)^\gamma \beta$ or $\alpha_i = \frac{i+d}{n+d}\beta$, for some pre-specified constants $0 < \gamma < 1$ and $d > 0$, with the β chosen as large as possible subject to Condition (4.12).

Recently, Gavrilov, Benjamini and Sarkar (2009) obtained a stepdown procedure controlling the FDR with independent p-values. We now derive a generalized version of this procedure providing a control of the k-FDR in the following theorem.

Theorem 4.4. *The step-down procedure with the critical values $\alpha_1 = \cdots = \alpha_k \leq \cdots \leq \alpha_n$ satisfying $\alpha_i/(1-\alpha_i) \leq i\beta/(n-i+1), i = 1, \ldots, n$, controls the k-FDR at level α for any fixed β satisfying*

$$\beta G_{k-1,i}(\alpha_{n+k-i-1}) \leq \alpha \text{ for all } k \leq i \leq n, \tag{4.19}$$

if the p-values are independent and $\hat{q}_i \sim U(0,1)$.

Proof.   From the proof of Theorem 4.3, we notice that for independent p-values

$$k\text{-}FDR \leq \sum_{i=1}^{n_0} \frac{\alpha_k}{k} P\left\{V_{n-1}^{(-i)} \geq k-1\right\} + \sum_{i=1}^{n_0}\sum_{r=k+1}^{n} P\left\{P_{(k)}^{(-i)} \leq \alpha_k, \cdots, \right.$$

$$\left. P_{(r-1)}^{(-i)} \leq \alpha_{r-1}, V_{n-1}^{(-i)} \geq k-1\right\}\left\{\frac{\alpha_r}{r} - \frac{\alpha_{r-1}}{r-1}\right\}$$

$$= \sum_{i=1}^{n_0}\sum_{r=k}^{n} \frac{\alpha_r}{r} P\left\{P_{(k)}^{(-i)} \leq \alpha_k, \cdots, P_{(r-1)}^{(-i)} \leq \alpha_{r-1}, P_{(r)}^{(-i)} > \alpha_r, \right.$$

$$\left. V_{n-1}^{(-i)} \geq k-1\right\}$$

$$\leq \beta \sum_{i=1}^{n_0}\sum_{r=k}^{n} \frac{1-\alpha_r}{n-r+1} P\left\{P_{(k)}^{(-i)} \leq \alpha_k, \cdots, P_{(r-1)}^{(-i)} \leq \alpha_{r-1}, P_{(r)}^{(-i)} > \alpha_r, \right.$$

$$\left. V_{n-1}^{(-i)} \geq k-1\right\}$$

$$= \beta \sum_{i=1}^{n_0}\sum_{r=k}^{n} \frac{1}{n-r+1} P\left\{P_{(k)}^{(-i)} \leq \alpha_k, \cdots, P_{(r-1)}^{(-i)} \leq \alpha_{r-1}, P_{(r)}^{(-i)} > \alpha_r, \right.$$

$$\left. \hat{q}_i > \alpha_r, V_{n-1}^{(-i)} \geq k-1\right\}$$

$$= \beta \sum_{i=1}^{n_0}\sum_{r=k}^{n} \frac{1}{n-r+1} P\left\{P_{(k)} \leq \alpha_k, \cdots, P_{(r-1)} \leq \alpha_{r-1}, P_{(r)} > \alpha_r, \right.$$

$$\left. \hat{q}_i > \alpha_r, V \geq k-1\right\}. \tag{4.20}$$

With A and U denoting, respectively, the total numbers of accepted and correctly accepted null hypotheses, we note that the last expression in (4.20) is $\beta E\{UI(V \geq k-1)/A \vee 1\}$, which is less than or equal to

$$\beta P\left\{V \geq k-1\right\} \leq \beta P\{\hat{R}_{n_0} \geq k-1\} \quad \text{(from Lemma 4.2)}$$

$$\leq \beta P\{\hat{q}_{(k-1)} \leq \alpha_{n_1+k-1}\}$$

$$= \beta G_{k-1,n_0}(\alpha_{n_1+k-1}).$$

This is controlled at α for any β satisfying (4.19). Thus, the theorem is proved. $\square$

Remark 4.4. Specifically, if we consider the critical constants $\alpha_i = i\beta/(n - i + 1 + i\beta)$, $i = k, \cdots, n$, as in Gavrilov, Benjamini and Sarkar (2009), where the constant β is determined subject to

$$\max_{k \leq n_0 \leq n} \left\{ \beta G_{k-1, n_0} \left(\frac{\beta(n - n_0 + k - 1)}{n_0 - k + 2 + \beta(n - n_0 + k - 1)} \right) \right\} = \alpha, \qquad (4.21)$$

the corresponding stepdown procedure will provide a control of the k-FDR at level α.

4.4. Numerical Studies

In this section, we present the results of a numerical study comparing the four different stepdown k-FDR procedures developed in Theorems 4.1-4.4 in terms of their critical values to gain an insight into their relative performance with respect to the number of discoveries. Let us denote the four sets of critical constants as $\alpha_i^{(j)}$, $i = k, \cdots, n$, with j referring to the jth procedure. For the first two procedures, the critical constants are defined in (4.4) and (4.10), respectively, and for the last two, the critical constants are $\alpha_i^{(3)} = (i \vee k)\beta/n$ and $\alpha_i^{(4)} = i\beta/(n - i + 1 + i\beta)$, where β is determined subject to (4.18) and (4.21), respectively. As the baseline method for comparison, we choose the stepdown procedure with the critical constants $\gamma_i = i\alpha/n, i = 1, \cdots, n$. This is the stepdown analog of the Benjamini-Hochberg (BH) stepup FDR procedure that, as Sarkar (2002) has shown, controls the FDR and hence the k-FDR under the conditions considered in Theorems 4.2-4.4, and would have been used by researchers without the knowledge of other stepdown k-FDR procedures. Considering $\zeta_i^{(j)} = \log_{10}\left(\alpha_i^{(j)}/\gamma_i \right), i = 1, \cdots, n$, for $j = 1, \cdots, 4$, we plot in Figure 4.1 the four sequences $\zeta_i^{(j)}, j = 1, \cdots, 4$, with $n = 500$, $k = 8$, and $\alpha = 0.05$.

As seen from Figure 4.1, the critical constants of the procedure in Theorem 4.1 (labeled RS) are all much less than those of the stepdown analog of the BH procedure (referred to as the stepdown BH in this article). For the procedure in Theorem 4.2 (labeled BL), the first few of its critical values are seen to be larger than those of the stepdown BH (labeled BH). The critical constants of the procedures in Theorems 4.3 and 4.4 (labeled SG and GBS respectively) are all uniformly larger than the corresponding critical values of the BH. Thus, there is a numerical evidence that the procedures in Theorems 4.3 and 4.4 are both more powerful than the stepdown BH, but the procedure in Theorem 4.1 is not. Since for a stepdown procedure, the power is mostly determined by some of its first critical values, the

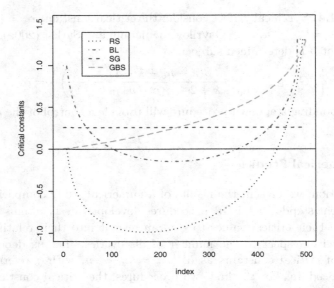

Fig. 4.1.   The logarithms with base 10 of ratios of critical constants of four stepdown k-FDR procedures with respect to that of the BH procedure for $n = 500, k = 8$, and $\alpha = 0.05$.

procedure in Theorem 4.2 may sometimes be more powerful than the step-down BH procedure.

We also compared the four stepdown k-FDR procedures with the step-down BH procedure in terms of their power. We simulated the average power, the expected proportion of false null hypotheses that are rejected, for each of these procedures. Figure 4.2 presents this power comparison. Each simulated power was obtained by (i) generating $n = 200$ independent normal random variables $N(\mu_i, 1)$, $i = 1, \cdots, n$ with n_1 of the 200 μ_i's being equal to $d = 2$ and the rest 0, (ii) applying the stepdown BH procedure and the four stepdown k-FDR procedures with $k = 4$ to the generated data to test $H_i : \mu_i = 0$ against $K_i : \mu_i > 0$ simultaneously for $i = 1, \ldots, 200$ at $\alpha = 0.05$, and (iii) repeating steps (i) and (ii) 1,000 times before observing the proportion of the n_1 false H_i's that are correctly declared significant.

As seen from Figure 4.2, the SG procedure in Theorem 4.3 is uniformly more powerful than the stepdown BH, with the power difference getting significantly higher with increasing number of false null hypotheses, while the GBS procedure in Theorem 4.4 is marginally more powerful than the stepdown BH, with the power difference getting significantly higher only

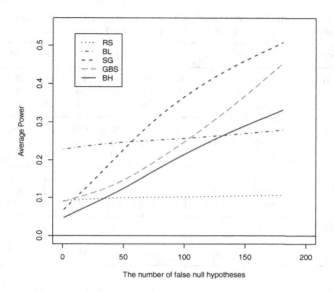

Fig. 4.2. Power of four stepdown k-FDR procedures in the case of independence with $n = 200, k = 4, d = 2$ and $\alpha = 0.05$.

after the number of false null hypotheses becomes moderately large. The BL procedure in Theorem 4.2 is the most powerful among these four stepdown procedures when the proportion of false null hypotheses is small. Even when the false proportion is moderately large, this is also more powerful than the stepdown BH. However, it loses its advantage over the stepdown BH when the proportion of false null hypotheses is very large. Finally, the RS procedure in Theorem 4.1 is less powerful than the stepdown BH, as we expected from Figure 4.1 showing the numerical comparisons of the critical constants of these procedures.

4.5. An Application to Gene Expression Data

Hereditary breast cancer is known to be associated with mutations in BRCA1 and BRCA2 proteins. Hedenfalk et al. (2001) report that a group of genes are differentially expressed between tumors with BRCA1 mutations and tumors with BRCA2 mutations. The data, which are publicly available from the web site

http://research.nhgri.nih.gov/microarray/NEJM_Supplement/,

Table 4.1. Numbers of differentially expressed genes for the data in Hedenfalk et al. (2001) using four stepdown k-FDR procedures.

procedure	level α	$k = 2$	5	8	10	15	20	30
RS	0.03	2	5	8	8	12	18	22
	0.05	3	8	11	16	20	24	32
	0.07	5	11	18	19	24	29	47
BL	0.03	8	34	73	82	120	150	191
	0.05	11	42	75	86	125	157	200
	0.07	11	47	76	91	129	159	203
SG	0.03	8	8	11	20	21	24	73
	0.05	74	75	76	76	76	82	94
	0.07	103	110	123	124	129	131	139
GBS	0.03	3	5	8	8	12	18	22
	0.05	73	73	73	73	73	73	73
	0.07	96	96	96	96	96	96	96

consist of 22 breast cancer samples, among which 7 are BRCA1 mutants, 8 are BRCA2 mutants, and 7 are sporadic (not used in this illustration). Expression levels in terms of florescent intensity ratios of a tumor sample to a common reference sample, are measured for $3,226$ genes using cDNA microarrays. If any gene has one ratio exceeding 20, then this gene is eliminated. Such preprocessing leaves $n = 3,170$ genes.

We tested each gene for differential expression between these two tumor types by using a two-sample t-test statistic. For each gene, the base 2 logarithmic transformation of the ratio was obtained before computing the two-sample t-test statistic based on the transformed data. A permutation method from Storey and Tibshirani (2003) with the permutation number $B = 2,000$ was then used to calculate the corresponding raw p-value. Finally, we applied to these raw p-values the stepdown BH and the four stepdown k-FDR procedures in Theorems 4.1-4.4.

At $\alpha = 0.03$, 0.05 and 0.07, the stepdown BH results in 3, 33 and 95 significant genes respectively, while those numbers for the present methods are presented in Table 4.1 for $k = 2, 5, 8, 10, 15, 20$ and 30, with the four procedures in Theorems 4.1-4.4 labeled RS, BL, SG and GBS respectively. As we can see from this table, the RS procedure in Theorem 4.1 generally detects less significant genes than the stepdown BH for moderate or large values of α. The BL procedure in Theorem 4.2 always detects more differentially expressed genes than the stepdown BH for slightly moderate values of k and small or moderate values of α. The SG procedure in Theorem 4.3 is seen to always detect more significant genes than the stepdown BH, while the GBS procedure in Theorem 4.4 detects almost the same number of differentially expressed genes as the stepdown BH except for moderate α.

4.6. Conclusions

We have presented a number of new stepdown k-FDR procedures in this article under different assumptions on the dependence structure of the p-values, generalizing some existing stepdown FDR procedures. These would be of use in situations where one is willing to tolerate at most $k - 1$ false rejections and is looking for a stepdown procedure controlling a powerful notion error rate than the k-FWER for exercising a control over at least k false rejections. Although any FDR stepdown procedure can also control the k-FDR, ours are more powerful than the corresponding FDR versions. Moreover, we offer better k-FDR stepdown procedures than the stepdwon analog of the BH stepup procedure, with its first $k-1$ critical values equal to the kth one, which would have been commonly used by researchers without knowing the existence of any other stepdown k-FDR procedure.

Acknowledgments

The research of Wenge Guo was supported by NSF Grant DMS-1006021 and the research of Sanat Sarkar was supported by NSF Grant DMS-0603868 and DMS-1006344.

References

1. Benjamini, Y. and Hochberg, Y. (1995). Controlling the false discovery rate: A practical and powerful approach to multiple testing. *J. Roy. Statist. Soc.* **B 57**, 289–300.
2. Benjamini, Y., Krieger, A. M. and Yekutieli, D. (2006). Adaptive linear step-up false discovery rate controlling procedures. *Biometrika* **93**, 491–507.
3. Benjamini, Y. and Liu, W. (1999). A step-down multiple hypotheses testing procedure that controls the false discovery rate under independence. *J. Statist. Plann. Inf.* **82**, 163–170.
4. Benjamini, Y. and Yekutieli, D. (2001). The control of the false discovery rate in multiple testing under dependency. *Ann. Statist.* **29**, 1165–1188.
5. Benjamini, Y. and Yekutieli, D. (2005). False discovery rate-adjusted multiple

confidence intervals for selected parameters. *J. Amer. Statist. Assoc.* **100**, 71–93.

6. Blanchard, G. and Roquain, E. (2008). Two simple sufficient conditions for FDR control. *Electron. J. Stat.* **2**, 963–992.

7. Dudoit, S., van der Laan, M. and Pollard, K. (2004). Multiple testing: Part I. Single-step procedures for control of general type I error rates. *Statist. App. Gen. Mol. Bio.* **3**, 1, Article 13.

8. Finner, H., Dickhaus, T. and Roters, M. (2007). Dependency and false discovery rate: Asymptotics. *Ann. Statist.* **35**, 1432–1455.

9. Finner, H., Dickhaus, T. and Roters, M. (2009). On the false discovery rate and an asymptotically optimal rejection curve. *Ann. Statist.* **37**, 596–618.

10. Gavrilov, Y., Benjamini, Y. and Sarkar, S. K. (2009). An adaptive step-down procedures with proven FDR control under independence. *Ann. Statist.* **37**, 619–629.

11. Genovese, C., Roeder, K. and Wasserman, L. (2008). False discovery control with p-value weighting. *Biometrika* **93**, 509-524.

12. Genovese, C. and Wasserman, L. (2002). Operarting characteristics and extensions of the false discovery rate procedure. *J. Roy. Statist. Soc.* **B 64**, 499–517.

13. Genovese, C. and Wasserman, L. (2004). A stochastic process approach to false discovery control. *Ann. Statist.* **32**, 1035–1061.

14. Guo, W. and Rao, M. B. (2010). On stepwise control of the generalized family wise error rates. *Electron. J. Stat.* **4**, 472–485.

15. Guo, W. and Romano, J. P. (2007). A generalized Sidak-Holm procedure and control of generalized error rates under independence. *Statist. App. Gen. Mol. Bio.* **6**, 1, Article 3.

16. Hedenfalk, I., Duggan, D., Chen, Y., Radmacher, M., Bittner, M., Simon, R., Meltzer, P., Gusterson, B., Esteller, M., Kallioniemi, OP, Wilfond, B., Borg, A. and Trent, J. (2001). Gene-expression profiles in hereditary breast cancer. *New Eng. J. Med.* **344**, 539–548.

17. Hochberg, Y. and Tamhane, A. C. (1987). *Multiple Comparison Procedures.* John Wiley, New York.

18. Hommel, G. and Hoffmann, T. (1987). Controlled uncertainty. In *Multiple Hypothesis Testing* (P. Bauer, G. Hommel and E. Sonnemann, Eds.), 154–161, Springer, Heidelberg.

19. Karlin, S. and Rinott, Y. (1980). Classes of orderings of measures and related correlation inequalities I: Multivariate totally positive distributions. *J. Mult. Anal.* **10**, 467–498.

20. Korn, E., Troendle, T., McShane, L. and Simon, R. (2004). Controlling the number of false discoveries: Application to high-dimensional genomic data. *J. Statist. Plann. Inf.* **124**, 279–398.

21. Lehmann, E. L. and Romano, J. P. (2005). Generalizations of the familywise error rate. *Ann. Statist.* **33**, 1138–1154.

22. Romano, J. P. and Shaikh, A. M. (2006a). Stepup procedures for control of generalizations of the familywise error rate. *Ann. Statist.* **34**, 1850–1873.

23. Romano, J. P. and Shaikh, A. M. (2006b). On stepdown control of the false

discovery proportion. In *The Second E.L. Lehamann Symposium - Optimality* (J. Rojo, Ed.), 40–61, IMS Lecture Notes - Monograph Series **49**.

24. Romano, J. P. and Wolf, M. (2005). Stepwise multiple testing as formalized data snooping. *Econometrica* **73**, 1237–1282.

25. Romano, J. P. and Wolf, M. (2007). Control of generalized error rates in multiple testing. *Ann. Statist.* **35**, 1378–1408.

26. Sarkar, S. K. (2002). Some results on false discovery rate in stepwise multiple testing procedures. *Ann. Statist.* **30**, 239–257.

27. Sarkar, S. K. (2004). FDR-controlling stepwise procedures and their false negatives rates. *J. Statist. Plann. Inf.* **125**, 119–137.

28. Sarkar, S. K. (2006). False discovery and false non-discovery rates in single-step multiple testing procedures. *Ann. Statist.* **34**, 394–415.

29. Sarkar, S. K. (2007). Stepup procedures controlling generalized FWER and generalized FDR. *Ann. Statist.* **35**, 2405–2420.

30. Sarkar, S. K. (2008a). Two-stage stepup procedures controlling FDR. *J. Statist. Plann. Inf.* **138**, 1072–1084.

31. Sarkar, S. K. (2008b). Generalizing Simes' test and Hochberg's stepup procedure. *Ann. Statist.* **36**, 337–363.

32. Sarkar, S. K. (2008c). On methods controlling the false discovery rate (with discussion). *Sankhyā* **A 70**, 135–168.

33. Sarkar, S. K. and Guo, W. (2009). On a generalized false discovery rate. *Ann. Statist.* **37**, 337–363.

34. Sarkar, S. K. and Guo, W. (2010). Procedures controlling generalized false discovery rate using bivariate distributions of the null p-values. *Statist. Sinica* **20**, 1227–1238.

35. Storey, J. D. (2002). A direct approach to false discovery rates. *J. Roy. Statist. Soc.* **B 64**, 479–498.

36. Storey, J. D. (2003). The positive false discovery rate: A Bayesian interpretation and the q-value. *Ann. Statist.* **31**, 2013–2035.

37. Storey, J. D., Taylor, J. E. and Siegmund, D. (2004). Strong control, conservative point estimation and simultanaeous conservative consistency of false discovery rates: A unified approach. *J. Roy. Statist. Soc.* **B 66**, 187–205.

38. Storey, J. D. and Tibshirani, R. (2003). Statistical significance for genomewide studies. *Proc. Nat. Acad. Sci.* **100**, 9440–9445.

39. van der Laan, M., Dudoit, S. and Pollard, K. (2004). Augmentation procedures for control of the generalized family-wise error rate and tail probabilities for the proportion of false positives. *Stat. App. Gen. Mol. Bio.* **3**, 1, Article 15.

Chapter 5

On Confidence Intervals for Expected Response in 2^n Factorial Experiments with Exponentially Distributed Response Variables

H. V. Kulkarni[1] and S. C. Patil[2]

[1] *Shivaji University, Kohlapur, India*
[2] *Padmashree Dr. D. Y. Patil A. C. S. College, India*

The exponential distribution is an important distribution in the analysis of lifetime data. Its one interesting characterization is that inter-occurrence times between the successive events of a Poisson process are independent exponentially distributed. In industrial designed-experiments setting, exponentially distributed response variable can be encountered, for example, when the response is inter-occurrence time between successive defective products when the defectives are produced according to a Poisson process. Traditionally in such a situation, the log-transformed response variable is analyzed using ordinary least squares (OLS). However, this approach has its own limitations and the generalized linear models (GLM) appear to be a good alternative. In the present work, a comparative study is attempted for analyzing 2^n factorial experiments for exponentially distributed response variable among GLM, methods based on Edgeworth expansion and a method based on exact distribution of a pivotal quantity. The comparison is based on theoretical considerations and an extensive simulation study related to the coverages and expected length of confidence intervals ($LOCI$) for the expected response. The method based on exact distribution of a pivotal quantity turns out to be the best among all the approaches. Results of the analysis of a real dataset agree with the theoretical findings.

Keywords: Confidence intervals, Edgeworth expansion, Exact method, Factorial experiments, Generalized linear models.

Contents

5.1. Introduction

Some industrial situations require the analyst to determine which of a large
number of controllable factors have an impact on a relevant response variable. Designed experiments provide a systematic and scientific approach for
assessing the extent and nature of influence of one or more controlled input
factors on the response variable under study. Designed experiments are
used to build a model that describes the relationship of a response variable
of interest with one or more independent input factors that have significant influence on the study variable. These models also help in studying
existence and extent of interactions among the input factors and to find
a combination of levels of input factors that yields best response. This
kind of analysis is helpful to improve the quality/ quantity of the final
product.

The Ordinary Least Square (OLS) technique is a widely used technique
for the analysis of such experiments where the analysis is carried out under
the assumptions that, the responses are independently and normally distributed variables with constant variance and the expected responses bear a
linear relationship with the input factor effects and interactions. However,
there are many situations, where the response variable is non-normal, have
nonconstant variance and the relationship between the response variable
with the input factors may be nonlinear. Lewis et al. (2001b) gave many
examples of designed experiments with non-normal responses. For example,
lifetime of a produced item, counts of defects in one unit of the final product
etc. For these variables, the responses may follow respectively a Gamma
and a Poisson distribution where, the above mentioned assumptions are
not valid. These responses may be related nonlinearly with the input factors and the variance is not constant but is a function of mean in each of
the above cases. Traditionally in such situations, a suitably transformed
response variable (most frequently a variance stabilizing transformation) is
analyzed using OLS and the results are transformed back to the original

scale. But this approach has many drawbacks, see for example Myers and Montgomery (1997).

Myers and Montgomery (1997) advocated the use of GLM in this context. Since GLM make use of the natural distribution of the response variable, they are expected to perform better than the traditional approaches. Myers and Montgomery (1997) used GLM to analyze some real data sets involving non-normal response variables. They observed that, for these data sets, models built using GLM perform better than those built with OLS and transformation approach with respect to response estimation and prediction. Lewis, Montgomery and Myers (2001a) attempted a simulation study on analysis of designed experiments involving non-normal response variable using GLM. They found that even for a small sample, performance of GLM is satisfactory with respect to the above mentioned criteria. However, they have not compared this performance with the above mentioned traditional approaches. Moreover, for each of the probability distribution of the response variable, their study is limited to only a pre-specified parameter combination. This does not seem to be adequate for assessing the overall performance of GLM.

Patil and Kulkarni (2011) attempted a comparative study among the three approaches, namely the ordinary least squares (OLS), log transformation (variance stabilizing transformation) and GLM approach for analyzing 2^n factorial experiments for exponentially distributed response variable. Their findings are based on theoretical as well as a broader simulation based study, covering a reasonable subset of the parameter space representing a variety of situations that could be encountered in practice. The study revealed that, the GLM approach performs the best among the three approaches with respect to interval estimation of expected responses. As a continuation of that work, in the present work we propose three new methods for the problem of interval estimation under consideration and attempt to compare the GLM approach with these methods. The proposed methods are based on Edgeworth expansion and exact distribution of a pivotal quantity.

The organization of material is as follows: the underlying design matrix and its properties are discussed in Section 5.2. Section 5.3 discusses analytical properties including expected lengths and coverage probabilities of CI based on GLM and three newly proposed methods. In Section 5.4 we compare among the methods studied in Section 5.3 based on simulation study. The method based on exact distribution of a pivotal quantity turns out to be the best among the above mentioned methods. Section 5.5 presents analysis of a real data set and Section 5.6 gives concluding remarks.

5.2. Nature of the Underlying Design Matrix

Throughout this paper, we consider a 2-level factorial experiment with $r(> 1)$ replications, where the number of parameters (m) is equal to the number of distinct treatment combinations (k). Note that, the underlying model can be viewed as a regression model with design matrix (cf. Myers and Montgomery, 1997),

$$X_{n \times k} = \begin{pmatrix} A \\ ... \\ A \end{pmatrix} \tag{5.1}$$

where $n = rk$ and A is a $k \times k$ Hadamard matrix (i.e. binary square matrix having elements ± 1 with mutually orthogonal columns). Note that, the matrix A corresponds to a single replicate of the underlying design matrix. The matrix being a Hadamard matrix, we have

$$A^{-1} = (1/k)A' \tag{5.2}$$

Furthermore, A can be taken as a symmetric matrix. In particular, design matrices of 2^n full factorial experiments and 2^{n-k} regular fractions where the two levels are denoted by $+1$ and -1 possess this structure. For example, the design matrix corresponding to a single replicate of a 2^2 experiment given by

$$A = \begin{pmatrix} 1 & 1 & 1 & 1 \\ 1 & 1 & -1 & -1 \\ 1 & -1 & 1 & -1 \\ 1 & -1 & -1 & 1 \end{pmatrix} \tag{5.3}$$

satisfies the above requirements. In a particular model, the linear component corresponds to $A\underline{\beta}$ where, the components of the vector of parameters $\underline{\beta}$ correspond to various main effects and interactions. For example, for 2^2 full factorial experiments, $\underline{\beta} = (\beta_0, \beta_1, \beta_2, \beta_{12})$ where β_i; $i = 1, 2$ represent the main effects, β_{12} represents the interaction effect and β_0 represents the overall mean effect.

From (5.1) and the assumptions made on the matrix A, it follows that

$$X'X = rkI_k, \quad XX' = \begin{pmatrix} kI_k & ... & kI_k \\ . & & . \\ kI_k & ... & kI_k \end{pmatrix} \tag{5.4}$$

where I_k is the identity matrix of order k.

In the next section we work out the theoretical details for the GLM based CI and CI based on some new proposed methods.

5.3. Some Analytical Methods for Obtaining CI for the Mean Response

Throughout this paper, we assume that the response variable Y is an exponentially distributed random variable. $Y_{ij}, i = 1, ..., k; j = 1, ..., r$ are observations on Y and $\underline{Y} = (Y_{11}, ..., Y_{kr})$. Let $\bar{Y}_{i.} = (1/r) \sum_{j=1}^{r} Y_{ij}$. μ_i denotes the expected response corresponding to the ith regressor vector x_i, which is i^{th} row of coefficient matrix X given in (5.1) corresponding to i^{th} treatment combination. In this section we give the confidence limits, length of confidence intervals ($LOCI$) around expected responses and coverage probability for the GLM approach and three newly suggested methods.

5.3.1. *GLM approach*

For GLM approach, the model is defined through a link function $\eta = X'\beta = g(\underline{\mu})$ where $g(.)$ is an appropriate monotonic and differentiable link function between the mean and the linear predictor η. The expected response is given by $E\underline{Y} = \underline{\mu} = g^{-1}(\eta)$ where g^{-1} is operated componentwise. For example, the identity link function produces the model $\mu = \eta = X'\beta$. Another link function is the log link, $\ln(\mu) = X'\underline{\beta}$ which produces the model $\mu = \exp(X'\underline{\beta})$. There exist two types of links, namely canonical and noncanonical link. The distribution of response variable Y is assumed to be a member of the exponential family. We refer to McCullagh and Nelder (1989) for more theoretical details of GLM.

Our response variable $\underline{Y}$ is exponentially distributed, which is a member of the exponential family. For such response variable, the canonical link selects $\eta_i = \mu_i^{-1}$, which produces $\mu_i = 1/X'_i\underline{\beta}$, which is known to be 'reciprocal link'. There are some drawbacks of this link, for example, certain estimates of $\underline{\beta}$ might lead to negative values of the estimated response, while the response values of exponentially distributed model are nonnegative. Therefore, we choose the noncanonical 'log link', which overcomes this drawback. For this link, the underlying model is

$$E(Y_{ij}) = \mu_i = \exp(X'_i\beta) \quad i = 1, ..., k, j = 1, ..., r. \quad (5.5)$$

Here x'_i, is i^{th} row of A in the design matrix given in equation (5.1). We denote this by $E(\underline{Y}) = \exp(X'\underline{\beta})$.

Parameter estimation in GLM is performed using the method of maximum likelihood estimation (MLE), which is in general an iterative process and may not yield closed form expressions of MLE However, it is important and interesting to note that, when the number of distinct treatment combinations (k) is equal to the number of parameters (m), simple closed form expressions of MLE are available. Note that, under the model (5.5), A^{-1} exists where, A is the coefficient matrix corresponding to a single replicate as defined in (5.3). Therefore, $\underline{\mu}$ (the vector of expected responses corresponding to a single replicate) is a one-one function and hence a reparameterization of $\underline{\beta}$, namely $\underline{\beta} = A^{-1}log(\underline{\mu})$. For an exponential distribution, MLE of the population mean is sample mean therefore, using invariance property of the MLE and (5.5) it follows that the estimated response is available in closed form, namely

$$\hat{Y}_{ij} = \hat{\mu}_i = \exp{(X_i' \hat{\underline{\beta}})} = \bar{Y}_{i.} \tag{5.6}$$

which is simply the mean of all observations receiving i^{th} treatment combination. (Note that, if m is less than k, A^{-1} does not exist and this closed form expression is not possible. In fact this is the reason we are assuming that $m = k$.) Let $\underline{\hat{Y}} = (\hat{Y}_{11}, \hat{Y}_{12}, ..., \hat{Y}_{kr}.)$

The CI for mean response μ_i is given by, $exp(X_i' \hat{\underline{\beta}} \pm Z_{\alpha/2} \hat{\sigma}_i)$ (cf. Myers and Montgomery, 1997) where, in the notation of Myers and Montgomery (1997), $\hat{\sigma}_i = (a(\phi))^{-1} \sqrt{x_i'(X'\Delta \hat{V} \Delta X)^{-1}x_i}$. $\hat{V}$ is the diagonal matrix whose i^{th} diagonal element is estimate of $Var(Y_i)$ and for our problem Δ is the diagonal matrix whose i^{th} diagonal element Δ_i is the derivative of $(-1/\mu_i)$ with respect to $X_i'\underline{\beta}$. For the design matrix (5.1) using (5.2) and (5.4), noting that for exponential response variable $a(\phi) = -1$, $\Delta_i = exp(-X_i' \hat{\underline{\beta}})$ and $Var(Y_i) = exp(2X_i' \hat{\underline{\beta}})$ this simplifies to $\hat{\sigma}_i = 1/\sqrt{r}$. Consequently the CI for μ_i is given by

$$\bar{Y}_{i.} exp(\pm Z_{\alpha/2}/\sqrt{r}) \tag{5.7}$$

The corresponding $LOCI$ is given by $\bar{Y}_{i.} (\exp(Z_{\alpha/2}/\sqrt{r}) - \exp(-Z_{\alpha/2}/\sqrt{r}))$. In particular, for $\alpha = 0.05$ and $r = 2$, this reduces to $\bar{Y}_{i.} 3.7484$. It is clear that, $LOCI$ in this case is directly proportional to $\bar{Y}_{i.}$ and the corresponding $E(LOCI)$ is given by

$$E(LOCI) = 3.7484\mu_i \tag{5.8}$$

In the sequel we discuss three new methods for obtaining CI for the expected response μ_i. The proposed methods are also based on the same statistic $\bar{Y}_i$. but make use of its other distributional aspects under exponential distribution.

5.3.2. *Edgeworth expansion based CI*

Let $Y_1, Y_2, ..., Y_n$ be i.i.d. observations with mean $\theta = EY_1$, variance $\sigma^2 = E(Y_1 - \theta)^2$, standardized skewness $\gamma = \sigma^{-3} E(Y_1 - \theta)^3$ and standardized kurtosis $\kappa = \sigma^{-4} E(Y_1 - \theta)^4 - 3$. Define polynomials

$$q_1(x) = \gamma(2x^2 + 1)/6$$

$$q_2(x) = x\{(1/12)\kappa(x^2 - 3) - (1/18)\gamma^2(x^4 + 2x^2 - 3) - (1/4)(x^2 + 3)\}$$

$$q_{21}(x) = q_1(x)q_1{}'(x) - (1/2)xq_1(x)^2 - q_2(x) \tag{5.9}$$

where $q_1{}'$ denotes derivative of q_1 with respect to x. Let $\hat{\theta}$, $\hat{\sigma}$, $\hat{\gamma}$, $\hat{\kappa}$, $\hat{q_1}$, $\hat{q_2}$, $\hat{q_{12}}$ denote the maximum likelihood estimates (MLE) of the respective quantities. Let $T = \sqrt{n}(\hat{\theta} - \theta)/\hat{\sigma}$ be the statistic with c.d.f. $K(x) = P(T \leq x)$. Then it can be shown that (cf. Hall, 1988, Section 3) $K(x)$ admits Edgeworth expansion given by

$$K(x) = \Phi(x) + n^{-1/2} q_1(x)\phi(x) + n^{-1} q_2(x)\phi(x) + O(n^{-3/2}) \tag{5.10}$$

where Φ and ϕ denote distribution and density functions of the standard normal distribution, respectively.

Furthermore, if y_α is the αth quantile of $K(x)$, i.e. $P[T \leq y_\alpha] = \alpha$, then inverse Cornish-Fisher expansion yields, $y_\alpha = z_\alpha - n^{-1/2} q_1(z_\alpha) + n^{-1} q_{21}(z_\alpha) + O(n^{-3/2})$, where z_α is such that $\Phi(z_\alpha) = \alpha$. It can be shown that, (cf. Hall, 1988, Section 3) sample versions replacing unknown parameters by their sample estimates (in our case MLEs) also admit similar expansions. In particular,

$$\hat{y_\alpha} = z_\alpha - n^{-1/2} \hat{q_1}(z_\alpha) + n^{-1} \hat{q_{21}}(z_\alpha) + O(n^{-3/2}) \tag{5.11}$$

The third order correct, $(1 - 2\alpha)100\%$ CI for θ based on these quantities (cf. Hall, 1988, Section 3) is

$$I(1 - 2\alpha) = (\hat{\theta}(\alpha), \hat{\theta}(1 - \alpha)) \tag{5.12}$$

where $\hat{\theta}(\alpha) = \hat{\theta} + n^{-1/2} \hat{\sigma}\hat{y_\alpha}$ and $\hat{\theta}(1 - \alpha) = \hat{\theta} + n^{-1/2} \hat{\sigma}\hat{y}_{1-\alpha}$.

To apply this method of obtaining CI in our case, note that GLM approach yielded the estimated mean response as the mean $\bar{Y}_{i.}$ of the observations receiving i^{th} treatment, which is MLE of μ_i. GLM approach uses asymptotic properties of MLE to obtain CI for μ_i. Here we attempt to build CI for μ_i based on the same statistic $\bar{Y}_{i.}$, making use of other aspects of its actual distribution. Here Y_{ij} are exponentially distributed which is a skewed and leptokurtic distribution. Edgeworth expansion based methods are designed to make correction for skewness and kurtosis. We propose here use of the CI (5.12) for the expected response μ_i.

In our case Y_{ij}, $\quad i = 1, ..., k, j = 1, ..., r$ are i.i.d. exponential (μ_i) variates for which $\hat{\theta} = \hat{\mu}_i = \bar{Y}_{i.}, \hat{\sigma} = \bar{Y}_{i.}, \gamma = 2, \kappa = 6$. Then using equations (5.9) and (5.11) the $(1 - 2\alpha)100\%$ CI based on Edgeworth expansion are given by equation (5.12) after omitting $O(n^{-3/2})$ terms. This yields the CI $(\bar{Y}_{i.} - r^{-1/2}\bar{Y}_{i.}y_\alpha, \bar{Y}_{i.} + r^{-1/2}\bar{Y}_{i.}y_{1-\alpha})$. The corresponding $LOCI$ is $LOCI = r^{-1/2}\bar{Y}_{i.}(y_{1-\alpha} - y_\alpha)$. In particular for $r = 2, \alpha = 0.025, y_\alpha = -9.181, y_{1-\alpha} = 5.0878$ and $LOCI = 10.0896\bar{Y}_{i.}$. Hence, the $E(LOCI)$ is

$$E(LOCI) = 10.0896\mu_i \tag{5.13}$$

5.3.3. *Shortest length CI*

If the distribution of $T = \sqrt{n}(\hat{\theta} - \theta)/\hat{\sigma}$ is known then we may choose v, w to minimize $v + w$ subject to

$$P(-w \leq T \leq v) = 1 - 2\alpha, 0 < \alpha < 1/2. \tag{5.14}$$

Then $(1 - 2\alpha)100\%$ CI for θ (cf. Hall, 1988) is

$$I_0 = (\hat{\theta} - n^{-1/2}\hat{\sigma}v, \hat{\theta} + n^{-1/2}\hat{\sigma}w). \tag{5.15}$$

It has same coverage as the equal tailed interval $(\hat{\theta}(\alpha), \hat{\theta}(1-\alpha))$, but usually has strictly shorter length than the one based on Edgeworth expansion. Suppose the Edgeworth expansion (5.10) holds. Let $\phi_i(x) = q_i(x)\phi(x)$ for $i \geq 1$, $\phi_0(x) = \Phi(x)$, $\phi_{ik}(x) = (\partial/\partial x)^k \phi_i(x)$, and $\Psi_{ik} = \phi_{ik}(z_{1-\alpha})$. Then it can be shown that (cf. Hall, 1988) the numbers v, w in (5.14) which minimize $v + w$ subject to (5.14) satisfy

$$v = z_{1-\alpha} + n^{-1/2}v_1 + n^{-1}v_2 + O(n^{-3/2}), w = z_{1-\alpha} - n^{-1/2}v_1 + n^{-1}v_2 + O(n^{-3/2}) \tag{5.16}$$

where $v_1 = -\Psi_{11}\Psi_{02}^{-1}$ and $v_2 = (\frac{1}{2}\Psi_{11}^2\Psi_{02}^{-1} - \Psi_{20})\Psi_{01}^{-1}$.

Further a third order correct approximation to (5.15) is $(\hat{\theta} - n^{-1/2}\hat{\sigma}\hat{v}, \hat{\theta} + n^{-1/2}\hat{\sigma}\hat{w})$ where $\hat{v}$ and $\hat{w}$ are obtained by replacing unknown parameters in v and w by their MLE. As explained earlier, in our case, $\hat{\theta} = \hat{\mu}_i = \bar{Y}_{i\cdot}, \hat{\sigma} = \bar{Y}_{i\cdot}$ and $n = r$. Therefore, the shortest length CI is given by $(\bar{Y}_{i\cdot} - r^{-1/2}\bar{Y}_{i\cdot}\hat{v}, \bar{Y}_{i\cdot} + r^{-1/2}\bar{Y}_{i\cdot}\hat{w})$ and the corresponding $LOCI = \bar{Y}_{i\cdot}.r^{-1/2}(\hat{w} + \hat{v})$. In particular, for our case $r = 2$, $\alpha = 0.025$, $\hat{v} = v = 3.8836$, $\hat{w} = w = 5.0202$, and this reduces to $LOCI = 6.2959\bar{Y}_{i\cdot}$. The corresponding $E(LOCI)$ is

$$E(LOCI) = 6.2959\mu_i \qquad (5.17)$$

5.3.4. *A method based on the quantiles of a pivotal quantity (Exact method)*

Since our response variable Y_{ij} is exponentially distributed variable, $T_i = \sum_{j=1}^{r} Y_{ij}$ follows a Gamma (μ_i, r) distribution and hence T_i/μ_i follows Gamma $(1, r)$ distribution. Thus, in this case $(1 - \alpha)100\%$ CI for T_i/μ_i is $(L(p), U(p))$ where, $L(p) = \zeta_p$ and $U(p) = \zeta_{1-(\alpha-p)}$, $0 \le p \le \alpha$. Here ζ_p denotes p^{th} quantile of Gamma $(1, r)$ distribution. These CI can be inverted to give CI for μ_i as, $(T_i/U(p), T_i/L(p))$ and therefore, $LOCI(p) = T_i(\frac{U(p)-L(p)}{U(p)L(p)}) = T_i h(p)$ where $h(p) = \frac{U(p)-L(p)}{U(p)L(p)}$. Note that, for given level α, p can be selected so as to minimize the length of the resulting CI i.e., $p^* = arg\min_p LOCI(p) = arg\min_p h(p)$

It can be seen that $h(p)$ is minimum at $p^* = 0.049692$ giving $h(p^*) = 2.72891$. This gives, CI for μ_i as,$(T_i/10.5308, T_i/0.3541)$ and the corresponding $LOCI = 2.7289T_i$ and $E(LOCI) = 2.7289E(T_i)$. As T_i follows Gamma(μ_i, r) distribution, $E(T_i) = \mu_i r$. In particular for $r = 2$ this gives

$$E(LOCI) = 5.4578\mu_i \qquad (5.18)$$

Note that, the $E(LOCI)$ given in (5.8), (5.13), (5.17) and (5.18) for all the above discussed methods depend only on the particular μ_i under consideration and not on other components of μ. Of course this is not surprising because all these methods make use of the same statistic $\bar{Y}_{i\cdot}$ for constructing the desired CI.

The $LOCI$ given by the GLM approach are smallest and those given by Edgeworth expansion based method are largest. Those given by exact method are little larger than those of GLM.

5.3.5. *Coverage probability for various methods*

i) Coverage probability for GLM approach

For GLM approach, the coverage probability for 95% CI is, $P(\bar{Y}_i. \exp(-Z_{\alpha/2}/\sqrt{r}) \leq \mu_i \leq \bar{Y}_i. \exp(Z_{\alpha/2}/\sqrt{r})) = P(r * exp(-1.96/\sqrt{r}) \leq \sum_{j=1}^r Y_{ij}/\mu_i \leq r * exp(1.96/\sqrt{r}))$. As Y_{ij} are exponentially distributed, $\sum_{j=1}^r Y_{ij}/\mu_i$ follows Gamma$(1,r)$ distribution. Therefore, the above coverage probability is $G(r * exp(1.96/\sqrt{r}), 1, r) - G(r * exp(-1.96/\sqrt{r}), 1, r)$, where $G(x, 1, r)$ is the cumulative distribution function at x for Gamma$(1, r)$ distribution. Note that, this does not depend on the value of μ_i. It can be seen that, this coverage probability is an increasing function of r and for $r = 2$ it is 0.9067. Thus for GLM approach, all coverages should be equal to 90.67% irrespective of the value of underlying μ.

ii) Coverage probability for Edgeworth expansion based CI

For Edgeworth expansion based CI, the coverage probability is, $P(\bar{Y}_i. + r^{-1/2}\bar{Y}_i.y_\alpha \leq \mu_i \leq \bar{Y}_i. + r^{-1/2}\bar{Y}_i.y_{1-\alpha}) = P(r/(1 + \frac{y_\alpha}{\sqrt{r}}) \leq \sum_{j=1}^r Y_{ij}/\mu_i \leq r/(1 + \frac{y_{1-\alpha}}{\sqrt{r}}))$. Recall that, $\bar{Y}_i. = (1/r)\sum_{j=1}^r Y_{ij}$, $y_\alpha = -9.181$, $y_{1-\alpha} = 5.0878$ and the distribution of $\sum_{j=1}^r Y_{ij}/\mu_i$ is Gamma$(1, r)$. In particular, for $r = 2$ and $\alpha = 0.025$, proceeding as in GLM, the coverage probability for Edgeworth expansion based CI is 0.9288. Note that, this also does not depend on the value of μ_i.

iii) Coverage probability for Shortest length CI

For shortest length CI, the coverage probability is, $P(\bar{Y}_i. - r^{-1/2}\bar{Y}_i.v \leq \mu_i \leq \bar{Y}_i. + r^{-1/2}\bar{Y}_i.w) = P(r/(1 - \frac{v}{\sqrt{r}}) \leq \sum_{j=1}^r Y_{ij}/\mu_i \leq r/(1 + \frac{w}{\sqrt{r}}))$. Recall that, $v = 3.8836$ and $w = 5.0202$ and the distribution of $\sum_{j=1}^r Y_{ij}/\mu_i$ is Gamma$(1, r)$. In particular, for $r = 2$, $\alpha = 0.025$ and applying the similar arguments used for the coverage probability for Edgeworth expansion based CI, the coverage probability for shortest length CI is 0.9275. As before, this also does not depend on the value of μ_i.

iv) Coverage probability for Exact method

For exact method, the coverage probability is, $P(T_i/U(p) \leq \mu_i \leq T_i/L(p))$. Note that, T_i/μ_i follows Gamma$(1, r)$ distribution, $L(p) = 0.3541$ and

$U(p) = 10.5308$. In particular, for $r = 2$, and $p = 0.049692$ the coverage probability is 0.95. Recall that, for exact method we have chosen $L(p)$ and $U(p)$ to minimize $LOCI(p)$ subject to $P(L(p) \leq T_i/\mu_i \leq U(p)) = 0.95$.

It follows that, the coverage probability of the exact method is equal to the confidence level (C.L.), that of the Edgeworth expansion based approaches is a little less than C.L. and that of GLM approach is smaller than C.L..

In the next section we assess the performance of above methods based on a simulation study. Although theoretical results are available for the methods discussed above, we also study simulation based performance of these methods for the sake of completeness.

5.4. Simulation Study

5.4.1. *Details of the simulation procedure*

The simulation study is carried out for a 2^2 full factorial experiment with two replications, where the response variable is exponentially distributed. Thus, we have $k = 2^2 = 4$ parameters, $r = 2$ replications and $n = rk = 8$ runs. Since in many real life situations particularly related to industry, it is very expensive to go for large number of replications, we consider only two replicates. The coefficient matrix is $X = \begin{pmatrix} A \\ A \end{pmatrix}$ where A is given in (5.3). The vector of responses $\underline{Y}$ is an eight dimensional vector.

For two replications, $E(\underline{Y}) = \underline{\mu} = (\underline{\mu}_1, \underline{\mu}_1)'$. The following 9 representative sets of $\underline{\mu}_1$ are selected for simulation. For each selected μ, we generated 5000 vectors of $\underline{Y}$ whose components are independent exponential variates with $E(\underline{Y}) = \underline{\mu}$. For each of these observation vectors $\underline{Y}$, $LOCI$ for each of the methods discussed in Section 5.3 are computed based on the procedures stated there. The box plots of averages of the 5000 $LOCI$ so obtained and the corresponding coverage probabilities (which are the proportions of the 5000 CIs, that cover the underlying μ_i) are plotted in Figure 5.1(a) and Figure 5.1(b) respectively.

5.4.2. *Observations based on simulation study*

The simulated $E(LOCI)$ and coverages for the methods of Section 5.3 conformed with the theoretical ones given in equations (5.8), (5.13), (5.17) and (5.18) and the theoretical coverages discussed in Subsection 5.3.5.

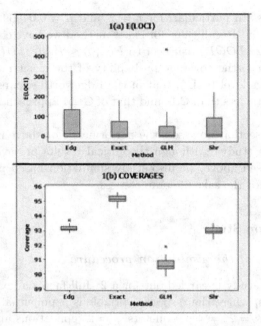

Fig. 5.1.   Box plots for GLM approach, Edgeworth expansion based CI, Shortest length
CI and Exact method.

5.5. Analysis of Real Data Set

We analyze a real data set extracted from Feigl and Zelen (1965), p. 830 so
as to fit our setup. The original data consists of 33 observations on three
variables namely WBC count; Survival time, Y (in weeks) in leukemia patients from the date of diagnosis; and presence or absence of a morphological
characteristics of white blood cells (AG), where AG is assigned value 1 and
-1 according as presence and absence of the morphological characteristics
respectively. The response variable Y which is survival time (in weeks) in
leukemia patients is assumed to follow an exponential distribution. Two
factors considered are $X_1 = WBC$ count taking values -1 (lower level) if
WBC $< 10,000$ and $+1$ otherwise and $X_2 =$ morphological characteristics
taking values as stated above. We have considered only a subset of the
data consisting of suitably chosen eight observations so as to fit the setup
considered in our study for illustration purpose. The corresponding design
matrix and response vector (Y) are given in Table 5.1. The CI and length
of CI $(LOCI)$ for expected survival time for each of the GLM approach,

Table 5.1. Design matrix and the Survival time of leukemia patients

X_1	X_2	$X_1 X_2$	Survival time of leukemia patients (Y)
1	1	1	4
1	-1	-1	108
-1	1	-1	65
-1	-1	1	65
1	1	1	43
1	-1	-1	5
-1	1	-1	7
-1	-1	1	100

Table 5.2. Confidence interval (CI) and *LOCI* for the Real Data set

Y	GLM		Edg		shr		Ext	
	CI	LOC	CI	LOC	CI	LOC	CI	LOC
4	(5.88,93.97)	88.1	(-129.06,108.04)	237.11	(-41.03,106.92)	159.88	(4.46,132.69)	128.23
108	(14.13,225.92)	211.8	(-310.30,259.77)	570.06	(-98.66,257.06)	384.38	(10.73,319.03)	308.30
65	(9.00,143.95)	134.9	(-197.71,165.51)	363.22	(-62.86,163.79)	244.92	(6.84,203.28)	196.44
65	(20.63,329.88)	309.2	(-453.09,379.30)	832.39	(-144.05,375.36)	561.27	(15.67,465.84)	450.17
43	(5.88,93.97)	88.1	(-129.06,108.04)	237.11	(-41.03,106.92)	159.88	(4.46,132.69)	128.23
5	(14.13,225.92)	211.8	(-310.30,259.77)	570.06	(-98.66,257.06)	384.38	(10.73,319.03)	308.30
7	(9.00,143.95)	134.9	(-197.71,165.51)	363.22	(-62.86,163.79)	244.92	(6.84,203.28)	196.44
100	(20.63,329.88)	309.2	(-453.09,379.30)	832.39	(-144.05,375.36)	561.27	(15.67,465.84)	450.17

Edgeworth expansion based CI, Shortest length CI and Exact method are tabulated in the Table 5.2 for this data set.

From Table 5.2 it is seen that, $E(LOCI)$ for GLM approach are uniformly smaller than that of remaining methods. For Edgeworth expansion based CI, the lower limits are taking negative values, which are undesirable. All these observations are consistent with the findings of the preceding sections. Thus, the real data set analysis also supports the use of Exact method and GLM approach.

5.6. Concluding Remarks

We have attempted a comparative study for analyzing 2^n factorial experiments for exponentially distributed response variable among the GLM approach and three new proposed methods.

GLM based CI have smaller expected lengths than all other methods but the coverage probabilities are much smaller than the desired level. The Edgeworth expansion based methods have a little improved coverage probabilities than GLM but still below the confidence level. The Exact method discussed in Section 5.3 attempts to reduce the $E(LOCI)$ than the Edgeworth expansion based methods, still maintaining the coverage probability to the desired level and thus outperforms all the above methods and is recommended for practical use.

Acknowledgment

H. V. Kulkarni was supported by the grants received by Government of India, Department of Science and Technology, India under the project Reference No.: SR/54/MS: 306/05.

References

1. Feigl, P. and Zelen, M. (1965). Estimation of exponential survival probabilities with concomitant information, *Biometrics* **21**, 826–838.
2. Hall, P. (1988). Theoretical Comparison of Bootstrap Confidence Intervals, *Ann. Statist.* **16**, 927–985.
3. Lewis, S., Montgomery, D. and Myers, R. (2001a). CI Coverage for Designed Experiments Analyzed with GLMs, *J. Qual. Tech.* **33**, 279–291.
4. Lewis, S., Montgomery, D. and Myers, R. (2001b). Examples of Designed Experiments with Nonnormal Responses,*J. Qual. Tech.* **33**, 265–278.
5. McCullagh, P. and Nelder, J. (1989). *Generalized Linear Models.* (2nd ed.), Chapman and Hall, London.
6. Myers, R. and Montgomery, D. (1997). A Tutorial on Generalized Linear Models, *J. Qual. Tech.* **29**, 274–290.
7. Patil, S. and Kulkarni, H. (2011). Fractorial Experiments with Exponentially Distributed Response Variable, *Appld. Math. Sc.* **5**, 459–476.

Chapter 6

Predictive Influence of Variables in a Linear Regression Model when the Moment Matrix is Singular

Md. Nurul Haque Mollah and S. K. Bhattacharjee

University of Rajshahi, Bangladesh

The choice of explanatory variables in a multiple linear regression model has been studied for Bayesian prediction. In presence of perfect multicollinearity, the problem is studied when the goal is to predict a vector of future responses. To test the significance of influence of variables for Bayesian prediction in the presence of perfect multicollinearity, Kullback-Leibler (K-L) divergence was decomposed into the discrepancies due to the location and scale parameters, respectively. However, one problem in that decomposition is that the derivation of the distributional form of discrepancy due to the location parameter as well as K-L divergence was very difficult. To overcome this problem, in this paper we consider the predictive influence of variables in a normal multiple linear regression model in the presence of perfect multicollinearity when the goal is to predict a single future response. Then we have derived the distributional results of both discrepancies due to the location and scale parameters, respectively and the test procedure is performed.

Keywords: Influence of variables and variable selection, Kullback-Leibler divergence, Multiple linear regression model, Perfect multicollinearity, Predictive distribution.

Contents

6.1. Introduction

The influence of explanatory variables in a multiple linear regression model has been studied in the classical approach for selecting the best regression model by Bhatterjee (2007). Partial F-tests, the multiple correlation coefficient R^2 and the C_p-plot technique are generally used for this purpose. Multicollinearity is a serious problem in estimating the regression parameters as a violation for the assumption of ordinary least square (OLS) method. Several methods have been proposed as a remedies of multicollinearity, where ridge regression is a widely used technique (Wedderburn, 2007). In Bayesian approach, influence of explanatory variables in a multiple linear regression model has been studied in absence of multicollinearity (Bhatterjee and Dunsmore, 1995; Bhatterjee, 2007; Lindley, 1968). A problem of the influence of variables in a logistic model in the predictive approach have been also considered by Bhatterjee and Dunsmore (1991). The predictive influence of variables in a normal multiple linear regression model in the presence of perfect multicollinearity when the goal was to predict a vector of future responses have been studied by Mollah and Bhattacharjee (2007). To test the significance of influence of variables in the presence of perfect multicollinearity, The K-L divergence (Kullback and Leibler, 1951) was decomposed into the discrepancies due to the location and scale parameters, respectively. However, one problem in that decomposition is that the derivation of the distributional form of discrepancy due to the location parameter as well as K-L divergence was very difficult. The aim of this paper is to assess the influence of explanatory variables in a normal multiple linear regression model for Bayesian prediction in the presence of perfect multicollinearity. To do this, an attempt is made to derive the predictive distribution of a single future response in the linear regression model using vague prior density. To assess the influence of explanatory variables, we formulate the K-L divergence between the predictive distributions for a single future response with and without a specified subset of the explanatory variables. To test the significance of influence of variables, we decompose the formulated K-L divergence into the discrepancies due to the location and scale parameters, respectively. Then we derive the distributional results of both discrepancies and the test procedure is performed. It should be noted here that a trial version of our current proposal was also discussed by Mollah and Bhattacharjee (2008).

In section 6.2 we discuss the methodology of the current proposal. A simple illustration using a little bit artificial dataset is given in section 6.3 and section 6.4 contains the conclusion.

6.2. Methodology for Bayesian Prediction in Presence of Perfect Multicollinearity

Let us consider the multiple linear regression model as

$$y = \beta_0 + \beta_1 x_1 + \beta_2 x_2 + ... + \beta_{k-1} x_{k-1} + u, \tag{6.1}$$

which can be written in matrix notation as

$$\boldsymbol{y} = X\boldsymbol{\beta} + \boldsymbol{u}, \tag{6.2}$$

where $\boldsymbol{y}$ is the n vector of responses of the dependent variable, X is an $n \times k$ matrix of observed values, $\boldsymbol{\beta}$ is a $k \times 1$ vector of unknown regression coefficients, $\mathbf{u}$ follows $N(0, \sigma^2.I)$. The predictive density for a future response y^f generated by the same model (6.2) as $y^f = \boldsymbol{x}^f\boldsymbol{\beta} + u^f$, where $\boldsymbol{x}^f = (1, x_1^f, x_2^f, ..., x_{k-1}^f)$ and $u^f \sim N(0, \sigma^2)$, is given by

$$p(y^f \mid \boldsymbol{x}^f, X, \boldsymbol{y}) = \int p(y^f \mid \boldsymbol{x}^f, \boldsymbol{\beta}, \sigma) p(\boldsymbol{\beta}, \sigma \mid X, \boldsymbol{y}) \partial\boldsymbol{\beta}\partial\sigma, \tag{6.3}$$

where $p(\boldsymbol{\beta}, \sigma \mid X, \boldsymbol{y})$ is the joint posterior density of $\boldsymbol{\beta}$ and σ, and

$$p(y^f \mid \boldsymbol{x}^f, \boldsymbol{\beta}, \sigma) \propto \frac{1}{\sigma} \exp\left\{ -\frac{1}{2\sigma^2}(y^f - \boldsymbol{x}^f\boldsymbol{\beta})^2 \right\}. \tag{6.4}$$

Let us suppose we want to measure the joint influence of any subset of size r explanatory variables which is denoted by $X^{(r)}$, where $r = 1, 2, ..., (k-1)$. For convenience and without loss of generality let us assume that $X^{(r)}$ contains the last r explanatory variables $X_{k-r}, X_{k-r+1}, ..., X_{k-1}$, where, it is assumed that the variables $X^{(r)}$ have not been observed. Let us construct the reduced linear regression model excluding variables $X^{(r)}$ as

$$y = \theta_0 + \theta_1 x_1 + \theta_2 x_2 + ... + \theta_{k-r-1} x_{k-r-1} + v. \tag{6.5}$$

In matrix notation, the model can be written as

$$\boldsymbol{y} = X_R\boldsymbol{\theta} + \boldsymbol{v}, \tag{6.6}$$

where $\boldsymbol{y}$ is the n vector of responses of the dependent variable as before, X_R is an $n \times (k - r)$ matrix of observed values, $\boldsymbol{\theta}$ is a $(k - r) \times 1$ vector of unknown regression coefficients, ν follows $N(0, \sigma_R^2.I)$. The predictive density for a future response y^f with the explanatory variables

$x_R^f = (1, x_1^f, x_2^f, ..., x_{k-r-1}^f)$ for model (6.6) is given by

$$p(y^f \mid x_R^f, X_R, y) = \int p(y^f \mid x_R^f, \theta, \sigma_R) p(\theta, \sigma_R \mid X_R, y) \partial\theta \partial\sigma_R, \quad (6.7)$$

where $p(\theta, \sigma_R \mid X_R, y)$ is the joint posterior density for θ and σ_R, and

$$p(y^f \mid x_R^f, \theta, \sigma_R) \propto \frac{1}{\sigma_R} \exp\left\{-\frac{1}{2\sigma_R^2}(y^f - x_R^f\theta)^2\right\}. \quad (6.8)$$

The K-L divergence $\mathscr{D}_{KL}$ (Kullback and Liebler, 1951) has been used as a measure of discrepancy between two predictive distributions (6.3) and (6.7) in some articles (Bhattacharjee and Dunsmore, 1995; Bhattacharjee, 2007; Mollah and Bhattacharjee, 2008). It is defined by

$$\mathscr{D}_{KL} = \int p(y^f \mid x_R^f, X_R, y) \log \frac{p(y^f \mid x_R^f, X_R, y)}{p(y^f \mid x^f, X, y)} dy^f. \quad (6.9)$$

In this paper, we also consider the K-L divergence as defined in (6.9) for the same purpose when the moment matrix $(X^T X)$ is singular, where the reduced moment matrix $(X_R^T X_R)$ may be singular or not. The moment matrix $X^T X$ will be singular when $n \times k$ matrix X is of rank q with $0 \le q < k$. This occurs, for example, when the observations of the explanatory variables satisfy an exact linear relation. This situation is commonly termed 'perfect multicollinearity'. Another example in which $n < k$, the moment matrix $(X^T X)$ becomes singular.

To derive the predictive density (6.3) when the moment matrix $(X^T X)$ is singular, let us first summarize the derivation of the joint posterior density $p(\beta, \sigma \mid X, y)$ of the model parameters β and σ. The regression model (6.2) is reparameterize as follows:

$$y = XP\gamma + u, \quad (6.10)$$

where $\gamma = P^T\beta$, a $k \times 1$ vector of parameters and P is a $k \times k$ orthogonal matrix such that

$$P^T X^T X P = \begin{bmatrix} D & 0 \\ 0 & 0 \end{bmatrix} = \text{diag}(\lambda_1, ..., \lambda_q, 0, ..., 0), \quad (6.11)$$

satisfying $\lambda_1 > ... > \lambda_q > 0$ with $D = \text{diag}(\lambda_1, ..., \lambda_q)$.

Then the "normal equations" for γ are

$$P^T X^T X P\gamma = P^T X^T y,$$
$$\implies \begin{bmatrix} D & 0 \\ 0 & 0 \end{bmatrix} \begin{bmatrix} \gamma_1 \\ \gamma_2 \end{bmatrix} = \begin{bmatrix} P_1^T X^T y \\ P_1^T X^T y \end{bmatrix} = \begin{bmatrix} P_1^T X^T y \\ 0 \end{bmatrix}, \quad (6.12)$$

where $\gamma^T = (\gamma_1^T : \gamma_2^T)$, with γ_1^T and γ_2^T are $q \times 1$ and $(k-q) \times 1$ vectors, respectively, and P_1 is a $k \times q$ submatrix of P given by $P = (P_1 : P_2)$. Also note that $P_2^T X^T = 0$ using equation (6.11). The complete solution of the normal equations in (6.12) is given by

$$\tilde{\gamma} = (P^T X^T X P)^* P^T X^T y + [I - (P^T X^T X P)^* P^T X^T X P] z, \quad (6.13)$$

where $(P^T X^T X P)^*$ denotes the generalized inverse (GI) of $P^T X^T X P$ and z is an arbitrary $k \times 1$ vector, with $z^T = (z_1^T : z_2^T)$. For convenience, we use the Moore-Penrose GI of $P^T X^T X P$ as follows:

$$(P^T X^T X P)^* = \begin{bmatrix} D^{-1} & 0 \\ 0 & 0 \end{bmatrix}. \quad (6.14)$$

Then equation (6.13) gives $\tilde{\gamma}_1 = D^{-1} P_1^T X^T y$ and $\tilde{\gamma}_2 = z_2$. Then we compute the vector $\beta = P\gamma$ of regression parameters as follows:

$$\tilde{\beta} = P\tilde{\gamma} = P(P^T X^T X P)^* P^T X^T y + [I - P(P^T X^T X P)^* P^T X^T X P] z,$$
$$= (X^T X)^* X^T y + [I - (X^T X)^* X^T X] Pz, \quad (6.15)$$

since $P(P^T X^T X P)^* P^T = (X^T X)^*$ is a GI of $X^T X$.

Now the likelihood function in terms of the parameters γ_1, γ_2 and σ based on the above discussion is given by

$$L(\gamma_1, \gamma_2, \sigma \mid y) \propto \frac{1}{\sigma^n} \exp\left\{ -\frac{1}{2\sigma^2} (y - XP\gamma)^T (y - XP\gamma) \right\}$$
$$\propto \frac{1}{\sigma^n} \exp\left\{ -\frac{1}{2\sigma^2} [a + (\gamma_1 - \tilde{\gamma}_1)^T D(\gamma_1 - \tilde{\gamma}_1)] \right\}, \quad (6.16)$$

which is independent of γ_2. Then we assume vague prior density for q elements of γ_1 and σ as follows:

$$p(\gamma_1, \sigma) \propto \frac{1}{\sigma}, \quad -\infty < \gamma_{1i} < \infty, i = 1, 2, ..., q \; 0 < \sigma < \infty. \quad (6.17)$$

Again we assume the prior density for $(k-q)$ elements of γ_2 given σ as follows:

$$p(\gamma_2 \mid \sigma) \propto \frac{1}{\sigma^{k-q}} \exp\left\{ -\frac{1}{2\sigma^2} (\gamma_2 - z_2)^T Q(\gamma_2 - z_2) \right\} \quad (6.18)$$

and assuming γ_2 is independent of γ_1 and Q is a nonsingular matrix. Combination of the prior pdf's in (6.17) and (6.18) with the likelihood function in (6.16) produces the posterior pdf for γ_1, γ_2 and σ as follows

$$p(\gamma_1, \gamma_2, \sigma | \boldsymbol{y}) \propto \frac{1}{\sigma^{n+k-q+1}} \exp \Big\{ - \frac{1}{2\sigma^2} [a + (\gamma_1 - \tilde{\gamma}_1)^T D (\gamma_1 - \tilde{\gamma}_1)$$
$$+ (\gamma_2 - \boldsymbol{z}_2)^T Q (\gamma_2 - \boldsymbol{z}_2) \Big\}. \tag{6.19}$$

On transforming from γ to β this posterior pdf becomes the joint posterior pdf for the regressions β and σ as follows:

$$p(\boldsymbol{\beta}, \sigma \mid X, \boldsymbol{y}) \propto \frac{1}{\sigma^{\nu+k+1}} \exp \Big\{ -\frac{1}{2\sigma^2} [a + (\boldsymbol{\beta} - \tilde{\boldsymbol{\beta}})^T (PFP^T)(\boldsymbol{\beta} - \tilde{\boldsymbol{\beta}})] \Big\}, \tag{6.20}$$

where

$$F = \begin{bmatrix} D & 0 \\ 0 & Q \end{bmatrix}$$

with $\nu = n - q$. In the current problem we assume $Q = \mathrm{diag}(\lambda^*_{q+1}, ..., \lambda^*_k)$ satisfying $\lambda_1 > ... > \lambda_q > \lambda^*_{q+1} > ... > \lambda^*_k > 0$, where λ^*_i's an arbitrary set of positive real numbers. A detail discussion about this posterior density can be found in Zelner (1971), pages 75-81.

Using (6.4) and (6.20) in (6.3), we obtain the predictive density for a future response y^f as

$$p(y^f \mid \boldsymbol{x}^f, X, \boldsymbol{y}) \propto \Big[1 + (\nu h)^{-1} (y^f - \boldsymbol{x}^f \tilde{\boldsymbol{\beta}})^2 \Big]^{-(\nu+1)/\nu}, \tag{6.21}$$

which follows univariate student t-distribution with mean $\boldsymbol{x}^f \tilde{\boldsymbol{\beta}}$ and variance $h = \tilde{\sigma}^2 [1 + \boldsymbol{x}^f (PFP^T)^{-1} \boldsymbol{x}^{f^T}]$, where $\tilde{\sigma}^2 = (\boldsymbol{y} - X\tilde{\boldsymbol{\beta}})^T (\boldsymbol{y} - X\tilde{\boldsymbol{\beta}})/\nu$. The derivation of the predictive density is given in appendix (A.1). Similarly, the predictive density (6.7) for y^f corresponding to the reduced model is obtained as

$$p(y^f \mid \boldsymbol{x}^f_R, X_R, \boldsymbol{y}) \propto \Big[1 + (\nu^* h^*)^{-1} (y^f - \boldsymbol{x}^f_R \tilde{\boldsymbol{\theta}})^2 \Big]^{-(\nu^*+1)/\nu^*}, \tag{6.22}$$

which follows univariate student t-distribution with mean $\boldsymbol{x}^f_R \tilde{\boldsymbol{\theta}}$ and variance $h^* = \tilde{\sigma}^2_R [1 + \boldsymbol{x}^f_R (QF_R Q^T)^{-1} \boldsymbol{x}^{f^T}_R]$, where $\nu^* = n - q^*$, $Q = (Q_1 : Q_2)$ is an orthogonal matrix such that $Q^T X_R{}^T X_R Q = \mathrm{diag}(d_1, ..., d_{q^*}, 0, ..., 0)$. Then $F_R = \mathrm{diag}(d_1, ..., d_{q^*}, d^*_{q^*+1}, ..., d^*_{k-r-1})$ satisfying $d_1 > ... > d^*_q > d^*_{q^*+1} > ... > d^*_{k-r-1} > 0$, where d^*_i's an arbitrary set of positive real numbers. Also note that $\tilde{\boldsymbol{\theta}} = Q\tilde{\gamma}^*$, where $\tilde{\gamma}^{*T} = (\tilde{\gamma}^{*T}_1 : \boldsymbol{z}^{*T}_2)$ with $\tilde{\gamma}^{*T}_1 = D_R^{-1} Q_1^T X_R{}^T \boldsymbol{y}$, where $D_R = \mathrm{diag}(d_1, ..., d_{q^*})$, $\boldsymbol{z}^*_2$ is an arbitrary vector of appropriate order and $\tilde{\sigma}^2_R = (\boldsymbol{y} - X_R \tilde{\boldsymbol{\theta}})^T (\boldsymbol{y} - X_R \tilde{\boldsymbol{\theta}})/\nu^*$.

An explicit expression for K-L divergence with Student's t-distributions is difficult to obtain. To overcome this problem, some authors, Jhonson and Geisser (1983) suggested the approximation of t-distribution to the normal distribution. Following them, some authors, Bhatterjee and Dunmore (1995), Mollah and Bhatterjee (2008) considered this approximation in their similar problems. The approximation of true model (likelihood function) to working model (quasi-likelihood function) also found in Wedderburn (1974) to overcome the problem of difficult data situation. Therefore, we approximate (6.21) and (6.22) to $N(\boldsymbol{x}^f\tilde{\boldsymbol{\beta}}, \; \delta^2)$ and $N(\boldsymbol{x}_R^f\tilde{\boldsymbol{\theta}}, \; \delta_R^2)$ respectively as the working model, where

$$\delta^2 = \frac{\nu}{\nu - 2}\tilde{\sigma}^2 \left[1 + \boldsymbol{x}^f(PFP^T)^{-1}\boldsymbol{x}^{fT}\right] \tag{6.23}$$

and

$$\delta_R{}^2 = \frac{\nu^*}{\nu^* - 2}\tilde{\sigma}_R^2 \left[1 + \boldsymbol{x}_R^f(QF_RQ^T)^{-1}\boldsymbol{x}_R^{fT}\right]. \tag{6.24}$$

Then the K-L divergence $\mathscr{D}_{\mathrm{KL}}$ is approximated as

$$\widehat{\mathscr{D}}_{\mathrm{KL}} = \frac{1}{2\delta^2}\left(\boldsymbol{x}^f\tilde{\boldsymbol{\beta}} - \boldsymbol{x}_R^f\tilde{\boldsymbol{\theta}}\right)^2 + \frac{1}{2}\left\{\frac{\delta_R{}^2}{\delta^2} - \log\frac{\delta_R{}^2}{\delta^2} - 1\right\} = L + S, \tag{6.25}$$

where $L = \frac{1}{2\delta^2}\left(\boldsymbol{x}^f\tilde{\boldsymbol{\beta}} - \boldsymbol{x}_R^f\tilde{\boldsymbol{\theta}}\right)^2$ is the discrepancy due to the difference of location parameters and $S = \frac{1}{2}\left\{\frac{\delta_R{}^2}{\delta^2} - \log\frac{\delta_R{}^2}{\delta^2} - 1\right\}$ is the discrepancy due to the difference of scale parameters. If the moment matrices (X^TX) and $(X_R{}^TX_R)$ are both non-singular, then $\tilde{\boldsymbol{\beta}} = (X^TX)^{-1}X^T\boldsymbol{y}$, $\tilde{\boldsymbol{\theta}} = (X_R{}^TX_R)^{-1}X_R{}^T\boldsymbol{y}$,

$$\delta^2 = \frac{(n - k - 1)}{(n - k - 3)}\tilde{\sigma}^2 \left[1 + \boldsymbol{x}^f(X^TX)^{-1}\boldsymbol{x}^{fT}\right] \tag{6.26}$$

and

$$\delta_R{}^2 = \frac{(n - k + r - 1)}{(n - k + r - 3)}\tilde{\sigma}_R^2 \left[1 + \boldsymbol{x}_R^f(X_R{}^TX_R)^{-1}\boldsymbol{x}_R^{fT}\right]. \tag{6.27}$$

The detail derivation of $\mathscr{D}_{\mathrm{KL}}$ is given in (A.2). The derivation of the distributional form of $\widehat{\mathscr{D}}_{\mathrm{KL}}$ is very difficult in our case as well as in the case of Bhattacharyee and Dunsmore (1995). So we have derived the distributional form of L and S, separately in the following theorems.

Theorem 6.1. *If both $X^T X$ and $X_R^T X_R$ are singular, then under the full model (6.1)*

$$g.L = \frac{g}{2\delta^2} \left(x^f \tilde{\beta} - x_R^f \tilde{\theta} \right)^2$$

is distributed as non-central $F'(1, n - q;\ \lambda)$ distribution with non-centrality parameter $\lambda = \frac{\mu^2}{\theta^2}$, where

$$\mu = x^f P_2 z_2 - x_R^f Q_2 z_2^* + \left(x^f P_1 D^{-1} P_1^T x^{f^T} - x_R^f Q_1 D_R^{-1} Q_1^T x_R^{f^T} \right) X\beta,$$

$$\theta^2 = \sigma \left[x^f P_1 D^{-1} P_1^T x^{f^T} + x_R^f Q_1 D_R^{-1} Q_1^T x_R^{f^T} - W_1 \right]$$

and g is a constant which is given by

$$g = \frac{2[\nu/(\nu - 2)] \left[1 + x^f (PFP^T)^{-1} x^{f^T} \right]}{x^f P_1 D^{-1} P_1^T x^{f^T} + x_R^f Q_1 D_R^{-1} Q_1^T x_R^{f^T} - W_1}$$

where $W_1 = 2 x^f P_1 D^{-1} P_1^T X^T X_R Q_1 D_R^{-1} Q_1^T x_R^{f^T}$. Under the reduced model (6.5) non-central F reduces to central F, if either $x = x^f$ and $x_R = x_R^f$ or $z_2 = 0$ and $z_2^* = 0$.

Theorem 6.2. *If $X^T X$ is singular and $X_R^T X_R$ is non-singular, then under the full model (6.1)*

$$g'.L = \frac{g'}{2\delta^2} \left(x^f \tilde{\beta} - x_R^f \tilde{\theta} \right)^2$$

is distributed as non-central $F'(1, n - q;\ \lambda')$ distribution with non-centrality parameter $\lambda' = \frac{\mu'}{\theta'^2}$, where

$$\mu' = x^f P_2 z_2 + \left(x^f P_1 D^{-1} P_1^T x^{f^T} - x_R^f (X_R^T X_R)^{-1} x_R^{f^T} \right) X\beta,$$

$$\theta'^2 = \sigma \left[x^f P_1 D^{-1} P_1^T x^{f^T} + x_R^f (X_R^T X_R)^{-1} x_R^{f^T} - W_2 \right]$$

and g' is a constant which is given by

$$g' = \frac{2[\nu/(n - k - 3)] \left[1 + x^f (PFP^T)^{-1} x^{f^T} \right]}{x^f P_1 D^{-1} P_1^T x^{f^T} + x_R^f (X_R^T X_R)^{-1} x_R^{f^T} - W_2},$$

where $W_2 = 2 x^f P_1 D^{-1} P_1^T X^T X_R (X_R^T X_R)^{-1} x_R^{f^T}$. Under the reduced model (6.5) non-central F reduces to central F, if either $x = x^f$ or $z_2 = 0$.

Theorem 6.3. *If both $X^T X$ and $X_R^T X_R$ are non-singular, then under the full model (6.1)*

$$g^o.L = \frac{g^o}{2\delta^2} \left(x^f \tilde{\beta} - x_R^f \tilde{\theta} \right)^2$$

is distributed as non-central $F'(1, n-q; \lambda^o)$ distribution with non-centrality parameter $\lambda^o = \frac{\mu^o}{\theta^{o2}}$, where

$$\mu^o = \left(x^f (X^T X)^{-1} x^{f^T} - x_R^f (X_R^T X_R)^{-1} x_R^{f^T} \right) X\beta,$$

$$\theta^{o2} = \sigma \left[x^f (X^T X)^{-1} x^{f^T} + x_R^f (X_R^T X_R)^{-1} x_R^{f^T} - W_3 \right]$$

and g^o is a constant which is given by

$$g^o = \frac{2[(n-k-1)/(n-k-3)] \left[1 + x^f (PFP^T)^{-1} x^{f^T} \right]}{x^f (X^T X)^{-1} x^{f^T} + x_R^f (X_R^T X_R)^{-1} x_R^{f^T} - W_3},$$

where $W_3 = 2 x^f (X^T X)^{-1} X^T X_R (X_R^T X_R)^{-1} x_R^{f^T}$. Under the reduced model (6.5) non-central F reduces to central F.

Theorem 6.4. *If both $X^T X$ and $X_R^T X_R$ are singular, then under the full model (6.1) $2S$ is distributed as*

$$G_1 + \frac{q - q^*}{n - q}.G_1.F'(q - q^*, n - q;\ \lambda_1')$$

$$- \log\left\{ G_1 + \frac{q - q^*}{n - q}.G_1.F'(q - q^*, n - q;\ \lambda_1') \right\} - 1,$$

where

$$\lambda_1' = \tilde{\gamma}_1^T \left[D - P_1^T X^T X_R Q_1 D_R^{-1} Q_1^T X_R^T X P_1 \right] \tilde{\gamma}_1 / \sigma^2$$

and

$$G_1 = \frac{(n - q - 2) \left[1 + x_R^f (QF_R Q^T)^{-1} x_R^{f^T} \right]}{(n - q^* - 2) \left[1 + x^f (PFP^T)^{-1} x^{f^T} \right]}.$$

Under the reduced model (6.5) non-central F' reduces to central F.

Theorem 6.5. *If $X^T X$ is singular and $X_R^T X_R$ is non-singular, then under the full model (6.1) $2S$ is distributed as*

$$G_2 + \frac{q - k + r - 1}{n - q}.G_2.F'(q - k + r - 1, n - q;\ \lambda_2') -$$

$$\log\left\{ G_2 + \frac{q - k + r - 1}{n - q}.G_2.F'(q - k + r - 1, n - q;\ \lambda_2') \right\} - 1,$$

where

$$\lambda_2' = \tilde{\gamma}_1^T \left[D - P_1^T X^T X_R (X_R^T X_R)^{-1} X_R^T X P_1 \right] \tilde{\gamma}_1 / \sigma^2$$

and

$$G_2 = \frac{(n - q - 2) \left[1 + \boldsymbol{x}_R^f (X_R^T X_R)^{-1} \boldsymbol{x}_R^{f\,T} \right]}{(n - k + r - 3) \left[1 + \boldsymbol{x}^f (PFP^T)^{-1} \boldsymbol{x}^{f\,T} \right]}.$$

Under the reduced model (6.5) non-central F reduces to central F.

Theorem 6.6. *If both $X^T X$ and $X_R^T X_R$ are non-singular, then under the full model (6.1) 2S is distributed as*

$$G_3 + \frac{r}{n - k - 1}.G_3.F'(r, n - k - 1;\ \lambda_3') -$$

$$\log \left\{ G_3 + \frac{r}{n - k - 1}.G_3.F'(r, n - k - 1;\ \lambda_3') \right\} - 1,$$

where

$$\lambda_3' = (X\boldsymbol{\beta})^T \left[X(X^T X)^{-1} X^T - X_R (X_R^T X_R)^{-1} X_R^T \right] X\boldsymbol{\beta} / \sigma^2$$

and

$$G_3 = \frac{(n - k - 3) \left[1 + \boldsymbol{x}_R^f (X_R^T X_R)^{-1} \boldsymbol{x}_R^{f\,T} \right]}{(n - k + r - 3) \left[1 + \boldsymbol{x}^f (PFP^T)^{-1} \boldsymbol{x}^{f\,T} \right]}.$$

Under the reduced model (6.5) non-central F reduces to central F.

The derivation of theorems 6.1 and 6.4 are discussed in appendixes (A.2.1) and (A.2.2), respectively. The derivation of theorems 6.2 and 6.5 are similar to derivation of theorems 6.1 and 6.4, respectively. The derivation of theorems 6.3 and 6.6 can be considered in Johnson and Geisser (1983). However, these results do not lead to a simple celebrative distribution. It is imperative to note that $\mathscr{D}_{\mathrm{KL}}$ depends on the future explanatory value $\boldsymbol{x}^f$, which therefore needs specification. Faced with a similar problem in assessing the influence of individual observations on predictive distributions, the same future values of the explanatory variables as in the data set has been considered in Johnson and Geisser (1983). Recalling the possible optimistic nature of results from such sample reuse, we prefer to try to face the incorporation of future values and so to tackle the problem in a different way.

The discrepancy measure has been evaluated for all possible subsets of the explanatory variables in Bhattacharjee and Dunsmore (1995). For each subset, values of $\mathscr{D}_{\mathrm{KL}}$ are derived for a sample of typical x^f values, chosen to lie within the region of previous experience satisfying

$$x^f (X^T X)^{-1} x^{f^T} \leq 2k/n. \qquad (6.28)$$

proposed in Hoaglin and Welsch (1978). A safe decision according to Huber's suggestion (Huber, 1980) is that x^f will lie within the region of previous experience if $\frac{2k}{n} \leq 0.2$. However, in our current problem, the moment matrix $(X^T X)$ is singular. So we used $(PFP^T)^{-1}$ instead of $(X^T X)^{-1}$ in equation (6.28) and follow Huber's suggestion to obtain future sample point x^f, where $(PFP^T)^{-1}$ is a G inverse of $(X^T X)$.

For assessing the influential explanatory variables in the regression model, we compute the discrepancies by the formulated K-L divergence (6.25) based on full model (6.1) and all possible reduced models (6.5). For comparison between two predictive distributions we consider median discrepancy as the cutoff-point. Typically the largest median discrepancy occur when the reduced model includes none of explanatory variables. A subset of variables is deleted if the sample median discrepancy between the corresponding two predictive distributions is less than 1% of the largest median discrepancy suggested in the literature (Bhatterjee and Dunsmore, 2008). Our proposal is to omit a subset of variables by testing the significance of discrepancies L and S due to the difference location parameters and the difference scale parameters, respectively. If interest lies in the difference between two predictive means, it is sufficient to consider location parameter 'L'. If interest lies in the differences between two predictive variances, it is sufficient to consider scale parameter 'S'. To test the significance of partial discrepancies between two predictive distributions (6.21) and (6.22) due to the location parameter 'L' and the scale parameter 'S', our proposal is to test the significance of sample median discrepancy for L and S based on the proposed theorems 6.1-6.3 and 6.4-6.6, respectively.

6.3. Illustration

To demonstrate the performance of the proposed method, we used a data set on five variables Y, X_1, X_2, X_3 and X_4 in arbitrary units given in Aitchison and Dunsmore (1975), exercise-1, page-25, where Y denotes the response variable, and X_1, X_2, X_3 and X_4 are the explanatory variables. To investigate the performance of the proposed method, we modified the values of

X_1 by the relation $X_1 = 5X_2$ so that the moment matrix $X^T X$ becomes as a singular matrix. The modified dataset is given in Table 1.

Table 1

Y	X_1	X_2	X_3	X_4
6.0	27.5	5.5	108	63
6.0	23.5	4.7	94	72
6.5	26.0	5.2	108	86
7.1	34.0	6.8	100	100
7.2	36.5	7.3	99	107
7.6	43.5	8.7	99	111
8.0	51.0	10.2	101	114
9.0	70.5	14.1	97	116
9.0	85.5	17.1	93	119
9.3	106.5	21.3	102	121

For illustration, first we consider given data (x) as future sample points x^f like, Johnson W. and Geisser, S. (1983). Then the discrepancy $\mathscr{D}_{KL}$ between the predictive density based on the full model and each of the predictive density based on all possible reduced models is computed. Figure 6.1(a) provides a summary of these estimates through Box plots. We see that the median discrepancy between the all and none cases is 13.59. If a single variable is deleted, then the smaller and negligible discrepancy occurs due to each of X_1, X_2 and X_3 variables, while the discrepancy due to deleting X_4 is not negligible using the rule of thumb described at the end of section 6.2. So variable X_4 is influential together with the variables X_1, X_2 and X_3. If two variables are deleted, the corresponding values of $\mathscr{D}_{KL}$ due to deleting two variables (X_1, X_3) or (X_2, X_3) are negligible, while the discrepancies due to deleting any other pairs including (X_1, X_2) are not negligible. Again, the discrepancies due to deleting any set of three variables are also not negligible at all using the same rule as before. Obviously, X_1 (X_2) is not influential with X_3 in presence of X_2 (X_1) with X_4 in the regression model. In other words, X_1 (X_2) is influential with X_4 in absence of X_2 (X_1) with X_3 in the regression model. Therefore, any one set of variables from (X_1, X_4) and (X_2, X_4) is influential for Bayesian prediction. To investigate the performance of the proposed method for random future data points, we consider 10000 future observations within the region of previous

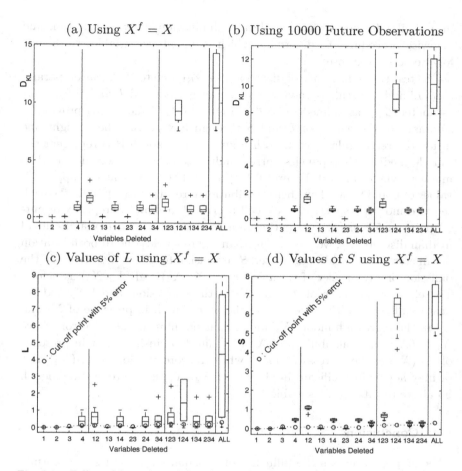

Fig. 6.1. Full model consists of 4 explanatory variables X_1, X_2, X_3 and X_4. (a) Box plots for the values of $\widehat{\mathscr{D}}_{\mathrm{KL}}$ with $X^f = X$ for all possible reduced models. (b) Box plots for values of $\widehat{\mathscr{D}}_{\mathrm{KL}}$ with 10000 future observations for all possible reduced models. (c) Box plots for values of location parameter L with $X^f = X$, where the marker style 'o' indicates cutoff-point that is upper 5% values of L and (d) Box plots for values of scale parameter S with $X^f = X$, where the marker style 'o' indicates cut-off point that is upper 5% values of S.

experience and the values of the discrepancy $\mathscr{D}_{\mathrm{KL}}$ (6.25) are computed corresponding to all possible reduced regression model. Figure 6.1(b) provides a summary of these values through Box plots. Comparing figures 6.1(a) and 6.1(b), we see that both cases produce almost similar results. We may in a similar fashion develop backward elimination, forward selection and stepwise procedures to determine which explanatory variables to include for predictive purpose.

To test the significance of discrepancies $\mathscr{D}_{\mathrm{KL}}$ due to the location parameter 'L' and the scale parameter 'S', we have separated L and S from the estimated $\mathscr{D}_{\mathrm{KL}}$ as defined in (6.25). Figures 6.1(c) and 6.1(d) provides a summary of the values of L and S through box plots. The straight line inside the rectangular box in both Figures 6.1(c) and 6.1(d) represent the sample median discrepancies corresponding to each reduced model. The marker style 'o' in both Figures 6.1(c) and 6.1(d) represent the upper 5% values of L and S based on their distributional results corresponding to each reduced model, respectively. We call it critical median discrepancy or cut-off point at 5% error. Comparing sample median discrepancy and critical median discrepancy, we see that median discrepancies due to both location parameter L and scale parameter S are insignificant at 5% error for the omission of any one set of variables from (X_1, X_3) and (X_2, X_3) as before. Therefore, in presence of X_1 and X_4 in the regression model X_2 and X_3 has no influence with 5% error, on the other hand, in presence of X_2 and X_4 in the regression model X_1 and X_3 has no influence at 5% error. Since $X_1 = 5X_2$, the contribution of X_1 (X_2) must be insignificant in presence of X_2 (X_1) in the regression model, while the contribution of X_1 (X_2) may or may not be insignificant in absence of X_2 (X_1) in the regression model. So above results are reasonable.

6.4. Conclusions

This chapter discusses the influence of explanatory variables in a normal multiple linear regression model for Bayesian prediction when the moment matrix is singular. To do this, we have derived the predictive distribution of a single future response in the linear regression model using vague prior density. To assess the influence of explanatory variables, we have formulated the K-L divergence between the predictive distributions for a single future response with and without a specified subset of the explanatory variables. To test the significance of influence of variables, we have decomposed the formulated K-L divergence into the discrepancies due to the location and scale parameters, respectively. Then we have derived the distributional results of both discrepancies and the test procedure is performed. The proposed method can be used to select the best regression model in presence of multicollinearity. Our future purpose is to compare the proposed method with other predictive model choice criteria's as discussed in Gelfand and Ghosh (1998).

References

1. Aitchison, J. and Dunsmore, I. R. (1975). *Statistical Prediction Analysis*. Cambridge University Press.
2. Bhattacharjee, S. K. and Dunsmore, I. R. (1991). Influence of Variables in a Logistic Model. *Biometrika* **78**, 851–856.
3. Bhattacharjee, S. K. and Dunsmore, I. R. (1995). The Predictive Influence of Variables in a Normal Regression Model. *JIOS* **16**, 01–08.
4. Bhattacharjee, S. K. (2007). The Predictive Influence of Variables in Multivariate Distribution. *J. Appld. Statist. Sc.* **16**.
5. Draper, N. R. and Smith, H. (1998). *Applied Regression Analysis*. New York, John Wiley & Sons.
6. Gelfand, A. E. and Ghosh, S. K. (1998). Model choice: A minimum posterior predictive loss approach. *Biometrika* **85**, 1–11.
7. Hoaglin, D. C. and Welsch, R. E (1978). The Hat Matrix in Regression and ANOVA. *Am. Statist.* **32**, 17–22.
8. Huber, P. J. (1980). *Robust Statistics*. John Wiley & Sons, New York.
9. Johnson, W. and Geisser, S. (1983). A Predictive View of the Detection and Characterization of Influential Observation in Regression Analysis. *J. Amer. Statist. Assoc.* **78**, 137–144.
10. Kullback, S. and Leibler, R. A. (1951). On Information and Sufficiency. *Ann. Math. Statist* **22**, 79–86.
11. Koutsoyiannies, A. (1977). *Theory of Econometrices*, 2nd edition. Macmillan Education Ltd, London.
12. Lindley, D. V. (1968). The Choice of Variable in Multiple Regression. *J. Roy. Statist. Soc.* **B 30**, 31–66.
13. Lindley, D. V. (1956). On a Measure of the Information Provided by an Experiment. *Ann. Math. Statist.* **27**, 986–1005.
14. Minami, M. and Eguchi, S. (2002). Robust Blind Source Separation by β-Divergence. *Neural Computn.* **14**, 1859–1886.
15. Mollah, M. N. H., Minami, M. and Eguchi, S. (2007). Robust Prewhitening for ICA by the Minimum β-Divergence Method and Its Application to FastICA. *Neural Proc. Lett.* **25**, 91–110.
16. Mollah, M. N. H. and Bhattacharjee, S. K. (2008). Predictive Influence of Variables in a Multivariate Distribution in Presence of Perfect Multicollinearity. *Commun. Statist.-Theory and Methods* **27**, 121–136.
17. Montgomery, D. C., Peck, E. A. and Vining, G. G. (2001). *Introduction to Linear Regression Analysis*, (Third Edition). New York, John Wiley & Sons.
18. Wedderburn, R. W. M. (1974). Quasi-Likelihood functions, generalized linear models, and the Gauss-Newton method. *Biometrika* **61**, 439–447.
19. Zelner, A. (1971). *An Introduction to Bayesian Inference in Econometrics*. New York, John Wiley & Sons, Inc.

A.1. Derivation of the Predictive Distribution when the Moment Matrix $(X^T X)$ is Singular

Using (6.4) and (6.19) in (6.3), we obtain the joint density for a future response y^f, $\boldsymbol{\beta}$ and σ as

$$p(y^f, \boldsymbol{\beta}, \sigma \mid \boldsymbol{x}^f, X, \boldsymbol{y}) \propto \frac{1}{\sigma^{n+p-q+2}} \times$$

$$\exp\left\{-\frac{1}{2\sigma^2}[a + (\boldsymbol{\beta} - \tilde{\boldsymbol{\beta}})^T PFP^T(\boldsymbol{\beta} - \tilde{\boldsymbol{\beta}}) + (y^f - \boldsymbol{x}^f\tilde{\boldsymbol{\beta}})^2]\right\}. \quad (A.1)$$

Then the marginal density of y^f and $\boldsymbol{\beta}$ is given by

$$p(y^f, \boldsymbol{\beta} \mid \boldsymbol{x}^f, X, \boldsymbol{y}) = \int_\sigma p(y^f, \boldsymbol{\beta}, \sigma \mid \boldsymbol{x}^f, X, \boldsymbol{y})d\sigma$$

$$\propto [a + (\boldsymbol{\beta} - \tilde{\boldsymbol{\beta}})^T PFP^T(\boldsymbol{\beta} - \tilde{\boldsymbol{\beta}}) + (y^f - \boldsymbol{x}^f\tilde{\boldsymbol{\beta}})^2]^{-\frac{n+p-q+1}{2}}.$$

Now
$$(\boldsymbol{\beta} - \tilde{\boldsymbol{\beta}})^T PFP^T(\boldsymbol{\beta} - \tilde{\boldsymbol{\beta}}) + (y^f - \boldsymbol{x}^f\tilde{\boldsymbol{\beta}})^2$$
$$= y^{f^2} + \tilde{\boldsymbol{\beta}}^T PFP^T\tilde{\boldsymbol{\beta}} - (PFP^T\tilde{\boldsymbol{\beta}} + \boldsymbol{x}^{f^T}y^f)M^{-1}(PFP^T\tilde{\boldsymbol{\beta}} + \boldsymbol{x}^{f^T}y^f)$$
$$[\boldsymbol{\beta} - M^{-1}(PFP^T\tilde{\boldsymbol{\beta}} + \boldsymbol{x}^{f^T}y^f)]^T M[\boldsymbol{\beta} - M^{-1}(PFP^T\tilde{\boldsymbol{\beta}} + \boldsymbol{x}^{f^T}y^f)],$$

where $M = (PFP^T + \boldsymbol{x}^{f^T}\boldsymbol{x}^f)$. Then the marginal or predictive density for y^f is given by

$$p(y^f \mid \boldsymbol{x}^f, X, \boldsymbol{y}) = \int_{\boldsymbol{\beta}} p(y^f, \boldsymbol{\beta} \mid \boldsymbol{x}^f, X, \boldsymbol{y})d\boldsymbol{\beta}$$

$$\propto \left[a + y^{f^2} + \tilde{\boldsymbol{\beta}}^T PFP^T\tilde{\boldsymbol{\beta}} - \right.$$

$$\left. (PFP^T\tilde{\boldsymbol{\beta}} + \boldsymbol{x}^{f^T}y^f)M^{-1}(PFP^T\tilde{\boldsymbol{\beta}} + \boldsymbol{x}^{f^T}y^f)\right]^{-\frac{n-q+1}{2}}$$

$$= \left[\nu\tilde{\sigma}^2 + (y^f - \boldsymbol{x}^f\tilde{\boldsymbol{\beta}})^T(1 - \boldsymbol{x}^f M^{-1}\boldsymbol{x}^{f^T})(y^f - \boldsymbol{x}^f\tilde{\boldsymbol{\beta}})\right]^{-\frac{n-q+1}{2}}$$

$$\propto \left[1 + h(y^f - \boldsymbol{x}^f\tilde{\boldsymbol{\beta}})^2\right]^{-(\nu+1)/2}$$

which follows univariate student t-distribution with mean $\boldsymbol{x}^f\tilde{\boldsymbol{\beta}}$ and variance $\frac{1}{\nu-2}h^{-1} = \frac{\nu\tilde{\sigma}^2}{\nu-2}(1 - \boldsymbol{x}^f M^{-1}\boldsymbol{x}^{f^T})^{-1} = \frac{\nu\tilde{\sigma}^2}{\nu-2}[1 + \boldsymbol{x}^f(PFP^T)^{-1}\boldsymbol{x}^{f^T}]$, where $\tilde{\sigma}^2 = a/\nu = (\boldsymbol{y} - X\tilde{\boldsymbol{\beta}})^T(\boldsymbol{y} - X\tilde{\boldsymbol{\beta}})/\nu$.

A.2. Formulation of K-L Divergence Based on Two Predictive Distributions (6.21) and (6.22)

Let $N(\boldsymbol{x}^f\tilde{\boldsymbol{\beta}}, \delta^2)$ and $N(\boldsymbol{x}_R^f\tilde{\boldsymbol{\theta}}, \delta_R^2)$ be two predictive densities denoted by $p(y^f)$ $q(y^f)$, respectively. Then the K-L divergence $\mathscr{D}_{\mathrm{KL}}$ is given by

$$
\begin{aligned}
\mathscr{D}_{\mathrm{KL}} &= \int q\left(y^f\right) \log \frac{q\left(y^f\right)}{p\left(y^f\right)} dy^f \\
&= \int q(y^f) \log \left[(2\pi\delta_R^2)^{-1/2} \exp\{-\frac{1}{2\delta_R^2}(y^f - \boldsymbol{x}_R^f\tilde{\boldsymbol{\theta}})^2\}\right] dy^f \\
&\quad - \int q(y^f) \log \left[(2\pi\delta^2)^{-1/2} \exp\{-\frac{1}{2\delta^2}(y^f - \boldsymbol{x}^f\tilde{\boldsymbol{\beta}})^2\}\right] dy^f \\
&= \log(2\pi\delta_R^2)^{-1/2} - \frac{1}{2} - \log(2\pi\delta^2)^{-1/2} + \frac{\delta_R^2}{2\delta^2} + \frac{1}{2\delta^2}(\boldsymbol{x}^f\tilde{\boldsymbol{\beta}} - \boldsymbol{x}_R^f\tilde{\boldsymbol{\theta}})^2 \\
&= \frac{1}{2\delta^2}(\boldsymbol{x}^f\tilde{\boldsymbol{\beta}} - \boldsymbol{x}_R^f\tilde{\boldsymbol{\theta}})^2 + \frac{1}{2}\left\{\frac{\delta_R^2}{\delta^2} - \log\frac{\delta_R^2}{\delta^2} - 1\right\} \\
&= L + S,
\end{aligned}
$$

where $L = \frac{1}{2\delta^2}(\boldsymbol{x}^f P\tilde{\gamma} - \boldsymbol{x}_R^f\tilde{\boldsymbol{\theta}})^2$ and $S = \frac{1}{2}\left\{\frac{\delta_R^2}{\delta^2} - \log\frac{\delta_R^2}{\delta^2} - 1\right\}$.

A.2.1. *Proof of Theorem 6.1*

We can write

$$
\begin{aligned}
\boldsymbol{x}^f\tilde{\boldsymbol{\beta}} - \boldsymbol{x}_R^f\tilde{\boldsymbol{\theta}} &= \boldsymbol{x}^f P\tilde{\gamma} - \boldsymbol{x}_R^f Q\tilde{\gamma}^* \\
&= \boldsymbol{x}^f(P_1\tilde{\gamma}_1 + P_2\boldsymbol{z}_2) - \boldsymbol{x}_R^f(Q_1\tilde{\gamma}_1^* + Q_2\boldsymbol{z}_2^*) \\
&= \boldsymbol{x}^f P_2\boldsymbol{z}_2 - \boldsymbol{x}_R^f Q_2\boldsymbol{z}_2^* + \\
&\quad \left(\boldsymbol{x}^f P_1 D^{-1} P_1^T X^T - \boldsymbol{x}_R^f Q_1 D_R^{-1} Q_1^T X_R^T\right) \boldsymbol{y}.
\end{aligned}
$$

Assuming that both $\boldsymbol{x}^f P_2$ and $\boldsymbol{x}_R^f Q_2$ reduces to zero vectors as both $\boldsymbol{x}P_2$ and $\boldsymbol{x}_R Q_2$ are zero vectors, where $\boldsymbol{x}$ and $\boldsymbol{x}_R$ are the row vectors of X and X_R as defined in (6.2) and (6.6), respectively. Hence under the full model, $(\boldsymbol{x}^f\tilde{\boldsymbol{\beta}} - \boldsymbol{x}_R^f\tilde{\boldsymbol{\theta}})$ follows $N(\mu, \eta^2)$, where

$$
\begin{aligned}
\mu &= \left(\boldsymbol{x}^f P_1 D^{-1} P_1^T X^T - \boldsymbol{x}_R^f Q_1 D_R^{-1} Q_1^T X_R^T\right) X\boldsymbol{\beta}, \\
&= \left(\boldsymbol{x}^f (PFP^T)^{-1} \boldsymbol{x}^{f^T} - \boldsymbol{x}_R^f (QF_R Q^T)^{-1} \boldsymbol{x}_R^{f^T}\right) X\boldsymbol{\beta},
\end{aligned}
$$

and

$$\eta^2 = \sigma \left[\boldsymbol{x}^f P_1 D^{-1} P_1^T \boldsymbol{x}^{f^T} + \boldsymbol{x}_R^f Q_1 D_R^{-1} Q_1^T \boldsymbol{x}_R^{f^T} - W_1 \right]$$

$$= \sigma \left[\boldsymbol{x}^f (PFP^T)^{-1} \boldsymbol{x}^{f^T} + \boldsymbol{x}_R^f (QF_R Q^T)^{-1} \boldsymbol{x}_R^{f^T} - W_1' \right],$$

where $W_1' = 2\boldsymbol{x}^f (PFP^T)^{-1} X^T X_R (QF_R Q^T)^{-1} \boldsymbol{x}_R^{f^T}$. Now we can write

$$L = \frac{1}{2\delta^2} \left(\boldsymbol{x}^f \tilde{\beta} - \boldsymbol{x}_R^f \tilde{\theta} \right)^2 = \frac{\left(\boldsymbol{x}^f \tilde{\beta} - \boldsymbol{x}_R^f \tilde{\theta} \right)^2}{\eta^2} \cdot \frac{\eta^2}{2\delta^2}.$$

Then

$$g.L = \frac{\left(\boldsymbol{x}^f \tilde{\beta} - \boldsymbol{x}_R^f \tilde{\theta} \right)^2 / \eta^2}{\tilde{\sigma}^2 / \sigma^2} = \frac{\chi'^2(1,\lambda)}{\chi^2(\nu)},$$

where $\chi^2(\nu)$ is the central chi-square distribution with ν degree of freedom and $\chi'^2(1,\lambda)$ is the non-central chi-square distribution with 1 degree of freedom and non-centrality parameter $\lambda = \frac{\mu^2}{\eta^2}$ and

$$g = \frac{2[\nu/(\nu-2)] \left[1 + \boldsymbol{x}^f (PFP^T)^{-1} \boldsymbol{x}^{f^T} \right]}{\boldsymbol{x}^f (PFP^T)^{-1} \boldsymbol{x}^{f^T} + \boldsymbol{x}_R^f (QF_R Q^T)^{-1} \boldsymbol{x}_R^{f^T} - W_1'}.$$

Now we will show that $\chi^2(\nu)$ and $\chi'^2(1,\lambda)$ are independent of each other. For this, we have $\left(\boldsymbol{x}^f \tilde{\beta} - \boldsymbol{x}_R^f \tilde{\theta} \right)^2 = \boldsymbol{y}^T A \boldsymbol{y}$ and $\nu \tilde{\sigma}^2 = \boldsymbol{y}^T B \boldsymbol{y}$, where

$$A = [X P_1 D^{-1} P_1^T \boldsymbol{x}^{f^T} \boldsymbol{x}^f P_1 D^{-1} P_1^T X^T + X_R Q_1 D_R^{-1} Q_1^T \boldsymbol{x}_R^{f^T} \boldsymbol{x}_R^f Q_1 D_R^{-1} Q_1^T X_R^T$$
$$- 2 X P_1 D^{-1} P_1^T \boldsymbol{x}^{f^T} \boldsymbol{x}_R^f Q_1 D_R^{-1} Q_1^T X_R^T] \text{ and}$$
$$B = (I - X P_1 D^{-1} P_1^T X^T).$$

Then it can be shown that $AB = 0$. Thus $\chi^2(\nu)$ and $\chi'^2(1,\lambda)$ are independent of each other. Hence

$$g.L = \frac{\chi'^2(1,\lambda)}{\chi^2(\nu)} = F'(1,\nu,\lambda),$$

which is non-central F with 1 and ν degrees of freedom.
Under the reduced model (6.6), the non-centrality parameter $\lambda = 0$, since

$$\mu = \left(\boldsymbol{x}^f P_1 D^{-1} P_1^T X^T - \boldsymbol{x}_R^f Q_1 D_R^{-1} Q_1^T X_R^T \right) X_R \boldsymbol{\theta} = 0.$$

Therefore, under the reduced model (6.6) non-central F reduces to central F.

A.2.2. *Proof of Theorem 6.4*

We have

$$2S = \frac{\delta_R^2}{\delta^2} - \log \frac{\delta_R^2}{\delta^2} - 1 = G_1 . \frac{\tilde{\sigma}_R^2}{\tilde{\sigma}^2} - \log \left\{ G_1 . \frac{\tilde{\sigma}_R^2}{\tilde{\sigma}^2} \right\} - 1$$

where $G_1 = \frac{n-q^*}{n-q} G$. We can write

$$\tilde{\sigma}^2 = (\boldsymbol{y} - X\tilde{\boldsymbol{\beta}})^T (\boldsymbol{y} - X\tilde{\boldsymbol{\beta}}) / \nu = (\boldsymbol{y}^T \boldsymbol{y} - \tilde{\gamma}_1^T P_1^T X^T \boldsymbol{y}) / \nu \text{ and}$$

$$\tilde{\sigma}_R^2 = (\boldsymbol{y} - X_R\tilde{\boldsymbol{\theta}})^T (\boldsymbol{y} - X_R\tilde{\boldsymbol{\theta}}) / \nu^* = (\boldsymbol{y}^T \boldsymbol{y} - \tilde{\gamma}_1^{*T} Q_1^T X_R^T \boldsymbol{y}) / \nu^*.$$

Let us denote

$$S_0 = \nu \tilde{\sigma}^2 = \boldsymbol{y}^T \boldsymbol{y} - \tilde{\gamma}_1^T P_1^T X^T \boldsymbol{y} \text{ and}$$

$$\begin{aligned} S_1 &= \nu^* \tilde{\sigma}_R^2 = \boldsymbol{y}^T \boldsymbol{y} - \tilde{\gamma}_1^{*T} Q_1^T X_R^T \boldsymbol{y} \\ &= (\boldsymbol{y}^T \boldsymbol{y} - \tilde{\gamma}_1^T P_1^T X^T \boldsymbol{y}) + (\tilde{\gamma}_1^T P_1^T X^T \boldsymbol{y} - \tilde{\gamma}_1^{*T} Q_1^T X_R^T \boldsymbol{y}) \\ &= S_0 + S_2, \end{aligned}$$

where

$$\begin{aligned} S_2 &= \tilde{\gamma}_1^T P_1^T X^T \boldsymbol{y} - \tilde{\gamma}_1^{*T} Q_1^T X_R^T \boldsymbol{y} \\ &= \boldsymbol{y}^T [X P_1 D^{-1} P_1^T X^T - X_R Q_1 D_R^{-1} Q_1^T X_R^T] \boldsymbol{y}. \end{aligned}$$

Again S_0 can be written as

$$S_0 = \boldsymbol{y}^T [I - X P_1 D^{-1} P_1^T X^T] \boldsymbol{y}.$$

Now it can be shown that both $[I - X P_1 D^{-1} P_1^T X^T]$ and $[X P_1 D^{-1} P_1^T X^T - X_R Q_1 D_R^{-1} Q_1^T X_R^T]$ are idempotent and their ranks are $(n-q)$ and $(q-q^*)$, respectively. Therefore $\frac{S_0}{\sigma^2}$ follows $\chi^2(n-q, \kappa_1)$ the central chi-square distribution with $(n-q)$ degrees of freedom under the full model, since the non-centrality parameter

$$\kappa_1 = (X\boldsymbol{\beta})^T [I - X P_1 D^{-1} P_1^T X^T] (X\boldsymbol{\beta}) / \sigma^2 = 0.$$

On the other hand, $\frac{S_2}{\sigma^2}$ follows $\chi'^2(q-q^*, \kappa_2)$ the non-central chi-square distribution with $(q-q^*)$ degrees of freedom under the full model, since the non-centrality parameter

$$\begin{aligned} \kappa_2 &= (X\boldsymbol{\beta})^T [X P_1 D^{-1} P_1^T X^T - X_R Q_1 D_R^{-1} Q_1^T X_R^T] (X\boldsymbol{\beta}) / \sigma^2 \\ &= (X\boldsymbol{\beta})^T \left[X(PFP^T)^{-1} X^T - X_R (Q_R F_R Q_R^T)^{-1} X_R^T \right] (X\boldsymbol{\beta}) / \sigma^2 \neq 0. \end{aligned}$$

Again since

$$[I - X P_1 D^{-1} P_1^T X^T][X P_1 D^{-1} P_1^T X^T - X_R Q_1 D_R^{-1} Q_1^T X_R^T](X\boldsymbol{\beta}) / \sigma^2 = 0.$$

The two quadratic forms S_0 and S_2 are independent of each other, and hence the two chi-square distributions $\chi^2(n-q, \kappa_1)$ and $\chi'^2(q-q^*, \kappa_2)$ are independent of each other. Therefore under the full model

$$\frac{S_2/(q-q^*)}{S_0/(n-q)} = \frac{\chi'^2(q-q^*, \kappa_2)/(q-q^*)}{\chi^2(n-q, \kappa_1)/(n-q)} = F'(q-q^*, n-q; \ \kappa_2)$$

which is the non-central F-distribution with $(q-q^*)$ and $(n-q)$ degrees of freedom and non-centrality parameter κ_2. If the reduced model is true, then

$$\kappa_2 = (X_R\boldsymbol{\theta})^T [X P_1 D^{-1} P_1^T X^T - X_R Q_1 D_R^{-1} Q_1^T X_R^T](X_R\boldsymbol{\theta})/\sigma^2 = 0.$$

Hence under the reduced model

$$\frac{S_2/(q-q^*)}{S_0/(n-q)} = \frac{\chi'^2(q-q^*, \kappa_2)/(q-q^*)}{\chi^2(n-q, \kappa_1)/(n-q)} = F(q-q^*, n-q)$$

which is the central F-distribution with $(q-q^*)$ and $(n-q)$ degrees of freedom.

Now we can write,

$$2S = \frac{(S_0+S_2)/(n-q^*)}{S_0/(n-q)}.G_1 - \log\left\{\frac{(S_0+S_2)/(n-q^*)}{S_0/(n-q)}.G_1\right\} - 1$$

$$= G + \frac{S_2}{S_0}.G - \log\left\{G + \frac{S_2}{S_0}.G\right\} - 1$$

$$= G + \frac{q-q^*}{n-q}.G.\frac{S_2/(q-q^*)}{S_0/(n-q)}.G$$

$$\quad - \log\left\{G + \frac{q-q^*}{n-q}.G.\frac{S_2/(q-q^*)}{S_0/(n-q)}.G\right\} - 1.$$

Hence under the full model, we may write

$$G + \frac{q-q^*}{n-q}.G.F'(q-q^*, n-q; \ \kappa_1)$$

$$\quad - \log\left\{G + \frac{q-q^*}{n-q}.G.F'(q-q^*, n-q; \ \kappa_1)\right\} - 1.$$

Under the reduced model, we have

$$G + \frac{q-q^*}{n-q}.G.F(q-q^*, n-q) - \log\left\{G + \frac{q-q^*}{n-q}.G.F(q-q^*, n-q)\right\} - 1.$$

Chapter 7

New Wrapped Distributions – Goodness of Fit

A. V. Dattatreya Rao[1], I. Ramabhadra Sarma[2]
and S. V. S. Girija[3]

[1]*Acharya Nagarjuna University, India*
[2]*K. L. University, India*
[3]*Hindu College, Guntur, India*

Dattatreya Rao et al. (2007) have introduced Wrapped versions of four well-known life testing models viz. Lognormal, Logistic, Weibull and Extreme-Value distributions and mentioned basic characteristics along with graphs of the above said distributions. Here an attempt is made on the lines of Dattatreya Rao et al. (2007) to derive some of the new circular models - Wrapped Half Logistic and Wrapped Binormal distributions by reducing a linear variable to its modulo 2π and using trigonometric moments. Goodness of fit for small samples using Kolmogorov-Smirnov (KS) type Kuiper one sample test [Mardia and Jupp, 2000; Fisher, 1993] and for large samples using χ^2-test were employed [Banks et al., 2000]. Three data sets each of sample sizes 10 and 100 respectively are taken for employing Kuiper and χ^2-tests.

Keywords: Circular distributions, Extreme-Value distribution, Goodness of fit, Wrapped Binormal distribution, Wrapped Half Logistic distribution.

Contents

7.1. Introduction

Circular distributions play a vital role in varied fields in various applications such as

(1) (a) Solving birds navigational problems (Kendall, 1974 ; Schimdt-Koeing, 1963)

   (b) Interpreting paleomagnetic currents (Jammalamadaka and Sen Gupta, 1972; Fuller et al., 1996)

   (c) Assessing variation in the onset of leukaemia (Hrushesky, 1994; Morgan, 1990)

   (d) Analyzing wind directions (Breckling, 1989)

   (e) Angle of knee flexion to assess the recovery of orthopaedic patients (Jammalamadaka et al., 1986)

   (f) Morphological Operators on the Unit Circle (Hanbury and Serra, 2001).

In continuous case the probability density function (pdf) $g(\theta)$ of a circular distribution exists and has the following basic properties

(1) $g(\theta) \geq 0, \forall \theta$ 1.1(a)

(2) $\int_0^{2\pi} g(\theta)d\theta = 1$ 1.1(b)

(3) $g(\theta) = g(\theta + 2k\pi)$ for any integer k (*i.e. g* is periodic)   1.1(c)

Broadly the following are the general methods of construction of a circular model:

- using geometrical considerations
- wrapping a linear distribution around a unit circle,
- characterizing properties such as maximum entropy, etc.
- transforming a bivariate linear r.v. to its directional component, the so called offset distributions
- one may start with a distribution on the real line $\mathbb{R}$ and apply a stereographic projection that identifies points x on $\mathbb{R}$ with those on the circumference of the circle, say θ,
- using the Toeplitz Hermitian Positive Definite (THPD) matrices,
- using the Rising Sun functions.

NOTATIONS

Standard notations as in Jammalamadaka and SenGupta (2001) are adopted here.

$f(x)$ and $F(x)$	- probability density function (pdf) and cumulative distribution function (cdf) respectively of the linear variate X.
$g(\theta)$ and $G(\theta)$	- pdf and cdf of the corresponding circular models.
X_w	- circular (wrapped) random variable corresponding to X.
$\varphi_X(t)$, $t \in \mathbb{R}$	- characteristic function on real line.
$\varphi_w(p)$, $p = 0, \pm 1, \pm 2, \ldots$	- characteristic function of the wrapped model.
α_p, β_p	- the pth trigonometric moments of the circular random variable θ and let $\rho_p = \sqrt{\alpha_p^2 + \beta_p^2}$ and $\mu_p^o = \tan^{-1}\left(\frac{\beta_p}{\alpha_p}\right)$, however, μ_1 and ρ_1 are generally denoted as μ and ρ to represent mean direction and resultant length. Also, μ is used as location parameter.
V_o	- circular variance.
σ_o	- circular standard mean deviation.
α_p^*, β_p^*	- central trigonometric moments.
γ_1^o	- skewness of circular distribution.
γ_2^o	- kurtosis of circular distribution.
σ	- scale parameter.
μ	- location parameter in linear case.
c	- shape parameter.
$\mathbb{R}$	- Set of Real numbers.
$\mathbb{N}$	- Set of Natural numbers.
$\mathbb{C}$	- Set of Complex numbers.
$\mathbb{Z}$	- Set of Integers.

7.2. Methodology of Wrapping

Modulo 2π reduction:

If X is a r.v. defined on $\mathbb{R}$, then the corresponding circular (wrapped) r.v. X_W is defined as

$$X_W \equiv X \, (mod \, 2\pi) \tag{7.1}$$

and is clearly a many valued function given by

$$X_w(\theta) = \{X(\theta + 2k\pi), k \in \mathbb{Z}\} \qquad (7.2)$$

The wrapped circular pdf $g(\theta)$ corresponding to the density function $f(x)$ of a linear r.v. X is obtained as,

$$g(\theta) = \sum_{k=-\infty}^{\infty} f(\theta + 2k\pi), \theta \in [0, 2\pi) \qquad (7.3)$$

It may be noted that the circular distribution is a probability distribution whose total probability is concentrated on the unit circle in the plane

$$\{(\cos\theta, \sin\theta)/0 \le \theta < 2\pi\}$$

which satisfies the properties 1.1(a) through 1.1(c).

The form of pdf of a wrapped circular model can be obtained through characteristic function of the linear r.v. X using trigonometric moments. Utilising inversion theorem of characteristic function, one can derive, circular models through trigonometric moments. These trigonometric moments can be obtained using Proposition (7.1) (Jammalamadaka and SenGupta, 2001, p.31) and Carslaw (1930). If α_p and β_p are the trigonometric moments and $\sum_{p=1}^{\infty} \left(\alpha_p^2 + \beta_p^2\right)$ is convergent then the random variable θ has a density g which is defined almost everywhere by

$$g(\theta) = \frac{1}{2\pi} \left[1 + 2\sum_{p=1}^{\infty}(\alpha_p \cos p\theta + \beta_p \sin p\theta)\right], \theta \in [0, 2\pi) \qquad (7.4)$$

The characteristic function and cdf of the resultant wrapped circular model are, $\varphi_\theta(p) = E(e^{ip\theta}) = \varphi(p), p = 0, \pm 1, \pm 2, \pm 3, \dots\dots$ and $G(\theta) = \int_0^\theta g(\theta')d\theta', \theta \in [0, 2\pi)$.

7.3. Characteristic Functions of Half Logistic and Binormal Distributions

Gravin and McClean (1997) have used Binormal model in certain applications of statistical process control and weather studies. Kantam and Rosaiah (1998) applied Half logistic distribution in acceptance sampling based on life tests.

An attempt is made here to derive the new circular models on the basis of Dattatreya Rao et al. (2007) and obtained the characteristic functions as they are required in the derivation of the said new circular models Wrapped Half Logistic distribution and Wrapped Binormal distribution.

To obtain trigonometric moments of Wrapped Half Logistic distribution and Wrapped Binormal distribution, one needs characteristic function of Half Logistic distribution and Binormal distribution respectively which are not available, hence are derived by us and are given as follows.

The characteristic function of Half Logistic distribution is

$$\phi_X(t) = 2e^{it\mu} \left[\left(\int_0^a \cos(t\sigma\, y)\frac{e^{-y}}{(1+e^{-y})^2} dy + \sum_{n=1}^{\infty} (-1)^n u_n \right) \right. \qquad (7.5)$$

$$\left. +i \left(\int_0^a \sin(t\sigma\, y)\frac{e^{-y}}{(1+e^{-y})^2} dy + \sum_{n=1}^{\infty} (-1)^n v_n \right) \right]$$

$$t \in \mathbb{R}, a > 0, \sigma > 0$$

where $u_n = \frac{ne^{-na}}{n^2+t^2\sigma^2} (n\cos(t\sigma\, a) - t\sigma\sin(t\sigma\, a))$, $v_n = \frac{ne^{-na}}{n^2+t^2\sigma^2} (t\sigma\cos(t\sigma\, a) + n\sin(t\sigma\, a))$

Now write $b_n = \int_0^a \cos(t\sigma\, y) \frac{e^{-y}}{(1+e^{-y})^2} dy + \sum_{n=1}^{\infty} (-1)^n u_n$, $c_n = \int_0^a \sin(t\sigma\, y) \frac{e^{-y}}{(1+e^{-y})^2} dy + \sum_{n=1}^{\infty} (-1)^n u_n$

Then $\phi_X(t) = 2\left(\cos(t\mu) + i\sin(t\mu)\right)\left(b_n + ic_n\right)$ \qquad (7.6)

$$= 2\left(b_n \cos(t\mu) - c_n \sin(t\mu)\right)$$

$$+2i\left(b_n \sin(t\mu) + c_n \cos(t\mu)\right)$$

$$t \in \mathbb{R}, a > 0, \sigma > 0$$

The characteristic function of wrapped Half Logistic distribution is

$$\phi_w(p) = 2\left(b_n \cos(p\,\mu) - c_n \sin(p\,\mu)\right) +$$

$$2i\left(b_n \sin(p\,\mu) + c_n \cos(p\,\mu)\right) \qquad (7.7)$$

Where

$$b_n = \int_0^a \cos(p\sigma\, y)\, \frac{e^{-y}}{(1 + e^{-y})^2} dy + \sum_{n=1}^{\infty} (-1)^{n-1} \frac{ne^{-na}}{n^2 + p^2\sigma^2}$$

$$(n \cos(p\sigma\, a) - p\sigma \sin(p\sigma\, a)) \qquad (7.8)$$

$$c_n = \int_0^a \sin(p\sigma\, y)\, \frac{e^{-y}}{(1 + e^{-y})^2} dy + \sum_{n=1}^{\infty} (-1)^{n-1} \frac{ne^{-na}}{n^2 + p^2\sigma^2}$$

$$(n \sin(p\sigma\, a) + p\sigma \cos(p\sigma\, a))$$

$$\text{for } p \in \mathbb{Z}$$

Since $a > 0$, it can be chosen arbitrarily. For the evaluation of b_n and c_n, we take $a = \frac{2\pi}{p\sigma}$ and are evaluated numerically using MATLAB. For this a we have

$$b_n = \int_0^{2\pi/p\sigma} \cos(p\sigma\, y)\, \frac{e^{-y}}{(1 + e^{-y})^2} dy + \sum_{n=1}^{\infty} (-1)^{n-1} \frac{n^2 e^{-na}}{n^2 + p^2\sigma^2}$$

$$c_n = \int_0^{2\pi/p\sigma} \sin(p\sigma\, y)\, \frac{e^{-y}}{(1 + e^{-y})^2} dy + \sum_{n=1}^{\infty} (-1)^{n-1} \frac{np\sigma\, e^{-na}}{n^2 + p^2\sigma^2},$$

$$\text{for } p \in \mathbb{Z} \text{ and } \sigma > 0$$

Thus obtaining the trigonometric moments, various characteristics of the Wrapped Half Logistic model fixing $\mu = 0.1$ and varying $\sigma = 0.25, 0.5, 1.0$ and 1.5 are computed borrowing expression from Mardia (1972) and are tabulated in Table 7.1.

Using the concepts on p.302 of Apostol (1986) the characteristic function of Binormal distribution is derived and presented as

$$\phi_X(t) = e^{it\mu} \left[\left(\frac{\sigma_1}{\sigma_1 + \sigma_2} e^{\frac{-t^2\sigma_1^2}{2}} + \frac{\sigma_2}{\sigma_1 + \sigma_2} e^{\frac{-t^2\sigma_2^2}{2}} \right) \right]$$

$$+ ie^{it\mu} \left(\frac{-2}{\sqrt{\pi}} \frac{\sigma_1}{\sigma_1 + \sigma_2} e^{\frac{-t^2\sigma_1^2}{2}} \right.$$

$$\sum_{n=1}^{\infty} \frac{\left(\frac{t\sigma_1}{\sqrt{2}} \right)^{2n-1}}{(2n-1)(n-1)!} + \frac{2}{\sqrt{\pi}} \frac{\sigma_2}{\sigma_1 + \sigma_2} e^{\frac{-t^2\sigma_2^2}{2}} \sum_{n=1}^{\infty} \frac{\left(\frac{t\sigma_2}{\sqrt{2}} \right)^{2n-1}}{(2n-1)(n-1)!} \right)$$

$$\sigma_1, \sigma_2 > 0, t \in \mathbb{R} \qquad (7.9)$$

Table 7.1. Characteristics of Wrapped Half Logistic distribution

Wrapped Half Logistic distribution		$\sigma=0.25$ $\mu=0.1$	$\sigma=0.5$ $\mu=0.1$	$\sigma=1$ $\mu=0.1$	$\sigma=1.5$ $\mu=0.1$
Trigono metric moments	α_1	0.8959	0.6559	0.2168	0.0234
	α_2	0.6559	0.2168	-0.0352	-0.0515
	β_1	0.3427	0.5496	0.5637	0.4117
	β_2	0.5496	0.5637	0.2916	0.1716
Resultant length ρ		0.9592	0.8557	0.6040	0.4123
circular variance V_o		0.0408	0.1443	0.3961	0.5877
Central Trigono metric moments	α_1^*	0.9592	0.8557	0.6039	0.4123
	α_2^*	0.8553	0.5929	-0.2216	-0.0706
	β_1^*	0	0	0	0
	β_2^*	-0.0285	-0.1148	0.1929	0.1647
Skewness γ_1^o		-3.4500	-2.0954	0.7739	0.3656
kurtosis γ_2^o		5.3022	2.7238	-2.2607	-0.2882
circular standard deviation σ_o		0.2887	0.5582	1.0043	1.3311
		0.5582	1.0043	1.5652	1.8543

The characteristic function of Wrapped Binormal distribution is

$$\phi_W(p) = e^{ip\mu}\left[\left(\frac{\sigma_1}{\sigma_1+\sigma_2}e^{\frac{-p^2\sigma_1^2}{2}} + \frac{\sigma_2}{\sigma_1+\sigma_2}e^{\frac{-p^2\sigma_2^2}{2}}\right)\right]$$

$$+ie^{ip\mu}\left(\frac{-2}{\sqrt{\pi}}\frac{\sigma_1}{\sigma_1+\sigma_2}e^{\frac{-p^2\sigma_1^2}{2}}\right)$$

$$\sum_{n=1}^{\infty}\frac{\left(\frac{p\sigma_1}{\sqrt{2}}\right)^{2n-1}}{(2n-1)(n-1)!} + \frac{2}{\sqrt{\pi}}\frac{\sigma_2}{\sigma_1+\sigma_2}e^{\frac{-p^2\sigma_2^2}{2}}\sum_{n=1}^{\infty}\frac{\left(\frac{p\sigma_2}{\sqrt{2}}\right)^{2n-1}}{(2n-1)(n-1)!}\right)$$

$$\text{for } p = 0, \pm 1, \pm 2, ... \tag{7.10}$$

$$\phi_W(p) = (a\cos(p\mu) - b\sin(p\mu)) + i(a\sin(p\mu) + b\cos(p\mu)) \tag{7.11}$$

where $a = \left(\frac{\sigma_1}{\sigma_1+\sigma_2}e^{\frac{-p^2\sigma_1^2}{2}} + \frac{\sigma_2}{\sigma_1+\sigma_2}e^{\frac{-p^2\sigma_2^2}{2}}\right)$

Table 7.2. Characteristics of Wrapped Binormal distribution

Wrapped Binormal distribution		$\sigma_1 = 0.5$ $\sigma_2 = 0.75$	$\sigma_1 = 0.5$ $\sigma_2 = 1.5$	$\sigma_1 = 0.5$ $\sigma_2 = 2.5$	$\sigma_1 = 0.5$ $\sigma_2 = 3.5$
Trigono metric moments	α_1	0.8059	0.4641	0.1837	0.1122
	α_2	0.4374	0.1600	0.1011	0.0758
	β_1	-0.4456	-0.5419 -	-0.3949	-0.2691
	β_2	-0.5914	0.3798	-0.2355	-0.1625
Resultant length ρ		0.9209	0.7135	0.4356	0.2915
circular variance V_o		0.0791	0.2865	0.5644	0.7085
Central Trigono metric moments	α_1^*	0.9209 0.7334	0.7135	0.4356	0.2915
	α_2^*		0.3507	0.1150	0.0621
	β_1^*	0	0	0	0
	β_2^*	0.0561	0.2165	0.2290	0.1682
Skewness γ_1^o		2.5206	1.4115	0.5401	0.2821
Kurtosis γ_2^o		2.2681	1.1157	0.2480	0.1094
Circular standard deviation σ_o		0.4059	0.8217	1.2893	1.5701

$$b = \left(\frac{-2}{\sqrt{\pi}} \frac{\sigma_1}{\sigma_1 + \sigma_2} e^{\frac{-p^2 \sigma_1^2}{2}} \sum_{n=1}^{\infty} \frac{\left(\frac{p\sigma_1}{\sqrt{2}}\right)^{2n-1}}{(2n-1)\,(n-1)!} + \frac{2}{\sqrt{\pi}} \frac{\sigma_2}{\sigma_1 + \sigma_2} e^{\frac{-p^2 \sigma_2^2}{2}} \right.$$

$$\left. \sum_{n=1}^{\infty} \frac{\left(\frac{p\sigma_2}{\sqrt{2}}\right)^{2n-1}}{(2n-1)\,(n-1)!} \right)$$

for $p = 0, \pm 1, \pm 2, \ldots$

The results are evaluated numerically using MATLAB. Thus obtaining the trigonometric moments, various characteristics of the Wrapped Binormal model fixing $\sigma_1 = 0.5$ and varying $\sigma_2 = 0.75, 1.5, 2.5$ and 3.5 are computed borrowing expressions from Mardia (1972) and are presented in Table 7.2.

7.4. New Circular Models

An attempt is made here to derive the new circular models on the basis of Dattatreya Rao et al. (2007) and making use of the trigonometric moments

Table 7.3. Half Logistic Distribution

Half logistic	Original Linear Model	New Circular Models
Pdf	$f(x; \mu, \sigma) = \frac{2}{\sigma} \left[\frac{e^{-\left(\frac{x-\mu}{\sigma}\right)}}{1+e^{-\left(\frac{x-\mu}{\sigma}\right)}} \right]^2$, $0 \leq \mu < \infty$, $\mu \leq x < \infty$, $\sigma > 0$ $= \frac{1}{2\sigma} \operatorname{Sech}^2 \left(\frac{x-\mu}{2\sigma}\right)$, $\sigma > 0$	**Modulo 2π Reduction** $g(\theta) = \sum_{k=-\infty}^{\infty} \frac{2}{\sigma} \left[\frac{e^{-\left(\frac{\theta+2k\pi-\mu}{\sigma}\right)}}{1+e^{-\left(\frac{\theta+2k\pi-\mu}{\sigma}\right)}} \right]^2$, $\theta \in [0, 2\pi)$, $\mu \in [0, 2\pi)$ and $\theta \geq \mu$, $\sigma > 0$ $= \sum_{k=0}^{\infty} \frac{1}{2\sigma} \operatorname{Sech}^2 \left(\frac{\theta+2k\pi-\mu}{2\sigma}\right)$, $\theta \in [0, 2\pi)$, $\mu \in [0, 2\pi)$ and $\theta \geq \mu$, $\sigma > 0$ **Trigonometric Moments**, for $p \in \mathbb{N}$ and $\sigma > 0$ $g(\theta) = \frac{1}{2\pi} \left[1 + 4 \sum_{p=1}^{\infty} (b_n \cos p(\theta - \mu) + c_n \sin p(\theta - \mu)) \right]$, $\theta \in [0, 2\pi)$, $\mu \in [0, 2\pi)$ and $\theta \geq \mu$, $\sigma > 0$ where $b_n = \int_0^{2\pi/p\sigma} \cos(p\sigma\, y) \frac{e^{-y}}{(1+e^{-y})^2}\, dy + \sum_{n=1}^{\infty} (-1)^{n-1} \frac{n^2 e^{-na}}{n^2 + p^2 \sigma^2}$ $c_n = \int_0^{2\pi/p\sigma} \sin(p\sigma\, y) \frac{e^{-y}}{(1+e^{-y})^2}\, dy + \sum_{n=1}^{\infty} (-1)^{n-1} \frac{np\sigma\, e^{-na}}{n^2 + p^2 \sigma^2}$

Table 7.4. Binormal Distribution

Binormal	Original Linear Model	New Circular Models
Pdf	$f(x)$ $=$ $\begin{cases} \sqrt{\dfrac{2}{\pi}}\dfrac{1}{\sigma_1+\sigma_2}\exp\left[\dfrac{-(x-\mu)^2}{2\sigma_1^2}\right], & x \le \mu \\[2ex] \sqrt{\dfrac{2}{\pi}}\dfrac{1}{\sigma_1+\sigma_2}\exp\left[\dfrac{-(x-\mu)^2}{2\sigma_2^2}\right], & x > \mu \end{cases}$ $\sigma_1, \sigma_2 > 0, x \in \mathbb{R}$	Modulo 2π Reduction for $\theta, \mu \in [0, 2\pi), \sigma_1, \sigma_2 > 0$ $g(\theta) = \begin{cases} \sum_{k=-\infty}^{\infty} \sqrt{\dfrac{2}{\pi}}\dfrac{1}{\sigma_1+\sigma_2}\exp\left[\dfrac{-(\theta+2k\pi-\mu)^2}{2\sigma_1^2}\right], & \theta \le \mu \\[2ex] \sum_{k=-\infty}^{\infty} \sqrt{\dfrac{2}{\pi}}\dfrac{1}{\sigma_1+\sigma_2}\exp\left[\dfrac{-(\theta+2k\pi-\mu)^2}{2\sigma_2^2}\right], & \theta > \mu \end{cases}$ Trigonometric Moments for $p = 0, \pm 1, \pm 2, \ldots$ $g(\theta) = \dfrac{1}{2\pi}\left[1 + 2\sum_{p=1}^{\infty}\left(a\cos p\left(\theta-\mu\right) + b\sin p\left(\theta-\mu\right)\right)\right]$ $\sigma_1, \sigma_2 > 0, \theta, \mu \in [0, 2\pi]$ $a = \dfrac{\sigma_1}{\sigma_1+\sigma_2}e^{\frac{-p^2\sigma_1^2}{2}} + \dfrac{\sigma_2}{\sigma_1+\sigma_2}e^{\frac{-p^2\sigma_2^2}{2}}$ $b = \dfrac{-2}{\sqrt{\pi}}\dfrac{\sigma_1}{\sigma_1+\sigma_2}e^{\frac{-p^2\sigma_1^2}{2}}\sum_{n=1}^{\infty}\dfrac{\left(\frac{p\sigma_1}{\sqrt{2}}\right)^{2n-1}}{(2n-1)(n-1)!} + \dfrac{2}{\sqrt{\pi}}\dfrac{\sigma_2}{\sigma_1+\sigma_2}e^{\frac{-p^2\sigma_2^2}{2}}$ $\sum_{n=1}^{\infty}\dfrac{\left(\frac{p\sigma_2}{\sqrt{2}}\right)^{2n-1}}{(2n-1)(n-1)!}$

obtained from the characteristic functions of the derived distributions mentioned in Tables 7.3 and 7.4.

7.5. Tests for Goodness of Fit

Random data from the proposed Wrapped distributions with specified values of parameters are drawn using Acceptance and Rejection method (Law and Kelton, 2003). Goodness of fit for small samples using Kolmogorov-Smirnov (KS) type Kuiper one sample test and for large samples using χ^2-test were employed, (Banks et al., 2000) methodology is adopted from the book mentioned above. Three data sets of sample sizes 10 and 100 respectively are taken for employing Kuiper's and χ^2-tests. The results of these tests are presented in Tables 7.5 and 7.6. Here it may be noted that the Kuiper's test is rotation invariant.

Table 7.5. Results of small sample goodness of fit of Kuiper's test sample size $n = 10$

Distribution	Data set	D_+	D_-	Kuiper's $V = \sqrt{n}(D_+ + D_-)$
Wrapped Half Logistic $\mu = 2.5, \sigma = 0.5$	1	0.1964	0.0892	0.9031
	2	0.1762	0.0879	0.8352
	3	0.2482	0.0878	1.0625
Wrapped Binormal $\mu = 2.5,$ $\sigma_1 = 0.5, \sigma_2 = 0.75$	1	0.2604	0.1000	1.1397
	2	0.3590	0.0827	1.3968
	3	0.2557	0.2417	1.5729

Table 7.6. Results of large sample ($n = 100$) goodness of fit for LOS $\alpha = 5\%$.

Distribution	Data set	χ^2-Statistic
Wrapped Half Logistic $\mu = 2.5, \sigma = 0.5$	1	17.6000
	2	17.2000
	3	18.6000
Wrapped Binormal $\mu = 2.5,$ $\sigma_1 = 0.5, \sigma_2 = 0.75$	1	15.8000
	2	17.2000
	3	16.5000
	3	14.9000

In all cases shown above, the results are not significant hence there is no evidence against the null hypothesis then the samples are drawn from the distribution assumed.
Since $\chi^2_{\alpha/2} < \chi^2 < \chi^2_{1-\alpha/2}$, we can conclude that the data sets drawn from the respective new circular models are introduced in this chapter.

Acknowledgements

Authors are thankful to Prof. S. Rao Jammalamadaka, University of California, Santa Barbara, USA and Ashis SenGupta, Applied Statistics Unit, Indian Statistical Institute, Kolkata, India for their encouragement while working on this project. Project was carried out as a part of FIP scheme of the third author.

References

1. Apostol, T. M. (1986). *Mathematical Analysis*. Addison – Wesley/Narosa.
2. Banks, J. Carson, J. S. and Nelson, B. L. (2000). *Discrete – Event System Simulation*. Prentice Hall of India Private Ltd., Delhi.
3. Breckling, J. (1989). *The Analysis of Directional Time Series: Applications to Wind Speed and Direction*. Springer Verlag, New York.
4. Carslaw, H. S. (1930). *Introduction to the Theory of Fourier's series and Integrals*. Dover, New York, 3rd edition.
5. Dattatreya Rao et al., A. V., Ramabhadra Sarma, I. and Girija, S. V. S. (2007). On Wrapped Version of Some Life Testing Models, *Communication in Statistics – Theory and Methods*, **36**, 2027–2035.
6. Fisher, N. I. (1993). *Statistical Analysis of Circular Data*, Cambridge University Press, Cambridge.
7. Fuller, M., Laj, C. and Herrero – Bervera, E. (1996). The reversal of the earth's magnetic field, *Amer. Sci.*, **84**, 552–561.
8. Gravin, A. and McClean, B. (1997). Convolution and sampling theory of the Binormal distribution as prerequisite to its application in statistical process control, *The Statistician*, **46**, 33–47.
9. Hanbury, A. and Serra, J. (2001). Morphological Operators on the Unit Circle, *IEEE Transactions on Image Processing*, **10**, 1842–1850.
10. Hrushesky, W. J. M. (1994). *Circadian Cancer Therapy*, CRC Press, Boca Raton.
11. Jammalamadaka, S. R. and SenGupta, A. (2001). *Topics in Circular Statistics*, World Scientific Press, Singapore.
12. Jammalamadaka, S. R. and Sengupta, S. (1972). Mathematical Techniques for Paleocurrent Analysis: Treatment of Directional Data, Journal of Intl. Assoc. Geol, **4**, 235–258.
13. Jammalamadaka, S. R. and Sarma, Y. R. K. (1986). Functional assessment

of knee and ankle during level walking, *Data Analysis in Life Science*, 21–54. Indian Statistical Institute, Calcutta, India.

14. Kantam, R. R. L. and Rosaiah, K. (1998). Half logistic distribution in acceptance sampling based on life tests, *Indian Association for Productivity Quality and Reliability (IAPQR) Transactions*, **23**, 117–125.

15. Kendall, D. G. (1974). Pole – Seeking Brownian motion and bird navigation, *J. Roy. Statist. Soc.*, **36**, 365–417.

16. Law, A. M. and Kelton, W. D. (2003). *Simulation Modeling and Analysis*, 2nd edition. McGraw–Hill Higher Education, Delhi.

17. Mardia, K. V. (1972). *Statistics of Directional Data*. Academic Press, New York.

18. Mardia, K. V. and Jupp, P. E. (2000). *Directional Statistics*. John Wiley, Chichester.

19. Morgan, E. (1990). *Chronobiology and Chronomedicine*. Peter Lang, Frankfurt.

20. Schmidt – Koeing, K. (1963). On the role of loft, the distance and site of release in Pigeon homing (the "cross – loft experiment"), *Biol. Bull.*, **125**, 154–164.

PART 3
SEMI-PARAMETRIC

Chapter 8

Non-Stationary Samples and Meta-Distribution

Dominique Guégan

University Paris 1 Panthéon-Sorbonne, France

In this chapter, we focus on the building of an invariant distribution function associated with a non-stationary sample. After discussing some specific problems encountered by non-stationarity inside samples like the "spurious" long memory effect, we build a sequence of stationary processes permitting us to define the concept of meta-distribution for a given non-stationary sample. We use this new approach to discuss some interesting econometric issues in a non-stationary setting, namely forecasting and risk management strategy.

Keywords: Copula, Non-stationarity, Risk management, SETAR processes, Switching processes.

Contents

8.1. Introduction

During decades time series modelling has focused on stationary linear models, and later also on stationary non-linear models. Recently the question of non-stationarity has arisen, and a sudden interest focuses on the modelling of non-stationary and/or non-linear time series.

An interesting type of questions has emerged when it has been observed that stationary non-linear processes exhibited empirical autocorrelation

function with a hyperbolic decreasing rate, although they are characterized by a theoretical short memory behavior (exponential decrease of the
autocorrelation function). This observation has highlighted the fact that
the behavior of some statistical tools under non-stationarity must be ques-
tioned. Indeed, the statistical tools we are using are meaningful only under
certain assumptions, the most crucial one being the stationarity. Hence,
the question arises what the statistical tools are telling us when used on
non-stationary data.

Thus, defining a correct framework in which we can analyse data sets is
fundamental before choosing the class of models we will use. Following this
idea, we develop here a framework permitting us to work in a stationary setting including the structural non-stationarities. We assume that a process
$(Y_t)_t$ is characterized by changes in the k-th order moments all along the
information set (corresponding in practice to the observed trajectory). This
corresponds to structural changes in financial time series causing the time
series over long intervals to deviate significantly from stationarity. This
means that we assume that non-global stationarity is verified for the sample. Then it is plausible that by relaxing the assumptions of stationarity in
an adequate way, we may obtain better fit followed by robust forecasts and
management theory. Doing that, we will see that we can get new insight to
approximate the unknown distribution function for complex non-stationary
processes.

Without using the whole sample, which is the source of non-stationarity,
we define a new way to analyse and model this information set dividing it in
subsamples on which stationarity is achieved. In this paper, our objectives
are twofold. First, we show that the non-stationarities observed on the
empirical moments pollute the theoretical properties of the statistics defined
inside a "global" stationary framework, and thus a new framework needs
to be developed. Second we propose a new way to study finite sample data
sets in presence of k-order non-stationarity.

As the non-stationarity affects nearly all the moments of a time series
when it is present, we first study the impact of the non-stationarity on a
non-linear transformation of the observed data set $(Y_t)_t$ considering $(Y_t^\delta)_t$,
for any $\delta \in R^+$, looking at its sample autocovariance function (ACF). We
exhibit the strange behavior of this ACF in presence of non stationarity
and illustrate it through several modellings. Then, in order to avoid misspecification using the sample ACF when we work with a practical point
of view, we propose to work with the distribution function. In case of
non-stationarity, this one also appears inadequate, justifying the necessity

to adopt a new strategy. Thus, we focus on the building of sequence of invariant distribution functions using the notion of homogeneity intervals introduced by Starica and Granger (2005). This methodology will conduce us to propose the notion of meta-distribution associated with a sample in presence of non-stationarity. This last notion lies on both the use of copula and sequence of homogeneity intervals characterized by invariant distribution functions.

Thanks to this new approach, we get new insights for robust forecastings, risk management theory and solutions of complex probabilistic problems.

The plan of the chapter is the following. In Section 8.2 we recall the notion of strict stationarity and we exhibit the specific behavior of the sample ACF for a $(Y_t^\delta)_t$ process in presence of non-stationarity leading to the creation of spurious behaviors that we describe through three different modellings. In Section 8.3 we specify a homogeneity test based on higher order cumulants and we show how the copula concept is usefull in presence of non-stationarity. The notion of meta-distribution is introduced. Some applications are proposed for econometricians and risk managers. In Section 8.4 we give some concluding remarks.

8.2. Empirical Evidence

A stochastic process is a sequence of random variables $(Y_t)_t$ defined on a probability space $(\Omega, \mathcal{A}, P)$. Then $\forall t$ fixed, Y_t is a function $Y_t(\cdot)$ on the space Ω, and $\forall \omega \in \Omega$ fixed, $Y_.(\omega)$ is a function on $\mathbb{Z}$. The functions $(Y_.(\omega))_{\omega \in \Omega}$ defined on $\mathbb{Z}$ are realizations of the process $(Y_t)_t$. A second order stochastic process $(Y_t)_t$ is such that, $\forall t$, $EY_t^2 < \infty$. For a second order stochastic process, the mean $\mu_t = EY_t$ exists $\forall t$ and so do the variance and the covariance. The covariance $\gamma(.,.)$ of a second order stochastic process $(Y_t)_t$ exists and is defined by

$$cov(Y_t, Y_{t+h}) = \gamma_Y(h, t) < \infty \ \forall h, \forall t \in \mathbb{Z}. \tag{8.1}$$

A stochastic process is completely known as soon as we know its probability distribution function. When several realizations of a process are available, the theory of stochastic processes can be used to study this distribution function. However, in most empirical problems, only a single realization is available. Each observation in a time series is a realization of each random variable of the process. Consequently, we have one realization of each random variable and inference is not possible. We have to restrict the properties of the process to carry out inference. To allow estimation, we need to restrict the process to be strictly stationary, because we work mainly with non-linear models.

Definition 8.1. A stochastic process $(Y_t)_t$ is strictly stationary if the joint distribution of $Y_{t_1}, Y_{t_2}, \cdots, Y_{t_p}$ is identical to that of $Y_{t_1+h}, Y_{t_2+h}, \cdots, Y_{t_p+h}$, for all h, where p is an arbitrary positive integer and $t_1, t_2, \cdots, t_p$ is a collection of k positive integers.

Strict stationarity means intuitively that the graphs over two equal-length time intervals of a realization of a time series should exhibit similar statistical characteristics. It means also that $(Y_{t_1}, \cdots, Y_{t_p})$ and $(Y_{t_1+h}, \cdots, Y_{t_p+h})$ have the same joint distribution for all positive integers h and p, and thus have the same k-th order moments. Therefore, strict stationarity requires that the distribution of $Y_{t_1}, Y_{t_2}, \cdots, Y_{t_p}$ is invariant under time shifts. In that case, we speak of global stationarity.

8.2.1. *Non-stationary stylized facts*

Even if we work always in a global stationary framework, a lot of non-stationarities are observed on real data sets. In this Section, we specify some of the non-stationarities which affect the major financial and economic data sets. These questions are the base of a lot of problems concerning the modelling of real data sets. Indeed, structural behaviors like volatility, jumps, explosions and seasonality provoke non-stationarity. Now specific transformations on the data sets like concatenation, aggregation or distortion are also at the origin of non-stationarity.

All these features imply that the property of global stationarity fails. Indeed, existence of volatility imposes that the variance depends on time. This latter one is generally modelled using time varying function. In presence of seasonality the covariance depends on time producing evidence of non stationarity. Existence of jumps produces several regimes inside data sets. These different regimes can characterize the level of the data or its volatility. Changes in mean or in variance affect the properties of the distribution function charaterizing the underlying process. Indeed, this distribution function cannot be invariant under time-shifts and thus a global stationarity cannot be assumed. Distorsion effects correspond to explosions that one cannot remove from any transformation. This behavior can also be viewed as a structural effect. Existence of explosions means that some higher order moments of the distribution function do not exist. Concatenated data sets used to produce specific behavior cannot have the same probability distribution function on the whole period as soon as it is a juxtaposition of several data sets. Aggregation of independent or weakly dependent random variables is a source of specific features. All these behaviors provoke the non existence of higher order moments or erractic behaviors of the sample ACF.

Until now, several authors tried to take into account these non-stationarities through models. The simple one consists of taking the square (or any transformation) of the data to model the conditional variance (Engle, 1982; Ding and Granger, 1996). Now, whatever the chosen methodology, the main tool for studying the data sets remains the use of the sample autocorrelation function. The sample ACF computed from $(Y_t^\delta)_t$, $\delta \in R^+$, presents inappropriate behaviors under non-stationarity. We exhibit below the asymptotic behavior of the sample autocorrelation function in that context and illustrate it through some modellings.

8.2.2. *Asymptotic behavior of the ACF of the* $(Y_t^\delta)_t$ *process*

In this section we highlight that $\gamma_Y(t, h)$ and $\tilde{\gamma}_Y(h)$ are different concepts in presence of non-stationary, where $\tilde{\gamma}_Y(h)$ is the sample covariance. We assume that we observe a non-stationarity sample $(Y_1, \cdots, Y_T)$, corresponding to an underlying process $(Y_t)_t$, from which we build $Y^\delta = (Y_1^\delta, Y_2^\delta, \cdots, Y_T^\delta)$, $\delta \in R^+$. We divide this last sample into r subsamples each consisting of distinct stationary processes with finite k-th order moments, $k \in N$, assuming existence of distinct stationary ergodic process on each of them. We denote $p_j \in R^+$, $j = 1, \cdots, r$ such that $p_1 + p_2 + \cdots + p_r = 1$. Here p_j is the proportion of observations from the jth subsample in the full sample. If we define now $q_j = p_1 + p_2 + \cdots + p_j$, $j = 1, \cdots r$, the whole sample is written as the union of r subsamples $Y^\delta = ((Y_1^{\delta(1)}, \cdots, Y_{Tq_1}^{\delta(1)}), \cdots, (Y_{Tq_{r-1}+1}^{\delta(r)}, \cdots, Y_{Tq_r}^{\delta(r)}))$.

The sample auto-covariance function for the series $(Y_t^\delta)_t$ is equal to

$$\tilde{\gamma}_{Y^\delta}(h) = \frac{1}{T} \sum_{t=1}^{T-h} (Y_t^\delta - \underline{Y}_T^\delta)(Y_{t+h}^\delta - \underline{Y}_T^\delta), \tag{8.2}$$

where $\underline{Y}_T^\delta$ is the sample mean of the process $(Y_t^\delta)_t$.

Proposition 8.1. *Let* $(Y_1^{\delta(i)}), \cdots, (Y_{Tq_i}^{\delta(i)})$, $i = 1, \cdots, r$ *be* r *subsamples from the sample* Y^δ, $\delta \in R^+$, *each subsample corresponding to a stationary* *distinct ergodic process with finite second order moments, whose sample covariance is equal to $\tilde{\gamma}_{(Y^{\delta(i)})}(h)$. Then the sample autocorrelation function $\tilde{\gamma}_{Y^\delta}$ of the sample Y^δ is such that*

$$\tilde{\gamma}_{Y^\delta}(h) \to \sum_{i=1}^{r} p_i \tilde{\gamma}_{(Y^{\delta(i)})}(h) + \sum_{1 \le i \le j \le r} p_i p_j (E(Y^{\delta(j)}) - E(Y^{\delta(i)}))^2, \quad (8.3)$$

as $T \to \infty$. The proof is given in the Appendix.

Under the property of stationary ergodicity, the ACF of each process has an exponential decay. Thus, the sample Y^δ has its sample ACF $\tilde{\gamma}_{Y^\delta}(h)$ that decays quickly for the first lags and then approach positive constants given by $\sum_{1 \le i \le j \le r} p_i p_j (E(Y^{(j)})^\delta - E(Y^{(i)})^\delta)^2$. Thus, in presence of non-stationarity, this last term explains the existence of persistence observed on the sample ACF when we compute it using the whole sample $(Y_t^\delta)_t$. When $\delta = 1$, this proposition permits to explain how shifts in the means could provoke a slow decay of the autocorrelation function - which can be associated to a long memory behavior - and the same behavior is observed for the variance as soon as we modelled it using Y_t^δ, for any δ. We refer to Guégan (2005) for a review on the long memory concepts. We illustrate now these previous facts using different modellings.

Let $(Y_t)_t$ be a two state Markov switching process

$$Y_t = \mu_{s_t} + \varepsilon_t. \quad (8.4)$$

The process $(s_t)_t$ is a Markov chain which permits the switch from one state to another according to the transition matrix P, whose elements are the fixed probabilities p_{ij} defined by $p_{ij} = P[s_t = j | s_{t-1} = i]$, $i, j = 1, 2$, $0 \le p_{ij} \le 1$ and $\sum_{j=1}^{2} p_{ij} = 1$, for all $i = 1, 2$. The process $(\varepsilon_t)_t$ is a strong white noise, independent to $(s_t)_t$. The process $(Y_t)_t$ switches from level μ_1 to level μ_2 with respect to the Markov chain. This model has been studied by Andèl (1993). The theoretical behavior of the autocorrelation function of such a model, under stationarity conditions, is similar to the one of an ARMA(1,1) process: its autocorrelation function decreases with an exponential rate towards zero for large h. Nevertheless respecting the stationary conditions, it is possible to exhibit sample ACFs which have a very slow decay. This effect is explained in that case by the behavior of the second term of the relationship (8.3) which stays always bounded. We exhibit in Figures 8.1–8.3 this kind of behaviors for some models (8.4), providing the autocorrelation function of some simulated series with $p_{11} = p_{22} = p$. Here n_{s_t} denotes the number of changes inside the states, and the sample size is equal to $T = 1000$.

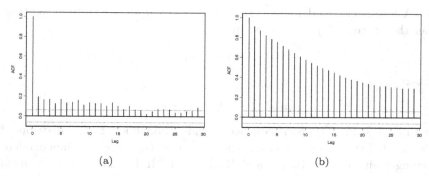

Fig. 8.1. Behaviors of the autocorrelation function of model (8.4): (a) $(\mu_1, \mu_2) = (0.5, -0.5)$ and $p = 0.99, n_{s_t} = 7$, (b) $(\mu_1, \mu_2) = (5, -5)$ and $p = 0.99, n_{s_t} = 7$

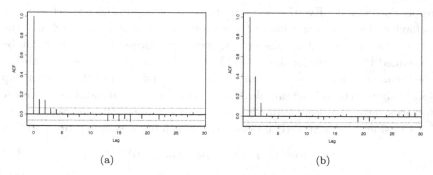

Fig. 8.2. Behaviors of the autocorrelation function of model (8.4): (a) $(\mu_1, \mu_2) = (0.5, -0.5)$ and $p = 0.75, n_{s_t} = 258$, (b) $(\mu_1, \mu_2) = (5, -5)$ and $p = 0.75, n_{s_t} = 258$

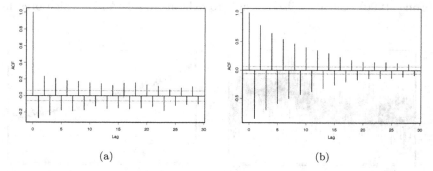

Fig. 8.3. Behaviors of the autocorrelation function of model (8.4): (a) $(\mu_1, \mu_2) = (0.5, -0.5)$ and $p = 0.05, n_{s_t} = 950$, (b) $(\mu_1, \mu_2) = (5, -5)$ and $p = 0.05, n_{s_t} = 950$

A StopBreak model also permits the switch from one state to another. Let the process $(Y_t)_t$ be defined by

$$Y_t = \mu_t + \varepsilon_t, \qquad (8.5)$$

where

$$\mu_t = (1 - \alpha\delta_t)\mu_{t-1} + \delta_t\eta_t, \qquad (8.6)$$

with $\alpha \in [0,2]$, $(\delta_t)_t$ is a sequence of independant identically distributed Bernoulli (λ) random variables and $(\varepsilon_t)_t$ and $(\eta_t)_t$ are two independent strong white noises (Breidt and Hsu, 2002). It is known that for fixed λ, this process - which models switches with breaks - has short memory behavior. This one is observed for long samples, but as soon as the jumps are rare relatively to sample size, the short memory behavior does not appear so evident. Even if the asymptotic theory describes a short memory behavior, a sample experiment for a short sample size looks much like the corresponding characteristics for long memory processes. This effect can be explained by the relationship (8.3). Indeed, for different values of α and λ, the means μ_1 and μ_2 are different, thus the second term of the relationship (8.3) is bounded and the sample ACF of model (8.5)-(8.6) does not decrease towards zero. We provide in Figure 8.4 an example of this behavior.

Consider now a SETAR process $(Y_t)_t$ which has a simple representation

$$Y_t = \mu_1 I(Y_{t-1} > 0) + \mu_2 I(Y_{t-1} \leq 0) + \varepsilon_t, \qquad (8.7)$$

where $I(.)$ is the indicator function, $(\varepsilon_t)_t$ being a strong white noise. This model permits to shift from the mean μ_1 to the mean μ_2 with respect to the value taken by Y_{t-1}. SETAR processes are known to be short memory (Tong, 1990). But it is also possible to exhibit sample ACFs which present

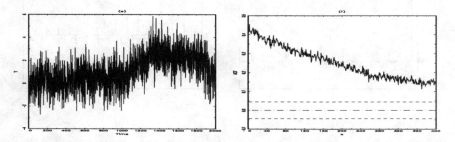

Fig. 8.4. The trajectory and ACF of the model (8.5)-(8.6) with $\lambda = 0.01$ and $\alpha = 0.9$, $T = 2000$, $\sigma_\varepsilon^2 = \sigma_\eta^2 = 1$.

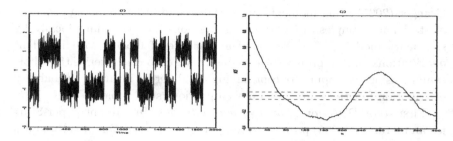

Fig. 8.5. Trajectory and ACF of the Threshold Auto-Regressive model defined by equation (8.7) with $T = 2000$, $\sigma_\epsilon^2 = 0.2$ and $\mu_0 = -\mu_1 = -1$.

slow decay. This slow decay can also be explained by the second term of the relationship (8.3), and also by the time spent in each state. We exhibit in Figure 8.5 an experiment corresponding to this fact.

Thus the use of the sample ACF creates confusion in the modelling of some data sets in presence of non-stationarity. This last statistical tool appears insufficient to characterize the behavior of any data set in that context. A new strategy needs to be developed as soon as the second order properties fail to give correct information for modelling. Moreover, it appears important to use the characteristics of the higher order moments or of the distribution function to solve this problem.

One way will be to test the invariance of the sample higher order moments and of the empirical distribution function all along the whole sample. With respect to the results of these tests, various strategies can be used. In presence of non-stationarity, we can use dynamical parameter models, consider models with a distribution function evolving in a dynamical way all along the whole sample or we can consider a sequence of stationary models. This means that we can define two strategies to study such a data set: the use of an unique distribution function with dynamic parameters or a set of several distribution functions invariant on each subsample.

8.3. A Meta-Distribution Function

This section concerns the discussion of several points of views in order to take into account non-stationarity and, to detect and model local stationarity. It is mainly a discussion of methodology.

The first problem is to detect non-stationarities or to test them. In a first insight, we consider an extension of the approach proposed by Starica and

Granger (2005) using moments up to the second order, building sequence of stationary samples, approximating the non-stationary data locally by stationary models. Then we build a sequence of models with invariant distributions and we propose a meta-distribution to characterize the whole sample, using the copula concept linking the different invariant probability distribution functions adjusted on each previous subsample.

First step: *Detection of homogeneity intervals.* In a recent paper Starica and Granger (2005) propose to test successively on different subsamples of a time series $(Y_t)_t$ the invariance of the spectral density. They propose a specific test and their strategy is the following. They consider a subset $(Y_{m_1}, \cdots, Y_{m_2})$, $\forall m_1, m_2 \in N$, on which they apply their test and build confidence intervals. Then they consider another subset $(Y_{m_2+1}, \cdots, Y_{m_2+p})$ for some $p \in N$. They apply again the test and verify if the value of the statistic belongs to the confidence interval previously built or not. If it belongs to the confidence interval, they continue with a new subset. If not, they consider $(Y_{m_1}, \cdots, Y_{m_2+p})$ as an interval of homogeneity and analyse the next subset $(Y_{m_2+p+1}, \cdots, Y_{m_2+2p})$ and define new confidence intervals from their statitic. At the end, they estimate a model on each homogeneity interval. They use these intervals to forecast.

The approach proposed by Starica and Granger (2005) is based on the spectral density which is built using the second order moments of a process. It is possible to extend this method by the use of empirical higher order moments. Using higher order moments is mainly justified by the fact that the moments up to the second order are non-stationary inside financial data sets. These higher order moments are estimated on the whole sample and on subsamples as before. Then we propose a test based on these higher order moments or their cumulants in order to obtain intervals of homogeneity. We compute the cumulants associated with the sample up to the k-th order, denoted c_k, and the spectral density of cumulant of order k, denoted $f_{c_k,Y}$. We define its estimate by $I_{c_k,Y,T}$. Consider the statistic

$$\tilde{T}_k(Y) = \sup_{\lambda \in [-\pi,\pi]} \left| \int_{[-\pi,\pi]^{k-1}} \left(\frac{I_{c_k,Y,T}(z)}{f_{c_k,Y}} - \frac{\tilde{c}_k}{c_k} \right) dz \right|, \qquad (8.8)$$

where $\tilde{c}_k$ is an estimate of c_k. It can be shown that, under the null that the cumulants of order k are invariant on the subsamples, this statistic $\tilde{T}_k(Y)$ converges in distribution to $\frac{(2\pi)^{k-1}}{c_k} B(\sum_{j=1}^{k-1} \lambda_j)$, when $T \to \infty$, $-\pi < \lambda < \pi$, and $B(.)$ is the Brownian bridge. Thanks to the knowledge of the critical values of this statistic, one can build homogeneity intervals, using moving windows. This statistic permits to use a more complete information from the data set in order to build homogeneity intervals.

Second step: *The use of copula concept.* When the homogeneity intervals are determined, we can associate to each subsample an invariant distribution function and compare it to the distribution function derived using the whole sample. Thus, we define a sequence of invariant distribution functions all along the sample.

We begin to recall the copula concept. Consider a general random vector $Z = (X, Y)^T$ and assume that it has a joint distribution function $F(x, y) = \mathbb{P}[X \leq x, Y \leq y]$ and that the random variables X and Y have continuous marginal distribution functions, denoted respectively by F_X and F_Y. It has been shown by Sklar (1959) that every 2-dimensional distribution function F with marginals F_X and F_Y can be written as $F(x, y) = C(F_X(x), F_Y(y))$ for an unique (because the marginals are continuous) function C that is known as the copula of F (this result is also true in the r-dimensional setting). Generally a copula will depend almost on one parameter, then we denote it by C_α and we have the following relationship:

$$F(x, y) = C_\alpha\big(F_X(x), F_Y(y)\big). \qquad (8.9)$$

Here, a copula C_α is a bivariate distribution with uniform marginals and it has the important property that it does not change under strictly increasing transformations of the random variables X and Y. Moreover, it makes sense to interpret C_α as the dependence structure of the vector Z. In the literature, this function has been called "dependence function", "uniform representation" and "copula". We keep this last denomination here. In Figure 8.6, we provide the graph of four different copulas using Gaussian (0,1) distributions as marginals.

Practically, to get the joint distribution F of the random vector $Z = (X, Y)^T$ given the marginals, we have to choose a copula that we apply to these marginals. There exists different families of copulas: the elliptical copulas, the Archimedean copulas, the meta-copulas (Joe, 1997). In order to choose the best copula adjusted for a pair of random variables, we need to estimate the parameters of the copula and to estimate the copulas, see Cherubini et al. (2004), and Caillault and Guégan (2005) for details on the procedures.

We have presented the method to adjust a copula in case of two processes. We can extend dynamically the adjustment for a sequence of r processes with invariant distribution functions $F_{Y^{(i)}}$, $i = 1, \cdots, r$. Thus, we will work step by step working with two subsamples at each step. This

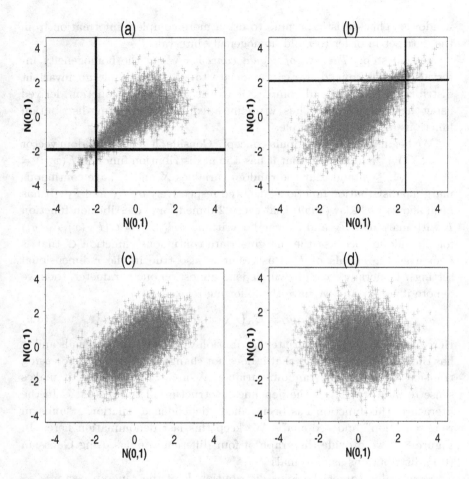

Fig. 8.6. Plots of 10000 samples of (a) Clayton, (b) Gumbel, (c) Frank and (d) Ali-Mikhail-Haq copulas with $\alpha = 0.5$, 2,5, 0.5 respectively.

will permit us to detect if the copula that we look for is the same all along the samples, or if it is the same but with different parameters or, if it changes with respect to the subsamples. In order to achieve this process, we can use nested tests, (Guégan, and Zhang, 2009) for applications.

Through the nested tests, we analyze the changes. If we admit that only the copula parameters change, we can apply the change-point analysis as in Dias and Embrechts (2004) to decide the change time. Moreover, considering that change-point tests have less power in case of "small" changes, we can assume that the parameters change according to a time-varying function of predetermined variables and test it. More tractably, we can decide the best copula on subsamples using the moving window, and then observe the changes.

Now we are going to use this copula concept in order to find the joint distribution function $F_Y = P[Y_1 \leq y_1, \cdots, Y_T \leq y_T]$ of a process $(Y_t)_t$ assuming that we observe $Y_1, \cdots, Y_T$, a sample of size T. The knowledge of this distribution function will permit us to do forecasting or to propose a risk management strategy for the data set of interest. If the process $(Y_t)_t$ is non-stationary, its distribution function F_Y is not invariant on the whole sample. In order to illustrate this fact, we provide an example of such a situation in Figure 8.7. In this figure, we have identified four homogeneity intervals characterized by changes in mean or in variance, and we have estimated different distribution functions on each subsample denoted $F_Y(1), F_Y(2), F_Y(3), F_Y(4)$. We now formalize this example.

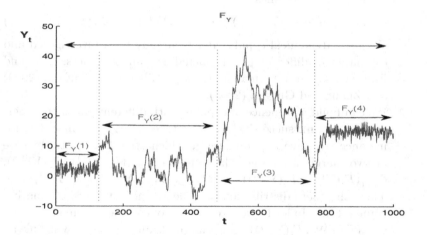

Fig. 8.7. Example of a sequence of invariant distribution functions

In order to find F_Y, we are going to determine on each homogeneity interval an invariant distribution function. These distribution functions can belong to the same class of distribution functions but with different parameters or can belong to different classes of distribution functions. To find them, we can use the Kolmogorov-Smirnov test, Q-Q plots or the χ^2 test.

Thus we get a sequence of r stationary subsamples $Y_1^{(i)}, \cdots, Y_{Tq_i}^{(i)}$, $i = 1, \cdots, r$, each characterized by an invariant distribution function $F_{Y^{(i)}}$, $i = 1, \cdots, r$. Using this sequence of invariant distribution functions, we can build the distribution function characterizing the whole sample, using a copula C_α linking this sequence of invariant distribution functions, such that

$$P(Y_1 \leq y_1, \cdots, Y_T \leq y_T) = C_\alpha(F_1(Y^{(1)}), \ldots, F_r(Y^{(r)})). \qquad (8.10)$$

The expression (8.10) provides a new way to characterize the joint distribution of the sample $Y_1, \cdots, Y_T$ using a sequence of invariant distributions. We call this distribution a meta-distribution and we will see that we can use it for different purposes.

There exists different ways to build this meta-distribution. Indeed, the expression (8.10) provides one approach but we can make it more complex.

(1) We can assume that the parameter α of the copula evolves in time so that (8.10) becomes

$$P(Y_1 \leq y_1, \cdots, Y_T \leq y_T) = C_{\alpha_t}(F_1(Y^{(1)}), \cdots, F_r(Y^{(r)})). \quad (8.11)$$

We get a dynamical copula. It has already been investigated and estimated in different papers including, among others, Dias and Embretchs (2004), Fermanian (2005), Caillault and Guégan (2009) and Zhang and Guégan (2009).

(2) We can build a sequence of copulas, with different parameters permitting to link successively two or three marginals, in order, for instance, to have a parameter α which stays constant. In case of two marginals, we get the sequence $C_{\alpha_{12}}(F_1(Y^{(1)}), F_2(Y^{(2)}))$, $C_{\alpha_{34}}(F_3(Y^{(3)}), F_4(Y^{(4)})), \ldots, C_{\alpha_{r,r-1}}(F_{r-1}(Y^{(r-1)}), F_r(Y^{(r)}))$. Then the meta-distribution will be defined, thanks to another copula C_β. Indeed, we are going to consider each expression $C_{\alpha_{ij}}(F_i(Y^{(i)}), F_j(Y^{(j)}))$ as a marginal, and then C_β will link all these marginals as

$$P(Y_1 \leq y_1, \cdots, Y_T \leq y_T) = C_\beta(C_{\alpha_{12}}, C_{\alpha_{34}}, \cdots, C_{\alpha_{r,r-1}}). \quad (8.12)$$

An estimation procedure has to be defined for this latter approach extending the previously cited works; see Guégan and Maugis (2011).

(3) Now we can use more than two marginals. In case of three marginals, we can use Archimedean copulas under specific restrictions (Joe, 1997). Then we define a copula C_ξ linking the marginals

$C_{\alpha_{ijk}}(F_i(Y^{(i)}), F_j(Y^{(j)}), F_k(Y^{(k)}))$ as

$$P(Y_1 \leq y_1, \cdots, Y_T \leq y_T) = C_\xi(C_{\alpha_{123}}, \ldots). \qquad (8.13)$$

(4) To work with more than two marginals, another interesting approach based on the vines can be conduced (Aas et al., 2009; Guégan and Maugis, 2011).

The methodology that we have proposed here permits to give new openings concerning several econometric issues.

(1) *To forecast with a non-stationary sample.*

- We can use one of the previous linking copulas C_α, C_{α_t}, C_β or C_ξ to get a suitable forecast for the process $(Y_t)_t$ assuming the knowledge of the whole information set $I_T = \sigma(Y_t, t < T)$. Then, for instance, we will compute $E_{C_\alpha}[Y_{t+h}|I_T]$.
- We can also decide to forecast using a smaller information set, based on one or several subsamples defined as homogeneity intervals:

    (a) If we consider the last homogeneity interval, then the predictor will be equal to $E_{F_Y^{(r)}}[Y_{t+h}|I_r]$, where $I_r = \sigma(Y_{Tq_{r-1}+1}^{(r)}, \cdots, Y_T^{(r)})$, and $F_Y^{(r)}$ is the marginal associated to this subset.

    (b) If we use the last two homogeneity intervals, we will compute the expectation under the copula linking the two marginals corresponding to each subsample and the information set will be the reunion of the two subsamples. Then we get $E_{C_\alpha}[Y_{t+h}|I_{r-1}\cup I_r]$, and C_α links $F_Y^{(r-1)}$ and $F_Y^{(r)}$.

    (c) If we look at the Figure 8.7, we can decide to use the intervals 1 and 4 to do forecast. In that case, we compute: $E_{C_\alpha}[Y_{t+h}|I(1)\cup I(4)]$ and C_α will link $F_Y^{(1)}$ and $F_Y^{(4)}$.

    We can expect that working with these approaches will provide better forecasts.

(2) *To compute a risk measure.* The same discussion as before can be done to compute for instance the classical Value-at-Risk mesure (VaR_α), which is simply the maximum loss that is exceeded over a specified period with a level of confidence $1 - \alpha$ for a given α. For

a random variable Y with distribution F_Y, it is defined by

$$F_Y(VaR_\alpha) = Pr[Y \le VaR_\alpha] = \alpha. \tag{8.14}$$

Then, to make this computation, we can decide to work with the distribution which appears more appropriate. This means that it could be one of the previous meta-distrbutions defined in (8.10), (8.11), (8.12) or (8.13). For instance, in case of our example (Figure 8.7), we can decide to use the two subsamples whose variablilities are more important to compute the VaR. Then the function F_Y will be the copula permitting to link $F_Y^{(2)}$ and $F_Y^{(3)}$, and in (8.14), we will use $F_Y = C_\alpha(F_Y^{(2)}, F_Y^{(3)})$. We can also use this methodology considering other risks measures (Guégan and Tarrant, 2010).

(3) *To determine the unconditional distribution of non-linear models.* Indeed, this approach could permit to solve this open problem. In particular, it could be possible to obtain the distribution function of any Markov switching model; this work will be the topic of a companion paper.

8.4. Conclusion

In this paper, we have discussed deeply the influence of non-stationarity on the stylized facts observed on the data sets and on specific statistics. For these statistics, a lack of robustness is observed in presence of non-stationarity, and this work emphasizes the fact that the theoretical concept of autocovariance function $\gamma_Y(t, h)$ and the concept of sample autocovariance $\tilde{\gamma}_Y(h)$ are totally different as soon as some specific features are detected in the sample. This evidence is illustrated through several examples. It is important to notice that this work does not concern asymptotic theory but discusses a new working framework to analyse non-stationary time series.

Then a new methodology is proposed in order to take the non-stationarity into account, building a sequence of invariant "homogeneity" intervals up to order four, and considering a new way to associate to a sample a joint distribution which can be used to compute risks or forecast. This new methodology opens the routes to solve a certain number of technical unsolved problems, as for instance the computation of the unconditional distribution of some non-linear processes.

Some extended research problems can also be considered; we cite some for illustration.

- The use of the change point theory could be used to verify the date at which we determine the beginning of an homogeneity interval. This could be a nice task. Indeed, most of the works concerning the change point theory concern detection of breaks in mean or in volatility. These works have to be reexamined taking into account the fact that breaks can provoke spurious long memory. Indeed, in the latter case, using the covariance matrix can be a problem in the sense that we cannot observe change point in the covariance matrix.

- The time spend in each state when we observe breaks is a challenging problem. We do not consider here the approach followed in the ACD models; we are interested to characterize the distribution function permitting the quantification of the time spend in a state. This last random variable is important in order to characterize the existence of states, and it can be connected to the creation of the long memory behavior. It is known that for a Markov switching process, this law is geometric depending on the transition probabilities. Deeper work is necessary to understand the role of this probability distribution in the creation of "spurious" long memory behavior. One way is to study the behavior of the autocovariance function when, for example, we assume that the time spend on a regime follows a specific law like the Poisson law or any classical continous law.

- The discussion of models taking into account sharp switches and time varying parameters. A theory has to be developed to answer to a lot of questions coming from practitioners. If the model proposed by Hyung and Franses (2005) appears interesting in that context, because it nests several related models by imposing certain parameter restrictions (Ar, Arfi, Stopbreak models for instance); more identification theory concerning this class of models need to be developed to understand how it can permit to give some answers to the problems discussed in this chapter.

- Another work concerns the test theory to detect "spurious" long memory begavior when this one is created by non-stationarity, some interesting references being Sibbersten and Kruse (2009), and Ohanassian, Russell and Tsay (2008). One way could be, using our approach to adjust an FI(d) on each interval, and to test the null that $d_1 = d_2 = \cdots = d_r = d$, where d is the value of the fractional differencing parameter obtained with the whole sample. Some preliminary empirical discussions on this approach have been done by Charffedine and Guégan (2008).

8.5. Appendix: Proof of Proposition 8.1

We develop the right hand side of the relationship (8.2):

$$\tilde{\gamma}_{Y^\delta}(h) = \frac{1}{T} \sum_{t=1}^{T-h} Y_t^\delta Y_{t+h}^\delta - \frac{\underline{Y}_T^\delta}{T} \sum_{t=1}^{T-h} (Y_t^\delta + Y_{t+h}^\delta) + \frac{1}{T} \sum_{t=1}^{T-h} \underline{Y}_T^{2\delta}. \tag{8.15}$$

Let

$$A = \frac{1}{T} \sum_{t=1}^{T-h} Y_t^\delta Y_{t+h}^\delta \tag{8.16}$$

and

$$B = -\frac{\underline{Y}_T^\delta}{T} \sum_{t=1}^{T-h} (Y_t^\delta + Y_{t+h}^\delta) + \frac{1}{T} \sum_{t=1}^{T-h} \underline{Y}_T^{2\delta}. \tag{8.17}$$

Thus $\tilde{\gamma}_{Y^\delta}(h) = A + B$. First, we compute A decomposing it on the r intervals as

$$A = \frac{1}{T} \sum_{i=1}^{r} \sum_{t=Tq_{i-1}+1}^{Tq_i-h} (Y_t^\delta)^{(i)} (Y_{t+h}^\delta)^{(i)}$$

$$+ \frac{1}{T} \sum_{i=1}^{r} [\sum_{t=Tq_i-h+1}^{Tq_{i+1}-h} (Y_t^\delta)^{(i)} (Y_{t+h}^\delta)^{(i)} + \cdots + \sum_{t=Tq_{r-1}-h+1}^{Tq_r-h} (Y_t^\delta)^{(i)} Y_{t+h}^{(r)}]. \tag{8.18}$$

Now we know that $cov((Y_t^\delta)^{(i)}, (Y_t^\delta)^{(j)}) = 0$ for all $i \neq j$ by building; thus

$$A = \frac{1}{T} \sum_{i=1}^{r} \sum_{t=Tq_{i-1}+1}^{Tq_i-h} (Y_t^\delta)^{(i)} (Y_{t+h}^\delta)^{(i)} + o(1). \tag{8.19}$$

We develop the term of the right hand of the previous relationship. Thus we get

$$\frac{1}{T} \sum_{i=1}^{r} \sum_{t=Tq_{i-1}+1}^{Tq_i-h} (Y_t^\delta)^{(i)} (Y_{t+h}^\delta)^{(i)} = \sum_{i=1}^{r} p_i \frac{1}{Tp_i} \sum_{t=Tq_{i-1}+1}^{Tq_i-h} (Y_t^\delta)^{(i)} (Y_{t+h}^\delta)^{(i)}$$

$$+ \sum_{i=1}^{r} p_i (E[(Y_t^\delta)^{(i)}])^2 - \sum_{i=1}^{r} p_i (E[(Y_t^\delta)^{(i)}])^2. \tag{8.20}$$

Thus

$$\frac{1}{T} \sum_{i=1}^{r} \sum_{t=Tq_{i-1}+1}^{Tq_i-h} (Y_t^\delta)^{(i)} (Y_{t+h}^\delta)^{(i)}$$

$$\rightarrow \sum_{i=1}^{r} p_i E[(Y_0^\delta)^{(i)} (Y_h^\delta)^{(i)}] - \sum_{i=1}^{r} p_i E[(Y_t^\delta)^{(i)}]^2 + \sum_{i=1}^{r} p_i (E[(Y_t^\delta)^{(i)}])^2$$

$$= \sum_{i=1}^{r} p_i (\gamma_{(Y^\delta)^{(i)}}(h) + (E[(Y_t^\delta)^{(i)}])^2). \qquad (8.21)$$

Thus

$$A \rightarrow \sum_{i=1}^{r} p_i \gamma_{(Y^\delta)^{(i)}}(h) + \sum_{i=1}^{r} p_i (E[(Y_t^\delta)^{(i)}])^2$$

in probability.

Now we compute B. Using the same remark as before, B can be simplified and we get

$$B = -\underline{Y}^{2\delta}{}_T + o(1). \qquad (8.22)$$

Now

$$-\underline{Y}^{2\delta}{}_T \rightarrow -(\sum_{i=1}^{r} p_i E[(Y^{\delta_t})^{(i)}])^2$$

$$= -\sum_{i=1}^{r} \sum_{j=1}^{r} p_i p_j E[(Y_t^\delta)^{(i)}] E[(Y_t^\delta)^{(j)}]$$

$$= -\sum_{i=1}^{r} (p_i E[(Y_t^\delta)^{(i)}])^2 - 2 \sum_{1 \leq i \leq j \leq r} p_i p_j E[(Y_t^\delta)^{(i)}] E[(Y_t^\delta)^{(j)}]. \quad (8.23)$$

Moreover $p_i = p_i^2 + p_i \sum_{j \neq i, j=1}^{r} p_j$. Thus

$$-(\underline{Y}^\delta)_T^2 \rightarrow -\sum_{i=1}^{r} p_i (E[(Y_t^\delta)^{(i)}])^2 + \sum_{1 \leq i \leq j \leq r} p_i p_j (E[(Y_t^\delta)^{(i)}] - E[(Y_t^\delta)^{(j)}])^2.$$

$$(8.24)$$

Then

$$B \rightarrow -\sum_{i=1}^{r} p_i (E[(Y_t^\delta)^{(i)}])^2 + \sum_{1 \leq i \leq j \leq r} p_i p_j (E[(Y_t^\delta)^{(i)}] - E[(Y_t^\delta)^{(j)}])^2. \quad (8.25)$$

Now using expressions found for A and B we get

$$(A+B) \to \sum_{i=1}^{r} p_i \gamma_{Y^{\delta(i)}}(h) + \sum_{i=1}^{r} p_i (E[(Y_t^\delta)^{(i)}])^2 - \sum_{i=1}^{r} p_i (E[(Y_t^\delta)^{(i)}])^2$$
$$+ \sum_{1 \le i \le j \le r} p_i p_j (E[(Y_t^\delta)^{(i)}] - E[(Y_t^\delta)^{(j)}])^2$$
$$= \sum_{i=1}^{r} p_i \gamma_{Y^{\delta(i)}}(h) + \sum_{1 \le i \le j \le r} p_i p_j (E[(Y_t^\delta)^{(i)}] - E[(Y_t^\delta)^{(j)}])^2. \quad (8.26)$$

Hence the proposition is proved.

Acknowledgments

The author wants to thank the participants to the different congresses or seminars where the ideas developed in this chapter have been presented, namely in the department of Mathematics in Singapore, in the University of Hong Kong, in the Business School of QUT, Brisbane, in the department of economics in Monash University, Melbourne and in the University of Adelaide, Australia, and also during the ISI Platinium Jubilee Conference ICSPRAR-08 in Kolkata. We particularly thank Professor L. Chen and Dr. Piotr Fryzlewicz for deep discussions, and Professor R. Wolff for his help in very carefully reading the manuscript.

References

1. Aas, K., Czado, C., Frigessi, A. and Bakken, H. (2009). Pair-copula constructions of multiple dependence. *Insurance: Mathematics and Economics* **44**, 182–198.
2. Andél, J. (1993). A Time Series Model with Suddenly Changing Parameters. *Journal of Time Series Analysis* **14**, 111–123.
3. Breidt, F. J. and Hsu, N. J. (2002). A class of nearly long-memory time series models. *International Journal of Forecasting* **18**, 265–281.
4. Charfeddine, L. and Guégan, D. (2008). Breaks or long memory behaviour: an empirical investigation. *Working paper WP-2008.22*, CES Paris 1 Pantheon-Sorbonne, France.
5. Caillault, C. and Guégan, D. (2005). Empirical Estimation of Tail Dependence Using Copulas: Application to Asian Markets. *Quantitative Finance* **5**, 489–501.
6. Cherubini, U., Luciano, E. and Vecchiato, W. (2004). *Copula methods in finance.* (Wiley Finance).

7. Dias, A. and Embrechts, P. (2004). Dynamic Copula Models for Multivariate High-Frequency Data in Finance. In *Risk Measures for the 21st Century* (G. Szegoe Ed.), Chapter 16, pp. 321-335, Wiley Finance Series.

8. Ding, Z. and Granger, C. W. J. (1996). Modelling volatility persistence of speculative returns: a new approach. *Journal of Econometrics* **73**, 185–215.

9. Engle, R. F. (1982). Autoregressive conditional heteroscedasticity with estimates of the variance of the United Kingdom. *Econometrica* **50**, 987–1007.

10. Fermanian, J. D. (2005). Goodness of fit tests for copulas. *Journal of multivariate analysis* **95**, 119–152.

11. Guégan, D. (2005). How can we define the concept of long memory? An econometric survey. *Econometric Reviews* **24**, 15–35.

12. Guégan, D. (2009). VaR computation in a Non-Stationary Setting. In *Model Risk Evaluation Handbook* (G. N. Gregoriou, C. Hopp and K. Wehn, Eds.) Chapter 15, McGraw Hill Cie.

13. Guégan, D. and Zhang, J. (2009). Change analysis of dynamic copula for measuring dependence in multivariate financial data. *Quantitative Finance* **8**, 1–20.

14. Guégan, D. and Maugis, P. A. (2008). Note on new prospects on vines. *Insurance Markets and Companies: Analyses and Actuarial Computations*, **1(1)**, 15–22.

15. Guégan, D. and Maugis, P. A. (2011). An econometric Study for Vine Copulas. *International Journal of Economics and finance* **2(1)**, 2–14.

16. Guégan, D. and Tarrant, W. (2010). On the necessity of five risk measures. Working paper of CES, Paris1 Pantheon-Sorbonne, France.

17. Hyung, N. and Franses, P. H. (2005). Forecasting time series with long memory and level shifts, *Journal of Forecasting* **24**, 1–16.

18. Joe, H. (1997). *Multivariate models and dependence concepts.* Chapman and Hall/CRC Press, London.

19. Ohanassian, A., Russell, J. R. and Tsay, R. S. (2008). True or Spurious Long Memory? A New Test?, *Journal of Business and Economic Statistics* **26**, 161–175.

20. Sibbertsen, P. and Kruse, R. (2009). Testing for a break in persistence under long-range dependencies, *Journal of Time Series Analysis* **30**, 263–285.

21. Sklar, A. (1959). Fonctions de répartition à *n* dimensions et leurs marges, *Publications de l'Institut de Statistique de l'Université de Paris* **8**, 229–231.

22. Starica, C. and Granger, C. W. G. (2005). Nonstationarities in stock returns, *The Review of Economics and Statistics* **87**, 3–18.

23. Tong, H. (1990). *Non-linear time series: a dynamical approach.* (Oxford Scientific Publications, Oxford).

Chapter 9

MDL Model Selection Criterion for Mixed Models with an Application to Spline Smoothing

Antti Liski[1] and Erkki P. Liski[2]

[1] *Tampere University of Technology, Finland*
[2] *University of Tampere, Finland*

For spline smoothing one can rewrite the smooth estimation as a linear mixed model (LMM) where the smoothing parameter appears as the ratio between the variance of the error terms and the variance of random effects. Smoothing methods that use basis functions with penalization can utilize maximum likelihood (ML) theory in the LMM framework. We introduce the minimum description length (MDL) model selection criterion for LMM's and propose an automatic data-based spline smoothing method based on the MDL criterion. Simulation study shows that the performance of MDL in spline smoothing is close to that of the Bayesian information criterion.

Keywords: Linear mixed models, Profile likelihood, Spline smoothing.

Contents

9.1. Introduction

This paper considers model selection for linear mixed models (LMM) using the MDL principle (Rissanen, 1996, 2000, 2007). Regression splines that use basis functions with penalization can be fit conveniently using the machinery of LMMs, and thereby borrow from a rich source of existing methodology (Brumback et al., 1999; Ruppert et al., 2003). The basis coefficients can be considered as random coefficients and the smoothing parameter as the ratio between variances of the error variables and random effects, respectively. In this article we present the MDL criterion under a LMM for choosing the number of knots, the amount of smoothing and the basis jointly. A simulation experiment was conducted to compare the performance of the MDL method with that of the corresponding techniques based on the Akaike information criterion (AIC), corrected AIC ($AICc$), and generalized crossvalidation (GCV).

The model known as the linear mixed model may be written as

$$y = \mathbf{X}\boldsymbol{\beta} + \mathbf{Z}\boldsymbol{b} + \boldsymbol{\varepsilon}, \quad \boldsymbol{b} \sim \mathrm{N}(\mathbf{0}, \phi^2 \mathbf{I}_m),$$
$$\boldsymbol{\varepsilon} \sim \mathrm{N}(\mathbf{0}, \sigma^2 \mathbf{I}_n), \quad \mathrm{Cov}(\boldsymbol{b}, \boldsymbol{\varepsilon}) = \mathbf{0}, \tag{9.1}$$

where $\mathbf{X}$ and $\mathbf{Z}$ are known $n \times p$ and $n \times m$ matrices, respectively, $\boldsymbol{b}$ is the $m \times 1$ vector of random effects that occur in the $n \times 1$ data vector $\boldsymbol{y}$ and $\boldsymbol{\beta}$ is the $p \times 1$ vector of unknown fixed effects parameters. Compared with the ordinary linear regression model, the difference is $\mathbf{Z}\boldsymbol{b}$, which may take various forms, thus creating a rich class of models. Then under these conditions we have

$$y \sim \mathrm{N}(\mathbf{X}\boldsymbol{\beta}, \sigma^2 \mathbf{V}) \tag{9.2}$$

and

$$y|b \sim \mathrm{N}(\mathbf{X}\boldsymbol{\beta} + \mathbf{Z}\boldsymbol{b}, \sigma^2 \mathbf{I}_n), \tag{9.3}$$

where $\mathbf{V} = \frac{1}{\alpha}\mathbf{Z}\mathbf{Z}' + \mathbf{I}_n$ for $\alpha = \sigma^2/\phi^2 > 0$. The parameter α is the ratio between the variance of the error variables ε_i, $1 \le i \le n$ and the variance of the random effects b_j, $1 \le j \le m$. The set of possible values for α is $[0, \infty]$. There are different types of LMMs, and various ways of classifying them. For these we refer to large literature on mixed models (see, e.g., Demidenko, 2004; Searle et al., 1992).

The paper is organized as follows. In Section 9.2, we consider likelihood estimation in LMMs. In Subsection 9.2.2 the estimates of the fixed effects and random effects parameters are presented as a function of the smoothing parameter. The MDL model selection criterion is introduced in Section 9.3, and it is applied to automatic scatterplot smoothing in Section 9.4. Section 9.5 presents simulation results.

9.2. Likelihood Estimation for Linear Mixed Models

In the LMM (9.1) the interest is either in the fixed effects parameter β, or also in the associated random effects b. If we focus only on the estimation of the vector of fixed effects β, then we have the linear model (9.2) and the vector of random effects b is a device for modelling the covariance structure for the response y. In many applications, the random effects themselves are of interest. In this case the choice of fixed versus random effects is a legitimate modelling choice.

Let $h(b; \sigma^2)$ denote the density function of the vector b of random effects, and $f(y|b; \beta, \sigma^2)$ the conditional density function of y given b. Then the joint density function of y and b is

$$
\begin{aligned}
&f(y|b; \beta, \sigma^2) h(b; \sigma^2) \\
&= \frac{1}{(2\pi\sigma^2)^{n/2}} \exp\left(-\frac{1}{2\sigma^2} \|y - X\beta - Zb\|^2\right) \\
&\quad \times \left(\frac{\alpha}{2\pi\sigma^2}\right)^{m/2} \exp\left(-\frac{\alpha}{2\sigma^2} \|b\|^2\right) \\
&= \frac{\alpha^{m/2}}{(2\pi\sigma^2)^{(n+m)/2}} \exp\left[-\frac{1}{2\sigma^2}(\|y - X\beta - Zb\|^2 + \alpha\|b\|^2)\right].
\end{aligned} \tag{9.4}
$$

The likelihood function for the model (9.1) is the density function (9.4) viewed as a function of the parameters β and σ^2 for fixed data y. Since the nonobservable vector b of random effects is part of the model, we integrate the joint density (9.4) with respect to b. The function

$$
L(\beta, \sigma^2; y) = \int f(y|b; \beta, \sigma^2) h(b; \sigma^2) \, db \tag{9.5}
$$

is the integrated likelihood function corresponding to the normal density $h(b; \sigma^2)$. The likelihood (9.5) takes the form (Pinheiro and Bates, 2000)

$$
L(\beta, \sigma^2; y) = \frac{1}{(2\pi\sigma^2)^{n/2}} \exp\left[-\frac{1}{2\sigma^2}(\|y - X\beta - Z\tilde{b}\|^2 + \alpha\|\tilde{b}\|^2)\right] |V|^{-1/2}, \tag{9.6}
$$

where $\tilde{b} = (Z'Z + \alpha I_m)^{-1} Z'(y - X\beta)$.

The vector denoted by $\tilde{b}$ in the function (9.6) can be thought of as a parameter vector just as β. The likelihood function (9.6) is used to determine the ML estimates of $\boldsymbol{\beta}$ and σ^2 as well as to estimate $\tilde{\boldsymbol{b}}$. Twice the logarithm of the likelihood function (9.6) is

$$2 \log[L(\boldsymbol{\beta}, \sigma^2)] = -n \log(\sigma^2) - \frac{1}{\sigma^2}\|\boldsymbol{y} - \mathbf{X}\boldsymbol{\beta} - \mathbf{Z}\tilde{\boldsymbol{b}}\|^2 - \alpha\|\tilde{\boldsymbol{b}}\|^2, \qquad (9.7)$$

where the unnecessary constants are omitted.

9.2.1. *Mixed model equations*

The function (9.7) can be considered as a penalized log-likelihood function. For a given α, the penalized maximum likelihood estimators for $\boldsymbol{\beta}$ and $\tilde{\boldsymbol{b}}$ from (9.7) are equivalent to the solution of the so-called mixed model equations (e.g., Searle et al., 1992)

$$\begin{pmatrix} \mathbf{X}'\mathbf{X} & \mathbf{X}'\mathbf{Z} \\ \mathbf{Z}'\mathbf{X} & \mathbf{Z}'\mathbf{Z} + \alpha\mathbf{I}_m \end{pmatrix} \begin{pmatrix} \hat{\boldsymbol{\beta}} \\ \hat{\boldsymbol{b}} \end{pmatrix} = \begin{pmatrix} \mathbf{X}'\boldsymbol{y} \\ \mathbf{Z}'\boldsymbol{y} \end{pmatrix}, \qquad (9.8)$$

which yield the estimates

$$\hat{\boldsymbol{\beta}} = (\mathbf{X}'\mathbf{V}^{-1}\mathbf{X})^{-1}\mathbf{X}'\mathbf{V}^{-1}\boldsymbol{y}, \qquad (9.9)$$

$$\hat{\boldsymbol{b}} = (\mathbf{Z}'\mathbf{Z} + \alpha\mathbf{I}_m)^{-1}\mathbf{Z}'(\boldsymbol{y} - \mathbf{X}\hat{\boldsymbol{\beta}}). \qquad (9.10)$$

The mixed model equations (9.8) refer to the LMM (9.1) which is an extension of the ordinary regression model. In Henderson (1963) it was shown that the derived estimates are indeed the best linear unbiased predictors (BLUP). In Robinson (1991), a wide ranging account of mixed model equations and BLUP are given with examples, applications and discussion.

Let $\boldsymbol{\delta} = (\boldsymbol{\beta}', \boldsymbol{b}')'$, $\mathbf{M} = (\mathbf{X}, \mathbf{Z})$ and $\mathbf{D} = \text{diag}(0, \ldots, 0, 1, \ldots, 1)$ a $(p + m) \times (p + m)$ diagonal matrix, whose first p diagonal elements are zero and the other m diagonal elements are 1. Then by (9.8)

$$\hat{\boldsymbol{\delta}} = \begin{pmatrix} \hat{\boldsymbol{\beta}} \\ \hat{\boldsymbol{b}} \end{pmatrix} = (\mathbf{M}'\mathbf{M} + \alpha\mathbf{D})^{-1}\mathbf{M}'\boldsymbol{y} \qquad (9.11)$$

and the ordinary least squares estimate

$$\tilde{\boldsymbol{\delta}} = (\mathbf{M}'\mathbf{M})^{-1}\mathbf{M}'\boldsymbol{y}$$

of $\boldsymbol{\delta}$ is obtained by putting $\alpha = 0$. Hence, for a given α, $\hat{\boldsymbol{\delta}}$ is the linear transformation

$$\hat{\boldsymbol{\delta}} = \mathbf{B}\tilde{\boldsymbol{\delta}} \qquad (9.12)$$

of $\tilde{\delta}$, where $\mathbf{B} = (\mathbf{M}'\mathbf{M} + \alpha\mathbf{D})^{-1}\mathbf{M}'\mathbf{M}$ is a shrinkage matrix whose eigenvalues lie in $[0, 1]$. Thus (9.12) is a ridge type estimator. Under the model (9.3)

$$\hat{\delta} \sim \mathrm{N}[\mathbf{B}\delta, \sigma^2 \mathbf{B}(\mathbf{M}'\mathbf{M})^{-1}\mathbf{B}]. \tag{9.13}$$

Maximizing the log-likelihood (9.7) with respect to σ^2 and inserting the estimators (9.9) and (9.10) for β and $\tilde{b}$ provide the estimate

$$\begin{aligned}
\hat{\sigma}^2 &= n^{-1}\|\boldsymbol{y} - \mathbf{X}\hat{\beta} - \mathbf{Z}\hat{\boldsymbol{b}}\|^2 = n^{-1}\|\boldsymbol{y} - \hat{\boldsymbol{y}}\|^2 \\
&= n^{-1}\boldsymbol{y}'(\mathbf{I} - \mathbf{H})^2\boldsymbol{y},
\end{aligned} \tag{9.14}$$

where the fitted values are

$$\hat{\boldsymbol{y}} = \mathbf{M}\hat{\delta} = \mathbf{H}\boldsymbol{y} \tag{9.15}$$

and the hat matrix $\mathbf{H}$ can be written as

$$\mathbf{H} = \mathbf{M}(\mathbf{M}'\mathbf{M} + \alpha\mathbf{D})^{-1}\mathbf{M}'. \tag{9.16}$$

Unlike for an ordinary linear regression model, $\mathbf{H}$ is not a projection matrix for $\alpha > 0$.

The conditional distribution of $\boldsymbol{y}|\boldsymbol{b}$ corresponding to (9.3) yields the normal density function

$$f(\boldsymbol{y}|\boldsymbol{b}; \beta, \sigma^2) = \frac{1}{(2\pi\sigma^2)^{n/2}} \exp(-\frac{1}{2\sigma^2}\|\boldsymbol{y} - \mathbf{M}\delta\|^2). \tag{9.17}$$

Here $\boldsymbol{b}$ is considered as a parameter vector just as β. The estimators for δ and σ^2 are given by (9.11) and (9.14), respectively.

9.2.2. *Profile likelihood estimates*

Note that $\hat{\beta}, \hat{\boldsymbol{b}}$ and $\hat{\sigma}^2$ are profile likelihood estimates depending on the value of α. The inverse of $\mathbf{V}$ can be written as follows

$$\mathbf{V}^{-1} = (\mathbf{I}_n + \frac{1}{\alpha}\mathbf{Z}\mathbf{Z}')^{-1} = \mathbf{I}_n - \mathbf{Z}(\alpha\mathbf{I}_m + \mathbf{Z}'\mathbf{Z})^{-1}\mathbf{Z}'. \tag{9.18}$$

If $\alpha \to 0$, then $\mathbf{V}^{-1} \to \mathbf{I}_n - \mathbf{Z}\mathbf{Z}^+$, where $\mathbf{Z}^+ = (\mathbf{Z}'\mathbf{Z})^{-1}\mathbf{Z}'$ is the Moore-Penrose inverse of $\mathbf{Z}$. Then using the above result, we conclude that $\hat{\beta}$ approaches to

$$\hat{\beta}_0 = [\mathbf{X}'(\mathbf{I}_n - \mathbf{Z}\mathbf{Z}^+)\mathbf{X}]^{-1}\mathbf{X}'(\mathbf{I}_n - \mathbf{Z}\mathbf{Z}^+)\boldsymbol{y}, \tag{9.19}$$

as $\alpha \to 0$. Similarly, it follows from (9.10) that

$$\hat{\boldsymbol{b}} \to \hat{\boldsymbol{b}}_0 = \mathbf{Z}^+(\boldsymbol{y} - \mathbf{X}\hat{\beta}_0) \quad \text{as} \quad \alpha \to 0. \tag{9.20}$$

Using (9.18), we have for $\mathbf{Z}'\mathbf{V}^{-1}$ the formula

$$\mathbf{Z}'(\mathbf{I}_n + \frac{1}{\alpha}\mathbf{Z}\mathbf{Z}')^{-1} = \mathbf{Z}' - \mathbf{Z}'\mathbf{Z}(\alpha\mathbf{I}_m + \mathbf{Z}'\mathbf{Z})^{-1}\mathbf{Z}'$$
$$= (\alpha\mathbf{I}_m + \mathbf{Z}'\mathbf{Z} - \mathbf{Z}'\mathbf{Z})(\alpha\mathbf{I}_m + \mathbf{Z}'\mathbf{Z})^{-1}\mathbf{Z}'$$
$$= \alpha(\alpha\mathbf{I}_m + \mathbf{Z}'\mathbf{Z})^{-1}\mathbf{Z}', \qquad (9.21)$$

which together with (9.10) implies

$$\hat{\boldsymbol{b}} = \frac{1}{\alpha}\mathbf{Z}'\mathbf{V}^{-1}(\boldsymbol{y} - \mathbf{X}\hat{\boldsymbol{\beta}}). \qquad (9.22)$$

Consequently, (9.22) is the conditional expectation $\mathrm{E}(\boldsymbol{b}|\boldsymbol{y})$ where $\boldsymbol{\beta}$ is replaced with $\hat{\boldsymbol{\beta}}$. Hence $\hat{\boldsymbol{b}} = \widehat{\mathrm{E}(\boldsymbol{b}|\boldsymbol{y})}$ is the ML estimate of the mean of $\boldsymbol{b}$ given a set of observations $\boldsymbol{y}$. If $\alpha \to \infty$, then clearly $\mathbf{V} \to \mathbf{I}_n$ and $\hat{\boldsymbol{b}} \to \mathbf{0}$. Thus clearly $\hat{\boldsymbol{\beta}}$ approaches to the ordinary least squares estimator

$$\hat{\boldsymbol{\beta}}_{OLS} = (\mathbf{X}'\mathbf{X})^{-1}\mathbf{X}'\boldsymbol{y} \quad \text{as} \quad \alpha \to \infty. \qquad (9.23)$$

9.3. Model Selection in Linear Mixed Models using MDL Criterion

9.3.1. *Model selection*

There is often uncertainty about which explanatory variables to use in $\mathbf{X}$, or how to select the matrix $\mathbf{Z}$. Typically we have a set of candidate models and the problem of model selection arises when one wants to decide which model to choose.

Let the variable η index the set of candidate models. We consider a set of conditional normal models corresponding to (9.3):

$$\boldsymbol{y}|\boldsymbol{b}_\eta \sim \mathrm{N}(\mathbf{X}_\eta\boldsymbol{\beta}_\eta + \mathbf{Z}_\eta\boldsymbol{b}_\eta, \sigma^2\mathbf{I}_n), \qquad (9.24)$$

where $\mathbf{X}_\eta$ and $\mathbf{Z}_\eta$ are $n \times p_\eta$ and $n \times m_\eta$ matrices, respectively, corresponding to the candidate model η. Here $\boldsymbol{\beta}_\eta$ and $\boldsymbol{b}_\eta$ are $n \times p_\eta$ and $n \times m_\eta$ parameter vectors for the model η. Note that the estimates $\hat{\boldsymbol{\beta}}_\eta$, $\hat{\boldsymbol{b}}_\eta$ and $\hat{\sigma}^2$ depend on the tuning parameter $\alpha \in [0, \infty]$. In this conditional framework we specify a model by giving the pair (η, α) and denote $\gamma = (\eta, \alpha)$.

9.3.2. *Normalized maximum likelihood*

Rissanen (1996) developed an MDL criterion based on the normalized maximum likelihood (NML) coding scheme (Barron et al., 1998). Assume that the response data are modelled with a set of density functions $f(\boldsymbol{y}; \gamma, \boldsymbol{\theta})$, where the parameter vector $\boldsymbol{\theta}$ varies within a specified parameter space. The NML function is defined by

$$\hat{f}(\boldsymbol{y};\gamma) = \frac{f(\boldsymbol{y};\gamma,\hat{\boldsymbol{\theta}})}{C(\gamma)}, \tag{9.25}$$

where $\hat{\boldsymbol{\theta}} = \hat{\boldsymbol{\theta}}(\boldsymbol{y})$ is the ML estimator of $\boldsymbol{\theta}$ and

$$C(\gamma) = \int f(\boldsymbol{x};\gamma,\hat{\boldsymbol{\theta}}(\boldsymbol{x}))\,\mathrm{d}\boldsymbol{x} \tag{9.26}$$

is the normalizing constant. The integral in (9.26) is taken over the sample space. Thus $\hat{f}(\boldsymbol{y};\gamma)$ defines a density function, provided that $C(\gamma)$ is bounded.

The expression

$$-\log \hat{f}(\boldsymbol{y};\gamma) = -\log f(\boldsymbol{y};\gamma,\hat{\boldsymbol{\theta}}) + \log C(\gamma) \tag{9.27}$$

is taken as the "shortest code length" for the data $\boldsymbol{y}$ that can be obtained with the model γ and it is called *the stochastic complexity* of $\boldsymbol{y}$, given γ (Rissanen, 1996). Here the estimate $\hat{\boldsymbol{\theta}} = (\hat{\boldsymbol{\beta}}, \hat{\boldsymbol{b}}, \hat{\sigma}^2)$ is given by (9.9), (9.10) and (9.14). The last term in the equation (9.27) is called *the parametric complexity*.

It is clear that the NML function (9.25) attains its maximum and the "code length" (9.27) its minimum for the same value of γ. According to the MDL principle we seek the value $\gamma = \hat{\gamma}$ that minimizes the stochastic complexity (9.27). In general, obtaining the $\hat{\gamma}$ may be computionally a very intensive task.

Here the NML density (9.25) is needed for the model (9.24). However, the normalizing constant (9.26) for the model (9.24) is not finite. Following Rissanen's renormalizing approach (Rissanen, 2000; Rissanen, 2007), data $\boldsymbol{y}$ is restricted to lie within a subset

$$\mathcal{Y}(s,R) = \{\boldsymbol{y} : \ \hat{\sigma}^2 \geq s, \ \hat{\boldsymbol{\delta}}'\mathbf{M}'\mathbf{M}\hat{\boldsymbol{\delta}} \leq nR\}, \tag{9.28}$$

where $s > 0$ and $R > 0$ are hyperparameters. Under the restriction (9.28) we have the NML density function

$$\hat{f}(\boldsymbol{y};\gamma,s,R) = f(\boldsymbol{y};\gamma,\hat{\boldsymbol{\theta}})/C(s,R), \tag{9.29}$$

where $\hat{\boldsymbol{\theta}} = (\hat{\boldsymbol{\delta}}, \hat{\sigma}^2)$. For the model (9.24) the numerator in (9.29) takes a simple form

$$f(\boldsymbol{y};\gamma,\hat{\boldsymbol{\theta}}) = (2\pi\hat{\sigma}^2 e)^{-\frac{n}{2}},$$

but the normalizing constant $C(s, R)$ will essentially depend on two hyper-parameters s and R.

The code length (9.27) corresponding to (9.29) is minimized by setting $s = \hat{s} = \hat{\sigma}^2$ and $R = \hat{R} = \hat{\delta}' \mathbf{M}' \mathbf{M} \hat{\delta}/n$, i.e. by maximizing the NML density (9.29) with respect to s and R under the restriction (9.28). The explicit formula of $C(s, R)$ is given in the Appendix (formula (9.39)). Since $\hat{f}(\boldsymbol{y}; \gamma, \hat{\sigma}^2(\boldsymbol{y}), \hat{R}(\boldsymbol{y}))$ of (9.29) is not a density function, we normalize it. To keep the normalizing constant finite, the sample space is restricted such that $\hat{\sigma}^2 \in [s_1, s_2]$ and $\hat{R} \in [R_1, R_2]$, where $0 < s_1 < s_2$ and $0 < R_1 < R_2$ are hyperparameters. The resulting NML function

$$\hat{f}(\boldsymbol{y}; \gamma) = \hat{f}(\boldsymbol{y}; \gamma, \hat{\sigma}^2(\boldsymbol{y}), \hat{R}(\boldsymbol{y}))/C(\gamma),$$

is a density function, where the normalizing constant $C(\gamma)$ depends on the hyperparameters. Athough the codelength will again depend on hyperparameters, they do not have essential effect on model selection. Derivation of the NML function for (9.24) under the LMM resembles that of the ordinary Gaussian linear regression (Rissanen, 2000; Rissanen, 2007).

9.3.3. *MDL criterion*

We are seeking models γ that minimize the "code length" $\log[1/\hat{f}(\boldsymbol{y}; \gamma)] = -\log \hat{f}(\boldsymbol{y}; \gamma)$. So, we define the selection criterion as $MDL(\gamma) = -2 \log \hat{f}(\boldsymbol{y}; \gamma)$, where the multiplier 2 is chosen just for convenience. For the model (9.24) under the LMM the MDL selection criterion takes the form

$$MDL(\gamma) = (n - d) \log \hat{\sigma}^2 + d \log \hat{R} - 2 \log \Gamma(\frac{n - d}{2}) - 2 \log \Gamma(\frac{d}{2}), \quad (9.30)$$

where $d = \operatorname{tr} \mathbf{H}$ defines the model's degrees of freedom and $\hat{R} = \|\hat{\boldsymbol{y}}\|^2/n$. Note that $p \leq d \leq p + m$, $d = p$, as $\alpha = 0$ and $d \to p + m$, as $\alpha \to \infty$.

If we apply Stirling's approximation

$$\Gamma(x + 1) \approx (2\pi)^{1/2}(x + 1)^{x+1/2} e^{-x-1}$$

to the Γ-functions in (9.30) and omit the unnecessary constants, the criterion (9.30) takes the form

$$MDL(\gamma) = (n - d) \log \frac{\hat{\sigma}^2}{n - d} + d \log \frac{\hat{R}}{d} + \log[d(n - d)].$$

The derivation of the criterion (9.30) is outlined in the Appendix. In the extreme cases $\alpha = 0$ and $\alpha \to \infty$, the criterion (9.30) reduces to the ordinary Gaussian regression with $p + m$ and p regressors, respectively.

9.4. Spline Smoothing using MDL Criterion

Suppose the smoothing model

$$y_i = r(x_i) + \sigma\varepsilon_i, \ i = 1, \ldots, n, \tag{9.31}$$

where y_i is the observation for the ith subject, x_i is a scalar covariate, $r(\cdot)$ is a smooth function giving the conditional mean of y_i given x_i and $\varepsilon_1, \ldots, \varepsilon_n$ are independent normally distributed error terms, i.e. $\varepsilon_i \sim N(0, 1)$. To pursue estimation, $r(\cdot)$ is replaced by a parametric regression spline model

$$r(x; \boldsymbol{\beta}, \boldsymbol{b}) = \beta_1 + \beta_2 x + \cdots + \beta_p x^{p-1} + \sum_{j=1}^{m} b_j z_j(x). \tag{9.32}$$

The first p terms are a $(p-1)$th order polynomial of x, covariates $z_1(x), \ldots, z_m(x)$ are elements of a smoothing basis, and $\boldsymbol{\beta} = (\beta_1, \ldots, \beta_p)'$ and $\boldsymbol{b} = (b_1, \ldots, b_m)'$ are unknown parameters. Then (9.32) can be written as

$$y_i = \boldsymbol{x}_i' \boldsymbol{\beta} + \boldsymbol{z}_i' \boldsymbol{b} + \sigma\varepsilon_i,$$

where $\boldsymbol{x}_i = (1, x_i, \ldots, x_i^{p-1})'$ and $\boldsymbol{z}_i = (z_1(x_i), \ldots, z_m(x_i))'$. Typically $\boldsymbol{x}_i$ is low-dimensional and $\boldsymbol{z}_i$ is high-dimensional basis linearly independent of $\boldsymbol{x}_i$. A convenient choice is to use the truncated power basis of degree $p-1$. Then the ith row of $\mathbf{Z}$ is $\boldsymbol{z}_i = ((x_i - \kappa_1)_+^{p-1}, \ldots, (x_i - \kappa_m)_+^{p-1})$ with x_+ as positive part, so that for any number x, x_+ is x if x is positive and is equal to 0 otherwise. The knots $\kappa_1, \ldots, \kappa_m$ are fixed values covering the range of $x_1, \ldots, x_n$.

The amount of smoothing is controlled by α, which is here referred to as a smoothing parameter. The fitted values for a spline regression are given by (9.15). In addition to the value of α, the degree of the regression spline and the number and location of knots must be specified. Here we adopt the procedure where the knots are located at "equally spaced" sample quantiles of $x_1, \ldots, x_n$. Thus the kth knot is the jth order statistic of $x_{(1)}, \ldots, x_{(n)}$ where j is $nk/(m+1)$ rounded to the nearest integer. As soon as the degree of the regression spline is specified, one has to fix the number of knots. It is often recommended to choose basis in a "generous" manner such that there are enough knots to fit features in the data (see, e.g., Ruppert, 2002). The relation between spline smoothing and mixed models in general has been discussed in Green and Silverman (1996), for example. Penalized spline estimation for smoothing was made popular in statistics by Eilers and Marx (1996).

In smoothing we control three modeling parameters: the degree of the regression spline $p-1$, the number of knots m and the smoothing parameter α. A model $\gamma = (p, m, \alpha)$ is specified by the triple where the values for the modeling parameters p, m and α should be determined in an optimal way. The choice of α has a profound influence on the fit. In fact, it was shown in Subsection 9.2.2 that α can be chosen to give any one of a spectrum of fits between the unconstrained regression spline fit and the least-squares polynomial fit. As $\alpha \to \infty$, the regression spline approach by (9.23) to a smooth polynomial. The case $\alpha = 0$ corresponds to the unconstrained case where the estimates of $\boldsymbol{\beta}$ and $\boldsymbol{b}$ are given by (9.19) and (9.20), repectively. A model estimator $\hat{\gamma}$ is obtained by minimizing the MDL selection criterion (9.30) with respect to model $\gamma = (p, m, \alpha)$, that is, with respect to parameters p, m and α, using numerical optimization routines.

9.5. Simulations

9.5.1. *Preamble*

In this section we give an outline of a simulation study which aims at the comparison of the performance of several model selection techniques in data based smoothing. Apart from smoothing, the number of knots is specified automatically. Along with the MDL criterion, we briefly review the performance of the model selection criteria $AICc$ (corrected AIC), BIC (Bayesian information criterion) and GCV (generalized cross-validation, see Ruppert et al., 2003).

In all investigated scenarios, we considered the model given in (9.32), where the x_i, $i = 1, \ldots, n$, were equally spaced on $[0, 1]$. Four diffrent regression functions $r(\cdot)$ were studied: the first, called "Logit", uses a logistic function

$$r(x) = 1/\{1 + \exp[-20(x - 0.5)]\}$$

and the second function "Bump" was

$$r(x) = x + 2\exp\{-16(x - 0.5)^2\}.$$

The third function "SpaHetj" is

$$r(x) = \sqrt{x(1 - x)} \sin\left(\frac{2\pi(1 + 2^{(9-4j)/5})}{x + 2^{(9-4j)/5}}\right),$$

where the parameter $j = 1, 2, \ldots$ controls spatial variation. The value $j = 1$ (SpaHet1) yields low spatial variation and larger values of j (eg. SpaHet3) imply greater spatial heterogeneity. The fourth function "Sinj"

$$r(x) = \sin(2\pi\theta), \ \theta = j$$

is a cyclic function, where the parameter θ controls the number of cycles. Ruppert (2002) used the above mentioned functions, among all, in his simulation study.

The knots were located at equally spaced sample quantiles, so that the number of knots determines the knot sequence. In this study, only the first degree $(p - 1 = 1)$, quadratic $(p - 1 = 2)$ and cubic splines $(p - 1 = 3)$ were considered. A model is specified by the triple $\gamma = (p, m, \alpha)$. For each combination of p and m the selection criterion was minimized with respect to α to determine the optimal model $\hat{\gamma}$. For each setting 500 datasets were simulated. The performance of the criteria were assessed by using the function $MSE(x)$ defined as the mean over the generated datasets of the squared error

$$SE(x; \hat{\gamma}) = [r(x; \hat{\gamma}) - r(x)]^2 \tag{9.33}$$

at the point x, $MASE$ defined as

$$MASE = \sum_{i=1}^{n} MSE(x_i)/n, \tag{9.34}$$

and the average squared error

$$ASE(\hat{\gamma}) = \sum_{i=1}^{n} SE(x_i; \hat{\gamma}) \tag{9.35}$$

of a model $\hat{\gamma}$ for a given dataset.

Along with the MDL criterion, also the criteria $AICc$, BIC and GCV were used to choose an appropriate spline smoothing model (see Ruppert et al., 2003) and the performance of these four criteria were compared. The corrected AIC criterion proposed in Hurvich et al. (1998) is given by

$$AICc(\gamma) = \log RSS(\gamma) + \frac{2[d(\gamma) + 1]}{n - d(\gamma) - 2},$$

where $d(\gamma) = tr\mathbf{H}(\gamma)$ and $RSS(\gamma)$ is the residual sum of squares. The criterion known as generalized cross-validation (GCV) is

$$GCV(\gamma) = \log RSS(\gamma)/[1 - d(\gamma)/n]^2.$$

The Bayesian information criterion (BIC) is given by

$$BIC(\gamma) = \log RSS(\gamma) + \frac{d(\gamma) \log n}{n}.$$

The model selection criteria are minimized numerically and the model $\hat{\gamma}$ that minimizes the criterion is selected. Bump, Logit, Sin3 and SpaHet3 functions were estimated using the spline model (9.32). An appropriate smoothing model was selected by using $AICc$, GCV, BIC and MDL criteria respectively, and the models were fitted to all simulated datasets. The performance of GCV is very close to that of $AICc$, but $AICc$ was uniformly better than GCV with respect to the $MASE$ criterion (9.34). Therefore the results for GCV are not reported in this paper.

9.5.2. *Results*

Inspection of Figure 9.1 shows that the average curves follow the true curves quite closely. However, in panel 2 each average curve tends to "straighten" the Logit function. It is obvious that the selected number of knots is not enough for the S-part of Logit function. Panel 1 shows that all criteria underestimate the bump part of the function, but underestimation is clearly greater when using MDL and BIC. These two criteria also react slower

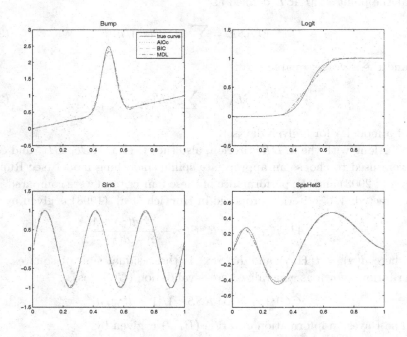

Fig. 9.1. The average fitted curves over 500 simulated data sets for the selection criteria $AICc, BIC$ and MDL when the sample size $n = 200$.

Table 9.1.   MASE and relative MASE for the criteria $AICc$, BIC and MDL.

function	n	degree	AICc	BIC	MDL
Bump	50	1	0.0284(1.33)	0.0303(1.39)	0.0319(1.38)
	100	1	0.0133(1.39)	0.0135(1.37)	0.0141(1.41)
	200	2	0.0078(1.28)	0.0082(1.28)	0.0083(1.29)
Logit	50	1	0.0116(1.65)	0.0115(1.62)	0.0119(1.65)
	100	1	0.0066(1.73)	0.0057(1.41)	0.0057(1.41)
	200	1	0.0036(1.67)	0.0033(1.40)	0.0033(1.41)
Sin3	50	2	0.0224(1.33)	0.0211(1.24)	0.0224(1.32)
	100	2	0.0116(1.32)	0.0107(1.19)	0.0109(1.21)
	200	3	0.0064(1.29)	0.0073(1.49)	0.0072(1.47)
Spahet3	50	3	0.0154(1.40)	0.0170(1.55)	0.0153(1.39)
	100	3	0.0079(1.38)	0.0083(1.43)	0.0078(1.34)
	200	3	0.0040(1.41)	0.0038(1.32)	0.0038(1.32)

when recovering from the bump at 0.6 in Panel 1. In panel 4 all criteria slightly underestimate the changes in SpaHet3 function.

In Table 9.1 the $MASE$ values are reported for $AICc$, BIC and MDL under various settings. The degree of the fitted spline model chosen by the four criteria varies from one to three. The degree reported in Table 9.1 is the most frequently selected one under a given setting. When computing the value of $MASE$ for a given criterion, say MDL, the model $\hat{\gamma} = (\hat{p}, \hat{m}, \hat{\alpha})$ for each data set is determined by minimizing the MDL criterion. Then $MASE$ is obtained as the average over the $ASE(\hat{\gamma})$ values. Besides $MASE$, also relative $MASE$ is reported. For computing the relative $MASE$, the minimum of the function $ASE(m) = ASE(\hat{p}, m, \hat{\alpha})$, say ASE^*, is determined with respect to the number of knots m for each data set. $MASE^*$ denotes the average of the ASE^* values over the generated datasets. Relative $MASE$ is defined as the ratio $MASE/MASE^*$. Clearly $ASE^* \leq ASE(\hat{\gamma})$ for each $\hat{\gamma}$, and consequently relative $MASE$ is not less than 1.

Inspection of the results in Table 9.1 shows that on the average the performance of MDL and BIC gets closer to each other as the sample size grows. This trend continues if we keep increasing the sample size n over 200. BIC and MDL do better than $AICc$ for Logit and Spahet3 with $n = 200$. A large value of relative $MASE$ indicates that the value of $MASE$ can be considerably decreased by choosing the number of knots optimally. In view of the relative $MASE$, BIC and MDL are closer to optimal knot selection than $AICc$, except in case of Bump and Sin3 (when $n = 200$). Most of the time BIC and MDL yield relative $MASE$'s very close to each other.

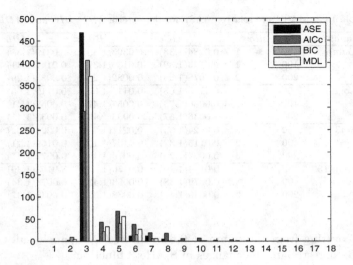

Fig. 9.2. Histograms of m as chosen by ASE, $AICc$, BIC and MDL criteria - *Spahet3*, $n = 200$.

Figure 9.2 displays histograms of the values of m chosen by the criteria $AICc$, MDL, BIC and ASE for SpaHet3 (500 datasets are generated). The ASE criterion uses the "oracle estimator" $\tilde{\gamma} = (\tilde{p}, \tilde{m}, \tilde{\alpha})$ that minimizes $ASE(\gamma)$. ASE chooses $m = 3$ in the vast majority of datasets. The behavior of BIC and MDL is closest to that of ASE. $AICc$ tends to choose larger values of m than BIC and MDL. The corresponding behavior remains also when data are generated from Logit and Bump (not reported here). All model selection criteria tend to choose less knots than ASE when data are generated from Sin3 function (Figure 9.3). BIC and MDL tend to choose even less knots than $AICc$. One can see that the knot selection behavior of BIC is close to that of MDL.

In Figure 9.4 the graphs of the function $MSE(x)$ are displayed for all criteria and functions under consideration. Again we observe that MDL and BIC are very close to each other. It is also evident that $AICc$ tends to react to function fluctuations more aggressively than BIC and MDL. MDL and BIC seem to need more observations than $AICc$ to detect sudden changes in a function. We may note that the absence of outliers seems to give some advantage to $AICc$ and GCV over BIC and MDL.

In Figure 9.5 the ASE values (9.35) of MDL are plotted against the ASE values of $AICc$ and BIC, respectively, when datasets are generated

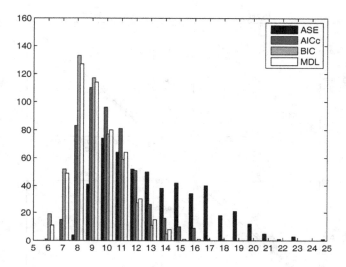

Fig. 9.3. Histograms of m as chosen by ASE, $AICc$, BIC and MDL criteria - *Sin3*, $n = 200$.

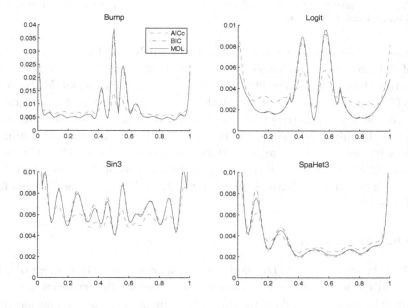

Fig. 9.4. $MSE(x)$ for each criterion as $n = 200$ (Degrees as in Table 9.1).

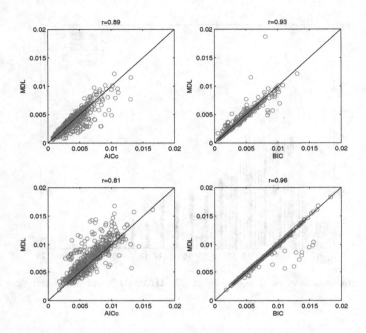

Fig. 9.5.   Scatterplots of the *ASE* values (9.35) for SpaHet3 (Panels 1 and 2) and Sin3 (Panels 3 and 4), $n = 200$.

from SpaHet3 (panels 1 and 2) and Sin3 (panels 3 and 4). In Panel 1 most of the *ASE* values are concentrated near the 45° line but *MDL* did clearly better than *AICc* more often ($r = 0.89$). In the scatterplot *MDL* versus *BIC* (panel 2) the values are nicely concentrated on the 45° line, except a couple of outliers ($r = 0.93$). In Panel 3 the majority of the ASE values lie on the upper side of the 45° line ($r = 0.81$) and the scatterplot in panel 4 again refers to the similar performance of *MDL* and *BIC* ($r = 0.96$). The *ASE* values lie very close to the 45° line except 11 cases where *BIC* fails. These outliers have an effect on the *MASE* value as can be seen from Table 9.1.

9.6. Conclusions

We have derived the *MDL* model selection criterion in the context of linear mixed models. It is an extension of the corresponding criterion known in linear regression. Spline smoothing is formulated as an estimation problem within the context of linear mixed models. Then an automatic *MDL* procedure for choosing the smoothing parameter, the number of knots and the smoothing basis is presented as a model selection problem. The performance of MDL is compared with the $AICc, BIC$ and GCV criteria. The simulation studies show that the results between the MDL approach and other methods are comparable in all cases. Furthermore, the performance of MDL is very close to that of BIC. No criterion dominates the other criteria uniformly. The MDL procedure outperforms the other methods in the case of SpaHet3 function.

9.7. Appendix

The estimator $\hat{\boldsymbol{\theta}} = (\hat{\boldsymbol{\delta}}, \hat{\sigma}^2)$ is a sufficient statistic for $\boldsymbol{\theta} = (\boldsymbol{\delta}, \sigma^2)$ under the model (9.17). By sufficiency the density (9.17) can be written as

$$f(\boldsymbol{y}; \boldsymbol{\theta}) = f(\boldsymbol{y}|\hat{\boldsymbol{\theta}})g(\hat{\boldsymbol{\theta}}; \boldsymbol{\theta}), \qquad (9.36)$$

where the conditional density $f(\boldsymbol{y}|\hat{\boldsymbol{\theta}})$ does not depend on the unknown parameter vector $\boldsymbol{\theta}$. The estimators $\hat{\boldsymbol{\delta}}$ and $\hat{\sigma}^2$ are not independent like in the ordinary linear regression, but we use the approximation

$$g(\hat{\boldsymbol{\delta}}, \hat{\sigma}^2; \boldsymbol{\delta}, \sigma^2) \approx g_1(\hat{\boldsymbol{\delta}}; \boldsymbol{\delta}, \sigma^2)g_2(\hat{\sigma}^2; \sigma^2), \qquad (9.37)$$

where $g_1(\hat{\boldsymbol{\delta}}; \boldsymbol{\delta}, \sigma^2)$ is the density function for the normal distribution (9.13).

The quadratic form $n\hat{\sigma}^2/\sigma^2 = \boldsymbol{y}'(\mathbf{I} - \mathbf{H})^2\boldsymbol{y}/\sigma^2$ does not follow a χ^2-distribution, since the matrix $\mathbf{H}$ in (9.16) is not idempotent (see, e.g., Searle, 1971). Let χ^2_ν denotes a gamma variable with parameters $\nu/2$ and 2. The simple Patnaik's two-moment approximation (Patnaik, 1949) consists of replacing the distribution of $Q = n\hat{\sigma}^2/\sigma^2$ by that of $c\chi^2_\nu$, where c and ν are chosen so that the first two moments of Q and $c\chi^2_\nu$ are the same, that is,

$$\mathrm{E}(Q) = \mathrm{E}(c\chi^2_\nu) \text{ and } \mathrm{Var}(Q) = \mathrm{Var}(c\chi^2_\nu).$$

Here ν can be fractional, and consequently χ^2_ν is not a proper χ^2-distribution.

However, instead of Patnaik's approximation we replace the distribution of Q by that of χ^2_d, which has a gamma distribution with parameters $d/2$ and 2 with $d = \mathrm{tr}\,\mathbf{H}$. This approximation gives results similar to Patnaik's approximation, but the derivation of the MDL criterion can be simplified. Now the approximate density g_2 of $\hat{\sigma}^2$ can be written as

$$g_2(\hat{\sigma}^2; \sigma^2) = \frac{n^{\frac{n-d}{2}}}{\Gamma(\frac{n-d}{2})2^{\frac{n-d}{2}}}(\hat{\sigma}^2/\sigma^2)^{\frac{n-d}{2}}(\hat{\sigma}^2)^{-1}e^{-\frac{n\hat{\sigma}^2}{2\sigma^2}}. \qquad (9.38)$$

Note that the function $\max_{\boldsymbol{\delta}} g_1(\hat{\boldsymbol{\delta}}; \boldsymbol{\delta}, \sigma^2) \equiv \tilde{g}_1(\sigma^2)$ depends on the parameter σ^2 only. We use the approximation $\tilde{g}_1(\sigma^2)g_2(\hat{\sigma}^2; \hat{\sigma}^2) \equiv \tilde{g}(\hat{\sigma}^2)$ to the function $g(\hat{\boldsymbol{\theta}}; \hat{\boldsymbol{\theta}})$. The function $\tilde{g}(\hat{\sigma}^2)$ can be written as

$$\tilde{g}(\hat{\sigma}^2) = A_{d,k}(\hat{\sigma}^2)^{-\frac{k}{2}-1}$$

where

$$A_{d,k} = \frac{|\mathbf{M}'\mathbf{M}|^{1/2}}{(\pi n)^{k/2}|\mathbf{B}|} \frac{(\frac{n}{2})^{\frac{k-d}{2}}}{\Gamma(\frac{n-d}{2})} (\frac{n}{2e})^{\frac{n}{2}}$$

and $k = p + m$.

Utilizing the factorization (9.36) and the above approximations we get the normalizing constant in (9.29) as follows

$$C(s, R) = \int_{\mathcal{T}(s,R)} \Big[\int_{\mathcal{Y}(\hat{\theta})} f(\boldsymbol{y}|\hat{\boldsymbol{\theta}}) \, \mathrm{d}\boldsymbol{y} \Big] \tilde{g}(\hat{\sigma}^2) \, \mathrm{d}\hat{\boldsymbol{\theta}}$$

$$= A_{d,k} \int_{s}^{\infty} (\hat{\sigma}^2)^{-\frac{k}{2}-1} \, \mathrm{d}\hat{\sigma}^2 \int_{\mathcal{D}(R)} \mathrm{d}\hat{\boldsymbol{\delta}}$$

$$= A_{d,k} V_k \frac{2}{k} \Big(\frac{R}{s} \Big)^{k/2}, \qquad (9.39)$$

where $\mathcal{T}(s, R) = \{\hat{\boldsymbol{\theta}} : \hat{\sigma}^2 \geq s, \hat{\boldsymbol{\delta}}'(\mathbf{B}')^{-1}\mathbf{M}'\mathbf{M}(\mathbf{B})^{-1}\hat{\boldsymbol{\delta}} \leq a_{n,d}R\}$ is the constrained estimation space. Integrating the inner integral in the first line of (9.39) over $\mathcal{Y}(\hat{\boldsymbol{\theta}}) = \{\boldsymbol{y} : \hat{\boldsymbol{\theta}} = \hat{\boldsymbol{\theta}}(\boldsymbol{y})\}$ for a fixed value of $\hat{\boldsymbol{\theta}}$ gives unity. In the last line of (9.39)

$$V_k R^{k/2} = \frac{(\pi n)^{k/2} R^{k/2} |\mathbf{B}|}{\frac{k}{2}\Gamma(\frac{d}{2})|\mathbf{M}'\mathbf{M}|^{1/2}(\frac{n}{2})^{\frac{k-d}{2}}}$$

is the volume of an ellipsoid $\mathcal{D}(R) = \{\hat{\boldsymbol{\delta}} : \hat{\boldsymbol{\delta}}'(\mathbf{B}')^{-1}\mathbf{M}'\mathbf{M}(\mathbf{B})^{-1}\hat{\boldsymbol{\delta}} \leq a_{n,d}R\}$ (Cramer, 1946), where $a_{n,d} = n^{k/2}\Gamma(k/2)/[(\frac{n}{2})^{\frac{k-d}{k}}\Gamma(d/2)]$. Note that $\mathcal{D}(R) = \{\hat{\boldsymbol{\delta}} : \tilde{\boldsymbol{\delta}}'\mathbf{M}'\mathbf{M}\tilde{\boldsymbol{\delta}} \leq a_{n,d}R\}$, since $\hat{\boldsymbol{\delta}} = \tilde{\boldsymbol{\delta}}\mathbf{B}$ by (9.12). By using Rissanen's renormalization technique (see Rissanen, 2007, p.115) we get rid of the two parameters R and s and obtain the MDL criterion (9.30).

References

1. Barron, A. R., Rissanen, J. and Yu, B. (1998). The MDL principle in modeling and coding. *Special Issue of Information Theory to Commemorate 50 Years of Information Theory* **44**, 2743–2760.

2. Brumback, B. A., Ruppert, D. and Wand, M. B. (1999). Comment on Shively, Kohn and Wood. *J. Amer. Statist. Assoc.* **94**, 794–797.

3. Cramér, H. (1946). *Mathematical Methods of Statistics*. Princeton, Princeton University Press.

4. Demidenko, E. (2004). *Mixed models*. Wiley.

5. Eilers, P. H. C. and Marx, B. D. (1996). Flexible smoothing with B-splines and penalties. *Statist. Sci.* **11**, 89–121.

6. Green, D. J. and Silverman B. W. (1996). *Nonparametric regression and generalized linear models*. Chapman & Hall, London.

7. Henderson, C. R. (1963). Genetic index and expected genetic advance. In *Statistical genetics and plant breeding* (W. D. Hanson and H. F. Robinson, Eds.), 141–163. National Academy of Research Counsil Publication No. 982, Washington, D. C.

8. Hurvich, C. M., Simonoff, J. S. and Tsai, C. (1998). Smoothing parameter selection in nonparametric regression using an improved Akaike information criterion. *J. Roy. Statist. Soc.* **B 60**, 271–293.

9. Patnaik, P. B. (1949). The non-central χ^2 and F-distributions and their applications. *Biometrika* **36**, 202–232.

10. Pinheiro, J. C. and Bates, B. M. (2000). *Mixed-Effects Models in S and S-PLUS*. New York, Springer.

11. Rissanen, J. (1996). Fisher Information and Stochastic Complexity. *IEEE Transactions on Information Theory* **42**, 40–47.

12. Rissanen, J. (2000). MDL Denoising. *IEEE Trans. on Information Theory* **46**, 2537–2543.

13. Rissanen, J. (2007). *Information and Complexity and in Statistical Modeling*. New York, Springer.

14. Robinson, G. K. (1991). That BLUP is a good thing: The estimation of random effects. *Statist. Sci.* **6**, 15–51.

15. Ruppert, D. (2002). Selecting the number of knots for penalized splines. *J. Comput. Graph. Statist.* **11**, 735–754.

16. Ruppert, D., Wand, M. P. and Carroll, R. J. (2003). *Semiparametric regression*. Wiley.

17. Searle, S. R. (1971). *Linear Models*. New York, Wiley.

18. Searle, S. R., Casella, G. and McCulloch, C. E. (1992). *Variance Components*. New York, Wiley.

Chapter 10

Digital Governance and Hotspot Geoinformatics with Continuous Fractional Response

G. P. Patil[1], S. W. Joshi[2] and R. E. Koli[3]

[1] *The Pennsylvania State University, USA*
[2] *Slippery Rock University of Pennsylvania, USA*
[3] *M. J. College, India*

The various impacts of modern communication and information technologies on society cannot be overstated and are reflected in worldwide emergence of digital governance. The purpose of digital governance is to empower the public through information access and analysis by enabling transparency, accuracy, and efficiency for societal good at large. In this context, development and applications of methodologies for geoinformatic hotspot analysis of spatial and temporal data are of utmost importance. The widely used SaTScan^{TM} software package, is applicable for Bernoulli, Poisson, ordinal, exponential, and normal response models. In this paper, we present a model that is suitable to analyze continuous fractional response data using the upper level set (ULS) scan statistic. The ULS scan statistic, an alternative to the circular scan statistic, uses arbitrarily shaped scanning windows. The circular/elliptic scan statistic uses the Euclidean distance as a sole basis to coalesce cells to form clusters, whereas the ULS scan statistic uses adjacency and the response ratio to combine cells to form candidate zones. The continuous fractional response model is presented after a brief review of the ULS scan statistic and its software implementation. The model is used for hotspot detection with forest cover data for Jalgaon district, Maharashtra State, India, as an illustrative example.

Keywords: Circular scan statistic, Hotspot detection, ULS tree, Upper level set scan statistic.

Contents

10.1. Introduction

The one dimensional version of the scan statistic has been thoroughly covered in Glaz and Balakrishnan (1999) and Glaz et al. (2001). A wide variety of methods have been proposed for modeling and analyzing geopsatial data (Cressie, 1991). More recently, the spatial scan statistic proposed by Kulldorff and Nagarwalla (1995) and Kulldorff (1997) provides a popular tool in the form of SaTScan^{TM} software system (Kulldorff et al., 1998) for detection and evaluation of disease clusters for discrete response data. A commercial system (Biomedware, 2001) is also available. Basic components of the scan statistic methodology are the topological structure under investigation, the probability distribution for modeling responses, and the scanning window. Suitable variations allow the scan statistic approach to be used for critical area analysis in various fields and for different types of response data. Patil and Taillie (2004) proposed the upper level set (ULS) scan statistic. Tango and Takahashi proposed methods and software for evaluating clusters of flexible shape using a scan-type statistic (Takahashi et al., 2004; Tango and Takahashi, 2005). Like SaTScan^{TM} their statistic also requires knowledge of location of centroids of cells. Patil et al. (2008a) report software implementation of the ULS scan statistic. The ULS scan statistic and its software implementation differ from the widely used SaTScan^{TM} system in three main respects:

(1) The ULS scan statistic uses an irregularly shaped scanning window unlike the circular/elliptic shaped window used by SaTScan^{TM}.
(2) The ULS scan statistic can be used to detect hotspots in any connected structure with the network topology whereas SaTScan^{TM} is applicable to geospatial regions only.
(3) The software provides an option of the use of the gamma distribution to model response data that are of continuous nature in addition to the binomial and Poisson models.

The second item in the above list seems to be quite significant in view of wide interest in hotspots in a network setting such as sensor networks

(Patil et al., 2008b). In addition to the responses that can be modeled using binomial, Poisson, and gamma distributions, there is a need for a model that can handle continuous fractional responses. In this paper, we present a model that can be used to analyze this kind of data for hotspot detection. The beta distribution is a natural choice for modeling continuous fractional data. But because of its lack of the additive property, it is not suitable for generating simulated replications of data which are essential for computing p-values. We propose a suitable transformation of the data so that the gamma distribution serves as a reasonably good approximate model. Software reported in Patil et al. (2008a) now has the capability to process continuous fractional data. We illustrate use of the software and viability of the proposed model to detect hotspots with forest cover data.

Section 10.2 introduces basic ideas underlying the scan statistic approach for hotspot detection. Sections 10.3 and 10.4 introduces the ULS scan statistic. The gamma model that forms the basis of the main topic of this paper is reviewed in Section 10.5. The continuous fractional model is discussed in Section 10.6. In Section 10.7 we illustrate the use of the model to detect forest cover hotspots in Jalgaon district, Maharashtra State, India.

10.2. Scan Statistics for Geospatial Hotspot Detection

We have the following:

R : A geographical connected region,

T : A set of 'cells' forming a partition or tessellation of R,

N : cardinality of T,

$n_1, n_2, \ldots, n_N$: 'sizes' of the N cells, and

$y_1, y_2, \ldots, y_N$: responses of interest for the N cells.

Here $y_1, y_2, \ldots, y_N$ are assumed to be a particular realization of independently distributed random variables $Y_1, Y_2, \ldots, Y_N$ that have distributions with a common form but with different parameter values that account for cell-to-cell response variation.

Interpretation of the size of a cell depends on the context. For example, if $Y_1, Y_2, \ldots, Y_N$ are binomial random variables, then $n_1, n_2, \ldots, n_N$ are respective numbers of trials. If $Y_1, Y_2, \ldots, Y_N$ have Poisson distribution, where, for $a = 1, 2, \ldots, N$, Y_a represents the number of events of a given type that occur at random in cell a with intensity λ_a, then $n_1, n_2, \ldots, n_N$ are areas of the N cells and $E[Y_a] = n_a \lambda_a$. In general, for $a = 1, 2, \ldots, N$,

y_a/n_a is the response rate or the intensity of the response for cell a. The spatial scan statistic seeks to identify "hotspots" or clusters of cells that have elevated response or, more precisely, elevated response rates compared with the rest of the region. Clearly, we are interested in the responses adjusted for cell sizes rather than in raw responses. It is possible that adjustment for some other characteristic such as gender or age is meaningful in some studies. Given a cluster of cells, C, the response rate for the cluster is the ratio:

$$\sum_C y_a / \sum_C n_a$$

where $\sum_C$ indicates summation over cells a belonging to the cluster C. This suggests that we assume that the parametrized family of distributions of $Y_1, Y_2, \ldots, Y_N$ is additive. In addition, a cluster of cells to be considered as a potential hotspot or a candidate hotspot needs to satisfy two geometrical properties:

(1) Cells within the cluster should be connected, that is, any two cells a_1, a_2 in the cluster should be adjacent to each other or there should be a sequence of cells $b_1, b_2, \ldots, b_k$, $k \geq 1$, all inside the cluster, such that a_1 is adjacent to b_1, a_2 is adjacent to b_k, and any two successive cells in the sequence are adjacent to each other. Such a cluster of connected cells will be called a zone. The set of all zones in R will be denoted by Ω. This requirement says that clusters without a common boundary with significantly elevated responses constitute distinct hotspots.

(2) The zone should not be excessively large so that the complement of the zone rather than the zone itself would constitute the background. This is achieved by limiting the search for hotspots to zones that do not comprise more than a certain threshold percentage, say, fifty percent of the entire region in size.

The process of hotspot detection then involves testing for each eligible zone in Ω the null hypothesis that its response rate is the same as that of the rest of the region, that is, the zone is not a hotspot, against the alternative hypothesis that its response rate is higher in comparison with that of the rest of the region. We conclude that there is no hotspot if the null hypothesis is not rejected for each eligible zone. This hypothesis testing model is formulated precisely as described below with the binomial response model used for illustration. Under the binomial response model, each Y_a is $Binomial(n_a, p_a)$, $1 \leq a \leq N$ and $Y_1, Y_2, \ldots, Y_N$ are independently distributed. Then the null hypothesis that there is no hotspot, that is, response rates for all cells are equal is stated as:

$$H_0 : p_1 = p_2 = \cdots = p_N = p_0, \text{ say,}$$

against the alternative hypothesis that there exists a non-empty zone $Z \in \Omega$ for which the response rate is higher than that for the rest of the region. Formally, the alternative hypothesis is:

H_1 : There is a non-empty zone $Z \in \Omega$ and values $0 \leq p_{nz}, p_z \leq 1$ such that

$$P_a = \begin{cases} p_z & \text{for all cells } a \text{ in } Z \\ p_{nz} & \text{for all cells } a \text{ in } R - Z \end{cases}$$

and $p_z > p_{nz}$.

The zone Z specified in the alternative hypothesis is an unknown parameter along with p_z and p_{nz}. Thus the full model involves three parameters: Z, p_z, and p_{nz} with $Z \in \Omega$ and H_0 implying $Z = \Phi$. For testing the null hypothesis using the likelihood ratio,

$$\lambda = L_{1,max}(Z, p_z, p_{nz})/L_{0,max}(p_0)$$

where

$L_{0,max}(p_0) = L_0(p_0)$ maximized over $0 \leq p_0 \leq 1$

$L_{1,max}(Z, p_z, p_{nz}) = L_1(Z, p_z, p_{nz})$ maximized over $0 \leq p_z, p_{nz} \leq 1$

$L_0(p_0) = \prod_{a \in T} b(y_a, n_a, p_0)$

$L_1(Z, p_z, p_{nz}) = \prod_{a \in Z} b(y_a; n_a, p_z) \prod_{a \in R-Z} b(y_a; n_a, p_{nz})$

$b(x; n, p) = n! p^x (1-p)^{n-x}/(x!(n-x)!)$

under the assumption of the binomial response model, we have $L_{0,max}(p_0) = L_0(\hat{p}_0)$ and $L_{1,max}(Z, p_z, p_{nz}) = L_1(Z, \hat{p}_z, \hat{p}_{nz})$ where $\hat{p}_0$, $\hat{p}_z$, and $\hat{p}_{nz}$ are the maximum likelihood estimates (MLEs) of p_0 under H_0 and of p_z and p_{nz} for a given zone Z under H_1, readily obtained as respective response ratios. Our objective is to maximize $L_1(Z, \hat{p}_z, \hat{p}_{nz})$ as Z varies over Ω, that is to compute the MLE of Z. If the ratio of the maximized $L_1(Z, \hat{p}_z, \hat{p}_{nz})/L_0(\hat{p}_0)$ is significantly high then MLE of Z is declared as a hotspot. However, Ω is generally so large that its size makes it impractical to maximize $L_1(Z, \hat{p}_z, \hat{p}_{nz})$ as Z varies over Ω by exhaustive search. One

approach to obtain an approximate solution to the maximization problem
is to replace the original parameter space Ω by a smaller, more tractable
subset Ω_0 of Ω, and maximize $L_1(Z, \hat{p}_z, \hat{p}_{nz})$ as Z varies over Ω_0 by exhaustive search. Success of this approach of reduction of the parameter space
depends on how well the reduced parameter space Ω_0 brackets the MLE
over full Ω.

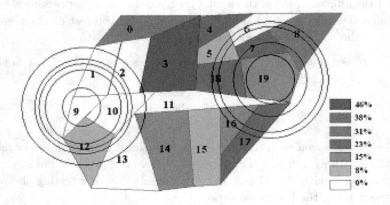

Fig. 10.1.   SaTScan's Circular Scan Window

Widely used SaTScan^{TM} software (Kulldorff, 2006a; 2006b) uses $\Omega_0 =$
$\Omega_{SaTScan}$ obtained as the set of zones covered by a collection of series of
expanding concentric circles with centers/ellipses at the centroid of each
cell. Figure 10.1 provides an example (Patil et al., 2006b) with a region
consisting of 20 cells numbered 0 through 19. The diagram shows only a few
concentric circles centered at centroids of cells 9 and 19. Entire cells whose
centroids are within the scan window form a candidate zone. The circular
scan window centered in a cell is expanded until zones of desired size are
formed. As it can be seen, SaTScan's reduced parameter space $\Omega_{SaTScan}$
is entirely dependent on geometrical proximity of cells without regard to
data and hence tends to produce zones that tend to be compact. It may
do a poor job of detecting actual hotspots that are not quite compact.
Below we review the ULS scan statistic, an alternative to the circular scan
statistic, as developed by Patil and Taillie (2004) that depends on the data
and takes care of connectedness of clusters using adjacency. It is based on
the concept of the upper level set (ULS) tree. For comparative nature of
the circular/elliptic scan statistic and the ULS scan statistic, the reader is
referred to Patil and Taillie (2004).

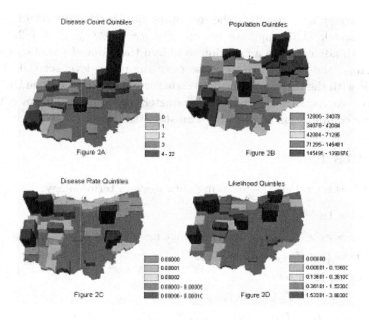

Fig. 10.2.   Ohio West Nile Cases, 2003

10.3. ULS Scan Statistic

We begin with an example to set the stage for formal definitions. Figure 10.2, 2A shows the cellular surface depicting number of human cases of the West Nile disease in different counties of the state of Ohio in 2003. To reduce complexity of the figure, we show quintiles instead of actual numbers of cases. The height of the surface over a county, which is proportional to the number of cases, y_a in the county, however, is not a true indicator of severity of the disease there. We must adjust the number of cases for the population size, n_a in each county. Figure 10.2, 2B shows the cellular surface depicting population of each county. Figure 10.2, 2C shows the cellular surface depicting the number of cases adjusted for population, that is, the rate $g_a = y_a/n_a$ for each county. It shows severity or intensity of the disease. The cellular surface in Figure 10.2, 2D shows likelihood for each county computed under the binomial model. Highest peaks in Figure 10.2, 2D together with some neighboring counties are most likely to emerge as hotspots. A careful study of the four figures will reveal that higher rates combined with higher populations tend to produce higher likelihood values, largely as a result of the monotone likelihood property of the binomial model. This suggests use of rates $g_a = y_a/n_a$, $a = 1, 2, \ldots, N$, along with adjacency as a criterion to obtain the reduced parameter space $\Omega_0 = \Omega_{\text{ULS}}$ and leads us to the concept of the upper level set (ULS) scan statistic with the corresponding data structure, the ULS tree. The ULS tree helps one to visualize the reduced parameter space and facilitates efficient adaptive computation of the ULS scan statistic.

10.4. ULS Tree

First, let us introduce the following notation and terminology:

$G = \{g_a | a \in T\}$,

$r_1 > r_2 > \cdots > r_m$ are all distinct members of G,

$T_i = \{a \in T | g_a = r_i\}, i = 1, 2, \ldots, m$,

upper level sets U_i

$\quad = T_1 \cup T_2 \cup \ldots \cup T_i$

$\quad = \{a \in T | g_a = r_i\}, r_i \in G, \text{for}, \ i = 1, 2, \ldots, m$,

$C_i = $ set of connected components of U_i, $i = 1, 2, \ldots, m$, and

$\Omega_{\text{ULS}} = C_1 \cup C_2 \cup \cdots \cup C_m$, the reduced parameter space.

The ULS tree $\mathcal{T} = \Omega_{\text{ULS}}$ is a tree whose nodes are members of Ω_{ULS}. Members of C_1 are leaf nodes. For $2 < i \leq m$, members of $C_i - C_{i-1}$ are level i nodes. To understand the parent-child relationship between nodes of the ULS tree, let Z be a node belonging to the ULS tree. Let $e = \min\{g_a | a \in Z\}$. Let $Z^* = \{a | a \in Z \text{ and } g_a = e\}$. Consider the set $Z - Z^*$. If $Z - Z^*$ is empty, then Z is a leaf node. If $Z - Z^*$ is not empty, then connected components of $Z - Z^*$ are children nodes of Z. The root of the ULS tree is $U_m = R$, the lowest level or level m node.

The ULS tree actually is an abstraction of the graph of the cellular surface like the one shown in Figure 10.2, 2C. Leaf nodes represent zones with local maxima. The root of the tree corresponds to the entire region.

Figure 10.3 shows development of the ULS tree for the data shown in Figure 10.1. The legend in Figure 10.1 shows cellular rates as percentages. Numbers shown in circles in Figure 10.3 are cell identification numbers. They have been sorted from top to bottom with the cell with the highest rate at the top. Cells at the same level have the same rates. Cells listed within pairs of braces show connected components at different levels.

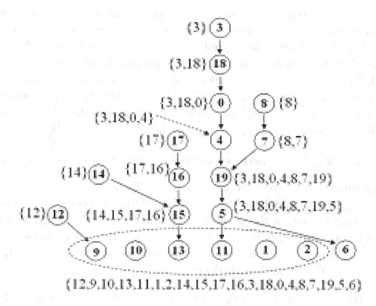

Fig. 10.3. Development of the ULS tree for data in Figure 10.1

One can think of synthesizing the ULS tree starting with leaf nodes and ending with the root node. To begin, connected components of T_1 are leaf nodes. Then given i, $1 < i \leq m$, if a connected component C in T_i has no cell that is adjacent to any cell in $T_1 \cup T_2 \cup \ldots \cup T_{i-1}$ then C forms a new leaf node. On the other hand, if C has cells that are adjacent to zones $Z_1, Z_2, \ldots, Z_k$ occurring at levels $1, 2, \ldots, i-1$, then the connected component of U_i containing $C \cup Z_1 \cup Z_2 \cup \ldots \cup Z_k$ is a new zone at level i and is the parent of zones $Z_1, Z_2, \ldots, Z_k$.

It is easy to see that the cardinality $|\Omega_{\mathrm{ULS}}| \leq N$, since

$$|\Omega_{\mathrm{ULS}}| = \sum_i |C_i| \text{ since } C_1, C_2, \ldots, C_m \text{ are all disjoint}$$

$$\leq \sum_i |T_i| \text{ since cells with rate } r_i \text{ are introduced in the ULS}$$

tree at level i and these cells need not be disconnected,

$$= |T|$$
$$= N$$

and that $|\Omega_{\mathrm{ULS}}| = N$ if and only if $m = N$.

To detect a hotspot we then compute the likelihood for each zone in Ω_{ULS} whose size does not exceed the desired maximum. Zones with high values of likelihood ratio are tested for statistical significance. However, since the data are not replicable, the standard likelihood ratio test cannot be used. Instead, distribution of the test statistic is obtained through place-Monte Carlo simulation (Dwass, 1957). The technique involves generating a large number of random samples from the conditional distribution under the null hypothesis of $Y_1, Y_2, \ldots, Y_N$ for the fixed value of $\sum_T Y_a =$ the actual value of $\sum_T y_a$, where $\sum_T$ indicates summation of objects over T. For each simulated sample, the maximized likelihood ratio statistic is evaluated using a new ULS tree for comparison with the observed likelihood ratio statistic to obtain the p-value of the observed test statistic. Small p-values favor rejection of the null hypothesis that there is no hotspot.

Two response models for discrete data that are commonly studied, implemented in software and applied are:

(1) **Binomial**: Here the size of cell a is a positive integer n_a and the cell response $Y_a \sim Binomial(n_a, p_a)$ where p_a, $0 < p_a < 1$, is the unknown cell characteristic hypothesized to measure the rate of the cell response. For this model, the conditional distribution of $(Y_1, Y_2, \ldots, Y_N)$, given $\sum_T y_a = y$ under the null hypothesis, is $(N-1)$-dimensional multivariate hypergeometric with parameters $(n_1, n_2, \ldots, n_N, y)$.

(2) **Poisson**: Here the size of cell a is a positive real number A_a and the cell response $Y_a \sim Poisson(\lambda_a A_a)$, where $\lambda_a > 0$ is the unknown cell characteristic hypothesized to measure the intensity of the cell response. For this model, the conditional distribution of $(Y_1, Y_2, \ldots, Y_N)$ given $\sum_T y_a = y$ under the null hypothesis is $(N-1)$-dimensional multinomial with parameters $(p_1, p_2, \ldots, p_N, y)$ where $p_a = A_a / \sum_T A_a$, $a = 1, 2, \ldots, N$.

In addition to the above two models, SaTScan^{TM} makes available also implementation of the exponential and normal response models. For continuous distributions such as gamma and lognormal, the exponential model has been shown to work well (Kulldorff, 2006b), but, in general, continuous response models have not received much attention. Patil and Taillie

(2004) discuss an approach to modeling continuous response distribution with gamma and lognormal as illustrations. In view of its relevance to the continuous fractional response model, a brief treatment of the gamma model is presented below.

10.5. Gamma Model

In order to parametrize the gamma model, Patil and Taillie (2004) are guided by two principles: (1) the mean response for a cell should be proportional to the cell size, and (2) the relative variability should decrease with the cell size. These are consistent with the behavior of the binomial and Poisson models. Patil et al. (2008a) discuss the gamma model further, and report software implementation of the gamma model with a case study.

If $Y \sim gamma(k, \beta)$, where k is the index parameter and β is the scale parameter, then

$E[Y] = k\beta$, $Var[Y] = k\beta^2$ and (coefficient of variation)$^2 = 1/k$.

Further, the gamma family is additive with respect to the index parameter. These observations suggest that the parameter k should be taken to be proportional to the cell size, which we will denote by A_a (A for area) of cell a, or $k_a = A_a/c$, c being an unknown parameter but with the same value for the entire region. Parameter β accounts for cell to cell response variability. Thus our hypothesis testing model for hotspot detection becomes:

$H_0 : c > 0$ is the same for all cells and $\beta_1 = \beta_2 = \cdots = \beta_N = \beta_0$, say, to be tested against the alternative hypothesis

$H_1 : c > 0$ is the same for all cells and there is a non-empty zone $Z \in \Omega$. and values $0 < \beta_{nz}, \beta_z$ such that

$$\beta_a = \begin{cases} \beta_z & \text{for all cells } a \text{ in } Z \\ \beta_{nz} & \text{for all cells } a \text{ in } R - Z \end{cases}$$

and $\beta_z > \beta_{nz}$.

The likelihood equation for c does not permit a closed form solution. But it can be solved efficiently using Newton-Raphson method (Patil et al., 2008a). As for β_0, β_z and β_{nz}, likelihood equations yield:

$$\beta_0 = c \left(\sum_R y_a / \sum_R A_a \right), \quad \beta_z = c(\sum_Z y_a / \sum_Z A_a),$$

and

$$\beta_{nz} = c \left(\sum_{nZ} y_a / \sum_{nZ} A_a \right)$$

where $\sum_R$, $\sum_Z$, and $\sum_{nZ}$ indicate summations extending over R, Z, and $R - Z$, respectively.

Under the null hypothesis, the conditional distribution of the vector $(Y_1, Y_2, \ldots, Y_N)$, given $\sum_T y_a = y$, is $N - 1$ dimensional Dirichlet with parameters $(k_1, k_2, \ldots, k_N)$, where $k_a = A_a/c$, c being an unknown parameter. We need to use this distribution to generate random copies of the data to compute the p-value of the candidate zone with the maximum likelihood. We approximate the distribution by substituting the MLE $\hat{c}_{Null}$ in place of c computed under the null hypothesis. A random copy is generated in two steps: (1) Generate $(X_1, X_2, \ldots, X_N)$ with X_a from $gamma(A_a/c, \beta)$, $a = 1, 2, \ldots, N$, with β and c replaced with their MLE's under the null hypothesis so that $(W_1, W_2, \ldots, W_N)$ is from $(N - 1)$-dimensional $Dirichlet(A_1/A, A_2/A, \ldots, A_N/A)$, where $W_a = X_a/(\sum_R X_a)$, $a = 1, 2, \ldots, N$, and $A = \sum_R A_a$. (2) Compute $Y_a = y * X_a/(\sum_R X_a)$, $a = 1, 2, \ldots, N$.

10.6. Continuous Fractional Response Model

The gamma model discussed in the previous section is applicable in hotspotting when continuous responses are positive valued and additive in nature, a situation that occurs quite frequently. Another situation with continuous responses that occurs frequently in practice is when they are between 0 and 1 when it seems plausible to postulate $Y \sim beta(\alpha, \beta)$ with the pdf

$$f_Y(y; \alpha, \beta) = \Gamma(\alpha + \beta)/(\Gamma(\alpha)\Gamma(\beta))y^{\alpha-1}(1 - y)^{\beta-1}, 0 < y < 1,$$

with $\alpha > 0$, $\beta > 0$, where Y represents a typical cell response. However, the beta family does not possess the additive property. Hence, to begin with, we propose the transformation:

$$X = Y/(1 - Y).$$

The random variable X has the beta distribution of the second kind (Patil et al. 1984) with the pdf

$$f_X(x; \alpha, \beta) = \frac{\Gamma(\alpha + \beta)}{\Gamma(\alpha)\Gamma(\beta)} x^{\alpha-1}/((1 + x)^{\alpha+\beta}), x > 0,$$

where $\alpha > 0, \beta > 0$. We note that this distribution also arises as a mixture of the gamma distributions $gamma(k, \alpha)$ on parameter k where $1/k \sim gamma(1, \beta)$ (Patil et al., 1984). In the absence of availability of an exact model with properties that are in conformation with our guiding principles, it appears reasonable to approximate the exact model, namely the mixture of the gamma distributions, with a straight gamma distribution that satisfies our criteria. Thus we propose to treat $Y/(1-Y)$ as $gamma(k, \beta)$.

In many situations with continuous fractional response the beta distribution of the first kind (Patil et al., 1984) rather than the standard beta may be more applicable. The beta distribution of the first kind with parameters $r, s, \alpha,$ and β has the pdf

$$f_Y(y; r, s, \alpha, \beta) = \Gamma(\alpha + \beta)/(\Gamma(\alpha)\Gamma(\beta))(y - r)^{\alpha-1}(s - y)^{\beta-1}/(s - r)^{\alpha+\beta-1},$$

for $r < x < s$, where $0 \leq r < s \leq 1, \alpha > 0, \beta > 0$.

The simple transformation, $Y' = (Y - r)/(s - r)$, takes us to the β scenario. However, r and s are typically unknown. Hence, to be able to deal with the beta distribution of the first kind using the technique developed for the standard beta distribution, one may use the transformation

$$Y' = (Y - \hat{r})/(\hat{s} - \hat{r}) \tag{10.1}$$

where $\hat{r}$ and $\hat{s}$ are reasonable estimates of r and s, respectively. For our purpose, we will use $y_{\min}$ and $y_{\max}$ for $\hat{r}$ and $\hat{s}$, respectively, where

$$y_{\min} = \min\{y_a | a \in T\}, \text{and}$$

$$y_{\max} = \max\{y_a | a \in T\}$$

mostly because of computational ease.

In Section 10.7 we will describe an application of the continuous fraction response model to Jalgaon district forest cover data using software implementation of the ULS scan statistic described in detail in Patil et al. (2008a). Its current version is able to handle the continuous fraction response model in addition to binomial, Poisson, and gamma models. Results reported in Section 10.7 indicate that application of the gamma model to approximate the beta model of the second kind is a viable technique to do hotspot detection with data, where the beta model is appropriate.

10.7. Jalgaon District Forest Cover

Table 10.1 shows Jalgaon district (Maharashtra) forest cover 2001-02 data by tehsil[a].

---

[a]Office of the Forest Conservator, Jalgaon and Yawal Subdivisions

Table 10.1.   Jalgaon District Forest Cover

Serial Number	Tehsil Name	Geographical Area (Hectares)	Forest Area (Hectares)	Forest Cover
0	Amalner	844.15	21.90	0.02594
1	Bhadgaon	484.53	78.49	0.16199
2	Bhusawal	413.38	29.60	0.07160
3	Bodvad	398.77	56.37	0.14136
4	Chalisgaon	1217.63	121.11	0.09946
5	Chopda	954.36	162.13	0.16988
6	Dharangaon	463.53	19.47	0.04200
7	Edlabad	646.11	132.57	0.20518
8	Erandol	511.03	24.26	0.04747
9	Jalgaon	825.07	142.68	0.17293
10	Jamner	1360.72	155.72	0.11444
11	Pachora	820.41	72.46	0.08832
12	Parola	791.21	98.59	0.12461
13	Raver	935.70	264.05	0.28220
14	Yawal	954.38	308.18	0.32291
Total		11620.98	1687.58	0.14522

We intend to determine if some cluster of tehsils can be considered as a hotspot based on the data in the last column of the table, using the ULS software (referred to as the 'program' hereafter) mentioned above. Figure 10.3 shows the contents of the input file to the program, as described below.

The program requires that each cell in the region be identified serially as 0, 1, 2, ... with one line of input for each cell in the data file in that order. The first entry in a given line is the cell identifier, the second entry is the 'size' of the cell, the third entry is the response. Remaining entries in the line are identifiers of cells that are adjacent to the cell mentioned at the beginning of the line. For the fractional continuous data as in the current case, the model to be used is specified by the user as 'beta'. For the beta model, each cell size needs to be input as 1 and the response is assumed to be between 0 and 1. The program automatically unitizes data using the transformation (10.1) above. However, it is necessary to adjust the data values 0 and 1 – considered as data values that are too small or too large to observe – so that computed probability densities are not zero. The program replaces the zero data value by $\varepsilon/2$ and the unit data value by $(1 + \upsilon)/2$, where ε is the smallest non-zero data value and υ is the largest non-unit data value after unitization. Unitized data values are further subjected to the transformation $y/(1 - y)$ before application of the gamma model. The program also allows the user to specify the threshold percentage to limit the size of a hotspot relative to the total size of the entire region. We ran

Table 10.2. Jalgaon District Forest Cover Hotspots

Threshold %	Member Count	Member Tehsils	p-value
10	1	14	0.180
20	3	7, 13, 14	0.040
30	3	7, 13, 14	0.049
40	3	7, 13, 14	0.083
50	3	7, 13, 14	0.106

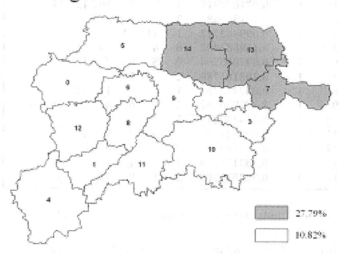

Jalgaon District Forest Cover

Fig. 10.4. The shaded area is a hotspot at 5% level for thresholds of 20% and 30%

the program five times with five threshold values of 10%, 20%, 30%, 40%, and 50%. Results of the five runs are summarized in Table 10.2. A map of Jalgaon district identifying each tehsil and the hotspot consisting of three tehsils as detected by the program are shown in Figure 10.4.

Incidentally, it may be worth noting that the zone consisting of tehsils 7, 13, and 14 happens to have the maximum likelihood value for threshold values of 20%, 30%, 40%, and 50%, however, with different p-values. This situation is explained by the fact that, when we increase the threshold, we increase the set of competing candidate zones and the maximum likelihood values occurring in the simulated samples exceed the likelihood of the top candidate zone as per actual data set more often. On the other hand, with the threshold of 10%, the p-value of the zone = {14} is greater than that of zone = {7, 13, 14} when the threshold is 0.20%. This is due to a greater probability of a high response over a small area purely by chance. We conclude that choice of the threshold is an important consideration in hotspot analysis from the point of view of the manager responsible for making practical decisions.

0	1	0.025943257	5	6	12			
1	1	0.161992034	4	8	11	12		
2	1	0.071604819	3	7	9	10	13	14
3	1	0.141359681	2	7	10			
4	1	0.099463712	1	11	12			
5	1	0.169883482	0	6	9	14		
6	1	0.042003754	0	5	8	9	12	
7	1	0.205181780	2	3	13			
8	1	0.047472751	1	6	9	11	12	
9	1	0.172930782	2	5	6	8	10	11	14
10	1	0.114439414	2	3	9	11		
11	1	0.088321693	1	4	8	9	10	
12	1	0.124606615	0	1	4	6	8	
13	1	0.282195148	2	7	14			
14	1	0.322911209	2	5	9	13		

Input File for Jalagaon Forest Cover Data

The three tehsils making up the hotspot in Figure 10.4 are Yawal, Raver, and Edlabad (now known as Muktainagar). All the three tehsils are located in the Satpuda mountain region and are known for their forests. Appropriately they have been identified as a hotspot. More importantly, the model presented in the paper asserts through the p-value the degree to which the tehsils stand out as forest covered areas within the district.

10.8. Conclusion

In the present age, when information is produced or can be generated in enormous quantities at blinding speed, when resources are in high demand, and when events occurring in an area have far reaching and quick consequences locally and globally, there is a need for development of tools to process information to isolate areas – geographical or otherwise in a network setting – of high activity or high intensity to produce early warnings, help design corrective actions, allocate resources equitably and efficiently, and so forth, through hotspot detection. Software tools for hotspot detection with discrete data have been available for some time. Here, we have presented a model based on the beta distribution to do hotspot detection with fractional data of continuous nature. Application of the model is illustrated through a real life example using the ULS scan statistic.

Acknowledgments

This material is based upon work supported by the United States National Science Foundation under Grant No. 0307010. Any opinions, findings, and conclusions or recommendations expressed in this material are those of the authors(s) and do not necessarily reflect the views of the agencies.

References

1. Biomedware (2001). *Software for the Environmental and Health Sciences.* Biomedware, Ann Arbor, MI.
2. Cressie, N. (1991). *Statistics for Spatial Data,* John Wiley and Sons, New York.
3. Glaz, J. and Balakrishnan, N. (1999). *Scan Statistics and Applications.* Birkhäuser, Boston, USA.
4. Glaz, J., Naus, J. and Wallenstein, S. (2001). *Scan Statistics.* Springer, New York, USA.
5. Kulldorff, M. (1997). A Spatial Scan Statistic. *Comm. Statist. Theory Methods* **26**, 1481–1496.
6. Kulldorff, M. (2006a). *SaTScan^{TM} v 7.0: Software for the spatial and space-time scan statistics.* Information Management Services Inc., Silver Spring, MD, USA
7. Kulldorff, M. (2006b). *SaTScan^{TM} User Guide for version 7.0.* http://www.satscan.org/
8. Kulldorff, M. and Nagarwalla, N. (1995). Spatial Disease Clusters: Detection and Inference. *Stat. Med.* **14**, 799–810.
9. Kulldorff, M., Rand, K., Gherman, G., Williams, G. and DeFrancesco, D. (1998). *SaTScan v 2.1: Software for the spatial and space-time scan statistics,* National Cancer Institute, Bethesda, MD, USA.
10. Patil, G. P. (2007). Statistical geoinformatics of geographic hotspot detection and multicriteria prioritization for monitoring, etiology, early warning and sustainable management for digital governance in agriculture, environment, and ecohealth. *J. Indian Soc. Agricultural Statist.* **61**, 132–146.
11. Patil, G. P., Boswell, M. T. and Ratnaparkhi, M. V. (1984). *Dictionary and Classified Bibliography of Statistical Distributions in Scientific Work.* Vol. 2:

Univariate Continuous Models, International Co-operative Publishing House, Burtonsville, MD, USA.

12. Patil, G. P. and Taillie, C. (2003). Geographic and network surveillance via scan statistics for critical area detection. *Statist. Sci.* **18**, 457–465.

13. Patil, G. P., Taillie, C. (2004). Upper level set scan statistic for detecting arbitrarily shaped hotspots. *Environ. Ecol. Stat.* **11**, 183–197.

14. Patil, G. P., Acharya, R., Glasmier, A., Myers, W., Phoha, S. and Rathbun, S. (2006a). Hotspot detection and prioritization geoinformatics for digital governance. In *Digital Government: Advanced Research and Case Studies* (H. Chen, L. Brandt, V. Gregg, R. Traunmuller, S. Dawes, E. Hovy, A. Macintosh, and C. Larson, Eds.), Springer, New York, USA.

15. Patil, G. P., Modarres, R., Myers, W. L. and Patankar, P. (2006b). Spatially Constrained Clustering and Upper Level Set Scan Hotspot Detection in Surveillance GeoInformatics. *Environ. Ecol. Stat.* **13**, 365–377.

16. Patil, G. P., Acharya, R., Myers, W., Phoha, S. and Zambre R. (2007). Hotspot Geoinformatics for detection, prioritization, and security. In *Encyclopedia of Geographical Information Science* (S. Shekhar and H. Xiong, Eds.), Springer Publishers.

17. Patil, G. P., Acharya, R. and Phoha, S. (2007). Digital governance, hotspot detection, and homeland security. In *Encyclopedia of Quantitative Risk Analysis*, Wiley, New York.

18. Patil, G. P., Acharya, R., Modarres, R., Myers, W. L. and Rathbun, S.L. (2007). Hotspot geoinformatics for digital government. In *Encyclopedia of Digital Government, Vol. II* (Ari-Veikko Anttiroiko and Matti Malkia, Eds.), 919.

19. Patil, G. P., Joshi, S. W. and Rathbun, S. L. (2007). Hotspot geoinformatics, environmental risk, and digital governance, In *Encyclopedia of Quantitative Risk Analysis*, Wiley, StateNew York, 927, Idea Group Reference, Hershey, PA, USA.

20. Patil, G.P., Joshi, S. W., Myers, W. L., and Koli, R. E. (2008a). ULS Scan Statistic for Hotspot Detection with Continuous Gamma Response, In *Scan Statistics: Methods and Applications* (J. Glaz, V. Pozdnyakov, and S. Wallenstein, Eds.), Birkhuser, Boston, MA.

21. Patil, G. P., Patil, V. D., Pawde, S. P., Phoha, S., Singhal, V., and Zambre, R. (2008b). Digital governance, hotspot geoinformatics, and sensor networks for monitoring, etiology, early warning, and sustainable management. In *Geoinformatics for Natural Resource Management* (P. K. Joshi, Ed.), Nova Science Publishers, New York (2009).

22. Takahashi, K, Yokoyama, T and Tango, T. (2004). FleXScan: Software for the flexible spatial scan statistic. *National Institute of Public Health, Japan.*

23. Tango, T and Takahashi, K. (2005). A flexibly shaped spatial scan statistic for detecting clusters, *International Journal of Health Geographics*, `http://www.i-healthgeographics.com/content/4/1/11`

Chapter 11

Bayesian Curve Registration of Functional Data

Z. Zhong[1], A. Majumdar[2] and R. L. Eubank[3]

[1] *Microsoft, USA*
[2,3] *Arizona State University, USA*

Functional data arise in numerous areas nowadays. When the functional responses evolve with respect to time, the subjects may experience events at different paces with the consequence that the sample curves are improperly aligned for inferential purposes. In particular, the sample mean function without alignment will fail to produce a satisfactory estimator of the true process mean function. In this article a new model for curve alignment or registration is developed from a Bayesian perspective. It incorporates nonparametric spline curve fitting methods with continuous Monte Carlo Markov chain (MCMC) techniques. The functional response curves are fit by nonparametric spline methods with their coefficients treated as random parameters. Similarly, the warping functions are modeled as random spline functions and random shift and amplitude coefficients are also included in the model formulation. An MCMC algorithm is created to estimate the parameters in the model. The performance of the proposed method is evaluated in an empirical study.

Keywords: MCMC technique, Nonparametric spline methods, Process mean function, Warping functions.

Contents

11.1. Introduction

Functional data analysis is a relatively new area in statistics. It concerns the study of observations on a function space valued random variable. Thus, in contrast to classical statistical themes, each individual observation represents a function rather than simply a scalar or vector value at a particular point. Functional data arise in many different fields, including economics, biology and signal processing. Techniques for the analysis of functional data are described in Ramsay and Silverman (1997) while data analysis case studies are presented in Ramsay and Silverman (2002).

The standard summary measures that are used in functional data analysis include the mean and covariance functions. After functional data are collected, each subject can typically be represented by a smooth function and these functions are, in turn, employed to construct summary measures. However, there are often systematic variations among those curves which have the consequence that direct averaging across curves will produce poor estimators of the true process average and covariance functions. In such cases the solution is to *register* or align the sample curves so that they only differ in amplitude and thereby produce a cross-sectional mean that gives a more satisfactory summary of the data. Without taking the phase variation into account, the cross-sectional mean always underestimates the amplitude of local maxima and overestimates the amplitude of local minima. This is problematic because local extreme are often the most important features of the process that is under study and their amplitude should be estimated accurately.

Various registration methods have been proposed in the literature for dealing with functional n data. Most of this work involves the use of warping functions. These are strictly monotone functions whose inverses transform observation time to a synchronized time scale where certain features on different curves occur simultaneously. A standard warping function model for functional data would assume that there an observed set of curve $x_1, \ldots, x_M$ stem from a model of the form

$$x_i(t) = \mu(h_i^{-1}(t)) + \epsilon_i(t), \qquad i = 1, \ldots, M$$

with $\mu(\cdot)$ the true mean function for the process, h_i^{-1} the inverse warping function for the ith subject and $\epsilon_i(\cdot)$ an error process. Registration for a sample of functional data can then be viewed as equivalent to estimating $h_1, \ldots, h_M$. Sakoe and Chiba (1978) provide an early example of the use of warping functions to align two curves. An extension of their approach that could be employed with a sample of functional data was developed by Gasser and Kneip (1995). Bayesian approaches to curve registration has been employed by McKeague (2005) and Alshabani et al. (2007) for dealing with problems of signature recognition and analysis of human movement curves, respectively. More in line with our proposed methodology is the work of Telesca and Inoue (2008) that will be discussed more fully at the end of Section 11.2.

The continuous monotone registration method of Ramsay and Li (1996) is a popular registration technique that is available from the FDA package in R. This method derives from a model where

$$z(t) = x_i\,[h_i(t)] + \epsilon_i(t),$$

with $x_i(\cdot)$ an observed sample curve and $z(\cdot)$ a fixed function that provides a template for the individual curves. The warping function is then estimated by the function h that minimizes

$$F_\lambda(z, x_i | h_i) = \int_0^{T_0} \{z(t) - x_i\,[h_i(t)]\}^2 dt + \lambda \int_0^{T_0} w_i^2(t) dt$$

with w_i representing the relative curvature: i.e., $w_i = D^2 h_i / D h_i$, with $D^2(\cdot)$ and $D(\cdot)$ being the second and first derivative, respectively. To implement the actual minimization the w_i are represented by linear combinations of B-spline basis functions. The choice of the function $z(\cdot)$ is problematic. In practice this is often taken to be the cross-sectional mean and the processes is applied in an iterative format to each of the sample curves while updating the cross-sectional mean at each step.

Our proposed approach has ties to the shape invariant models (SIM) proposed by Lawton et al. (1972) In particular, we pursue a direction that is similar to that of Brumback and Lindstrom (2004) in a SIM context. Their model can be written as

$$x_{ij} = \theta_{i1}\mu(h^{-1}(t_{ij}, \phi_i), \beta) + \theta_{i4} + \epsilon_{ij}, \quad i = 1, \ldots, M, \quad j = 1, \ldots, n_i,$$

where μ is a common shape function dependent on parameters and h is a common parametric form for the individual warping functions. Here θ_{i1}, θ_{i4} and the ϕ_i are random, curve-specific quantities and the ϵ_{ij} are random errors. The h_i^{-1} were then modelled as monotone splines with random coefficients and μ was treated as a cubic spline. The resulting framework can be viewed as producing a nonlinear mixed effects model and estimation methodology was developed from that perspective.

Our goal in the remainder of this paper is development of an automatic, data driven method for registration of functional data. In the next section we construct a registration model from a Bayesian perspective. Using this framework we then derive the desired automatic estimation algorithm. Section 11.3 describes the results of a small empirical study that examines the performance of our approach relative to continuous monotone registration. We conclude in Section 11.4 with a summary of our principal results and a discussion of future areas of investigation.

11.2. Bayesian Approach

In this section we lay out our Bayesian framework for data registration. The basic premise is that we observe responses x_{ij} being obtained at time ordinates t_{ij} for the ith subject/experimental unit with $j = 1, \ldots, n_i$ representing the readings for a particular subject. The x_{ij} then satisfy

$$x_{ij} = a_{sh_i} + a_{sc_i}\mu\{h_i^{-1}(t_{ij})\} + \epsilon_{ij}, \quad i = 1, \ldots, M, \quad j = 1, \ldots, n_i, \quad (11.1)$$

where the ϵ_{ij} are normal random errors, a_{sh_i} and a_{sc_i} are shift and scale parameters, respectively, $\mu(\cdot)$ is a common mean function and $h_i(\cdot)$ is the warping function for the ith subject. To simplify notation we will use $\mathbf{x}_i$ and $\mathbf{t}_i$ to indicate the responses and t ordinates for the ith subject and let $N = \sum_{i=1}^{M} n_i$ in what follows. Then, for any function f we use $f(\mathbf{t}_i)$ to indicate an $n_i \times 1$ vector with elements $(f(t_{i1}), \ldots, f(t_{in_i}))$.

In the next subsection we give a detailed development of the model we will employ. The basic premise behind our formulation is relatively simple. The idea is to approximate all unknown functions by linear combinations of B-splines and the resulting set of coefficients are then estimated from the data. However, there are complications that arise due to the monotonicity restrictions that must be treated with special care in order to produce a satisfactory approach. Section 11.2.2 then discusses the specific Monte Carlo Markov chain algorithms that we will employ while Section 11.2.3 gives the details behind our estimation algorithm.

11.2.1. *Model formulation*

A fundamental tool for our approach is the B-spline basis that will be used for modeling. Following developments in de Boor (1978), we can give a recursive definition for these functions. Specifically, let $0 < \xi_1 < \ldots < \xi_k < 1$ be a set of interior knots and define $2m$ additional knots as $\xi_{-(m-1)}, \ldots, \xi_{-1} = \xi_0 = 0$, $\xi_{k+1}, \ldots, \xi_{k+m} = T$ for some specified upper limit T. The B-splines of order m with knots at $\xi_1, \ldots, \xi_k$ are then defined recursively by

$$B_{i,m}(t) = \frac{t - \xi_i}{\xi_{i+m-1} - \xi_i} B_{i,m-1}(t) + \frac{\xi_{i+m} - t}{\xi_{i+m} - \xi_{i+1}} B_{i+1,m-1}(t), \qquad (11.2)$$

for $i = -(m-1), \ldots, k$, with

$$B_{i,1}(t) = \begin{cases} 1, t \in [\xi_i, \xi_{i+1}), \\ 0, \text{ otherwise,} \end{cases} \qquad (11.3)$$

being the initial step in the recursion. In using the recursion formula we need to employ the property that a B-spline of order r corresponding to an $r+1$ coincident knot is zero. With this convention, equation (11.2) provides both a definition and computational formula for the basis functions.

We will represent the warping functions as linear combinations of B-splines. Since warping functions will be monotone increasing, we need to enforce this property on the corresponding B-splines. Specifically, since the derivative of a B-spline is also a B-spline of one lower degree, monotonicity is obtained through an inequality constraint on the coefficients. Accordingly, we now take the ith warping function to be

$$h_i^{-1}(\mathbf{t}_i) = \mathbf{U}(\mathbf{t}_i)\boldsymbol{\phi}_i, \qquad (11.4)$$

for $i = 1, \ldots, M$, with $\mathbf{U}(\mathbf{t}_i)$ representing the quadratic B-spline basis with D uniformly spaced knots evaluated at time $\mathbf{t}_i$ and $\boldsymbol{\phi}_i$ the corresponding coefficient vector whose elements must be strictly increasing. So, $\mathbf{U}(\mathbf{t}_i)$ is a $n_i \times (D+3)$ matrix and $\boldsymbol{\phi}_i$ a vector of dimension $D+3$.

Similar to the developments for warping functions, we model the common mean function μ with a cubic spline: i.e.,

$$\mu(\cdot) = \mathbf{B}(\cdot)^T \boldsymbol{\beta},$$

with $\mathbf{B}(\cdot)$ the vector of B-spline basis functions for the common shape function and $\boldsymbol{\beta}$ a corresponding coefficient vector. Here $\mathbf{B}(\cdot)$ is a $(Q+4)$-vector with Q being the number of knots. The knots are taken to be equally spaced and $\boldsymbol{\beta}$ is a vector of dimension $Q+4$.

Upon inserting the models for the warping and mean function into (11.1), we obtain

$$\mathbf{x} = \text{diag}\{\mathbf{1_{n_1}}, \ldots, \mathbf{1_{n_M}}\}\mathbf{a_{sh}} + \text{diag}\{\mathbf{B}[\mathbf{U}(\mathbf{t}_1)\boldsymbol{\phi}_1]\boldsymbol{\beta}, \ldots, \mathbf{B}[\mathbf{U}(\mathbf{t}_M)\boldsymbol{\phi}_M]\boldsymbol{\beta}\}\mathbf{a_{sc}} + \boldsymbol{\epsilon}, \qquad (11.5)$$

where

$$\mathbf{x} = (x_{1,1}, \ldots, x_{1,n_1}, \ldots, x_{M,1}, \ldots, x_{M,n_M})^T,$$

$\mathbf{a_{sh}} = (a_{sh_1}, \ldots, a_{sh_M})^T, \mathbf{a_{sc}} = (a_{sc_1}, \ldots, a_{sc_M})^T$, diag$\{A_1, \ldots, A_M\}$ denotes a diagonal matrix with diagonal elements being matrices $A_1, \ldots, A_M$ and

$$\epsilon = (\epsilon_{1,1}, \ldots, \epsilon_{1,n_1}, \ldots, \epsilon_{M,1}, \ldots, \epsilon_{M,n_M})^T \sim \mathcal{N}(\mathbf{0}, \sigma^2 \mathbf{I})$$

with $\mathbf{I}$ the $N \times N$ identity matrix. For each observation x_{ij}, we can write (11.5) as

$$x_{ij} = a_{sh_i} + a_{sc_i}\{\mathbf{B}[\mathbf{U}(\mathbf{t}_i)\boldsymbol{\phi}_i]\}_j\beta + \epsilon_{ij}, \quad i = 1, \ldots, M, j = 1, \ldots, n_i,$$

with $\{\mathbf{B}[\mathbf{U}(\mathbf{t}_i)\boldsymbol{\phi}_i]\}_j$ being the jth row of $\mathbf{B}[\mathbf{U}(\mathbf{t}_i)\boldsymbol{\phi}_i]$. For easy interpretation, we will denote $\{\mathbf{B}[\mathbf{U}(\mathbf{t}_i)\boldsymbol{\phi}_i]\}_j$ as $\mathbf{g}_{ij}^T$ in the future. The qth column of $\mathbf{g}_{ij}^T$ is the qth B-spline basis function evaluated at $\mathbf{U}(\mathbf{t}_i)\boldsymbol{\phi}_i$: i.e.,

$$(\mathbf{g}_{ij}^T)_q = B_q[\mathbf{U}(\mathbf{t}_i)\boldsymbol{\phi}_i], \tag{11.6}$$

With this notational convention, our model can be expressed as

$$x_{ij} = a_{sh_i} + a_{sc_i}\mathbf{g}_{ij}^T\beta + \epsilon_{ij}, \quad i = 1, \ldots, M, \ j = 1, \ldots, n_i$$

or, in matrix form, as

$$\mathbf{x} = \text{diag}(\mathbf{1_{n_1}}, \ldots, \mathbf{1_{n_M}})\mathbf{a_{sh}} + \text{diag}(\mathbf{g}^T\boldsymbol{\beta})\mathbf{a_{sc}} + \epsilon$$

with diag$(\mathbf{g}^T\boldsymbol{\beta}) = $ diag$\{\mathbf{B}[\mathbf{U}(\mathbf{t_1})\boldsymbol{\phi}_1]\boldsymbol{\beta}, \ldots, \mathbf{B}[\mathbf{U}(\mathbf{t}_M)\boldsymbol{\phi}_M]\boldsymbol{\beta}\}$.

Before we begin our analysis, we put some restrictions on the model. Without loss we restrict all the curves to have a common start and end time: i.e., $t_{i,1} = 0$ and $t_{i,n_i} = T$ for all $i = 1, \ldots, M$. Since we want the real start/end time and the warped start/end time to coincide, we then restrict the time transformations to match at both the start points and the endpoints. That is, $h_i(t_1) = 0$ and $h_i(t_{n_i}) = T$ for all $i = 1, \ldots, M$. Hence, in our setting we fix $\phi_{i,1} = 0$ and $\phi_{i,D+3} = T$. This allows us to focus on the time differences among curves and makes the estimation less complicated.

As noted above, to ensure that the warping function is monotone increasing, we constrain its coefficients $\boldsymbol{\phi}_i = (\phi_{i,1}, \ldots, \phi_{i,D+3})^T$ to be strictly increasing in that $\phi_{i,1} < \phi_{i,2} < \cdots < \phi_{i,D+3}$. Thus, let $\Phi = (\boldsymbol{\phi}_1, \ldots, \boldsymbol{\phi}_M)^T$. As in Brumback and Lindstrom (2004) the monotone increasing constraint on Φ will then be enforced by modeling Φ with a Jupp inverse transformation that will be discussed subsequently.

For identifiability reasons, we force the average time transformation to be fixed at the identity transformation. That is, the mean of the $\boldsymbol{\phi}_i$ is the vector of identity spline coefficient. Specifically, the parameters ϕ_i are constrained so that

$$\frac{1}{M} \sum_{i=1}^{M} \sum_{j=1}^{D+3} U_j(t)\phi_{i,j} = t.$$

This is equivalent to

$$\sum_{j=1}^{D+3} U_j(t)\{\frac{1}{M} \sum_{i=1}^{M} \phi_{i,j}\} = t. \tag{11.7}$$

Let

$$s_j = \frac{1}{M} \sum_{i=1}^{M} \phi_{i,j}$$

and define

$$\mathbf{s} = (s_1, \ldots, s_{D+3})^T.$$

This vector can then be obtained through the recursion

$$s_1 = 0,$$
$$s_{q+1} = \frac{\gamma_{q+r} - \gamma_{q+1}}{r-1} + s_q, \quad q = 1, \ldots, D+2. \tag{11.8}$$

In (11.8), γ_i, $i = 1, \ldots, D+6$ are the knots with domain [0, T]. There are D interior knots and 2 boundary knots. The boundary knots are repeated 3 times for the purpose of applying the B-spline functions. That is, $(\gamma_1, \ldots, \gamma_{D+6})^T = (0, 0, 0, \gamma_4, \ldots, \gamma_{D+3}, T, T, T)^T$ and $0 < \gamma_4 < \gamma_{D+2} < \cdots < \gamma_{D+3} < T$. In what follows we will use $\mathbf{S}$ to denote the $M \times (D+3)$ matrix with all rows equal to $\mathbf{s}^T$.

Application of the Jupp transformation (Jupp, 1978) can be accomplished as follows. Let $\mathbf{v}$ be an ordered vector of length Q: that is, $\mathbf{v} = (v_1, \ldots, v_Q)^T$ with $v_1 < v_2 < \cdots < v_Q$. Define the Jupp transformation as $\text{Jupp}(\mathbf{v}) = \mathbf{w} = (w_1, \ldots, w_Q)^T$, where

$$w_q = \begin{cases} v_q & q = 1, Q, \\ \log(\frac{v_{q+1} - v_q}{v_q - v_{q-1}}) & q = 2, \ldots, Q-1. \end{cases}$$

It follows from this that the inverse of the Jupp function $\mathbf{w} = \text{Jupp}^{-1}(\mathbf{v})$ will be a vector of increasing elements with $w_1 = v_1$ and $w_Q = v_Q$. It can be calculated via the following steps:

(1) Define $a_1 = 1$, $a_q = \exp(\sum_{k=2}^{q} v_k)$, for $q = 2, \ldots, Q-1$;

(2) Let $c_k = (v_Q - v_1)a_k/(\sum_{q=1}^{Q-1} a_q)$, for $k = 1, \ldots, Q-1$;

(3) $w_1 = v_1$, $w_Q = v_Q$, and $w_j = v_1 + \sum_{k=1}^{j-1} c_k$, for $j = 2, \ldots, Q-1$.

It is easy to verify that for a matrix A we will have $\mathrm{Jupp}(\mathrm{Jupp}^{-1}(A)) = A$.

We now define the Φ matrix by

$$\Phi = \mathrm{Jupp}^{-1}([\mathbf{0} \quad \mathbf{Z}^T\mathbf{P} \quad \mathbf{0}] + \mathrm{Jupp}(\mathbf{S})) \tag{11.9}$$

with $\mathbf{0}$ a vector whose elements are all equal to 0, $\mathbf{P}$ a matrix of unconstrained parameters and $\mathbf{Z}$ an $(M-1) \times M$, full row-rank matrix determined by $\mathbf{Z}^T\mathbf{Z} = \mathbf{W}$ with

$$\mathbf{W} \equiv -\frac{1}{M-1}\mathbf{J}_M + \frac{M}{M-1}\mathbf{I}_M.$$

Here $\mathbf{J}_M$ is an $M \times M$ matrix with all entries 1, and $\mathbf{I}_M$ is an $M \times M$ identity matrix.

Since we know that the first column of Φ is equal to $(0, \ldots, 0)^T$ and the last column of Φ is equal to $(T, \ldots, T)^T$, there are only $D+1$ columns of "free parameters" in Φ. Since $\sum_{i=1}^{M} \phi_i \doteq M\mathbf{s}$, the last row of Φ, Φ_M can be expressed in terms of the first $M-1$ rows. So that

$$\Phi_M \doteq M\mathbf{s} - \sum_{i=1}^{M-1} \phi_i$$

The row dimension of Φ is therefore equal to $M-1$ and we can define a $(M-1) \times (D+1)$ matrix, $\tilde{\Phi}$, of free parameters by taking the Φ matrix and deleting its first column, last column and last row to obtain

$$\tilde{\Phi} = \begin{pmatrix} \phi_{12} & \cdots & \phi_{1,D+2} \\ \vdots & \ddots & \vdots \\ \phi_{M-1,2} & \cdots & \phi_{M-1,D+2} \end{pmatrix}. \tag{11.10}$$

The matrix in (11.10) will be the focus of subsequent developments using MCMC technology.

We will need to work with transformations between $\tilde{\Phi}$ and $\mathbf{P}$. To facilitate that, define $\tilde{Z}$ as the matrix Z by deleting its last column. So $\tilde{Z}$ is a $(M-1) \times (M-1)$ square matrix of full rank. Similarly, let $\tilde{S}$ be a $(M-1) \times (D+1)$ matrix obtained by deleting the first and last column and the last row of $\mathbf{S}$. Since the matrix $\tilde{Z}$ is invertible, we can now transform between the variable $\mathbf{P}$ and variable $\tilde{\Phi}$. Specifically, $\mathbf{P}$ can be determined in terms of $\tilde{\Phi}$ from the relation

$$\mathbf{P} = (\tilde{Z}^T)^{-1}(\mathrm{Jupp}(\tilde{\Phi}) - \mathrm{Jupp}(\tilde{\mathbf{S}}))$$

After the transformations on the parameters, we now have three parameters β, $\mathbf{P}$ and σ^2 to estimate in our model. Since our samples are chosen randomly and the effects come from some random variables, it is reasonable to model these parameters as random effects. Accordingly, we propose to estimate them from a Bayesian perspective and will use Monte Carlo Markov Chain methods to realize the values of the parameters.

11.2.2. *MCMC Algorithm*

Two of the most commonly used classes of MCMC algorithms are the Gibbs Sampler and Metropolis Hastings algorithm (e.g., sections 10 and 7 of Robert and Casella, 2004). The Gibbs sampler can be used when the posterior distribution of the parameter of interest is of some specific form we know such as a normal distribution. When this is not the case the Metropolis-Hastings algorithm comes into play. The Metropolis-Hastings algorithm is used to generate a Markov chain that uses an acceptance/rejection rule to converge to the specified target distribution. It can be shown that the chain generated by the Metropolis-Hastings algorithm is a Markov chain and converges to the stationary distribution.

We will also apply the Langevin-Hastings Hybrid algorithm in our work. The Langevin-Hastings algorithm (Møller and Waagepetersen, 2004) revises the Metropolis-Hastings algorithm using a truncated proposal density $Q(\gamma|\gamma^*)$ for γ a parameter vector. Usually the truncated proposal density is chosen to be a normal distribution wherein

$$Q(\gamma^*|\gamma) \sim \mathcal{N}\left(\gamma + \frac{\delta}{2} \bigtriangledown (\gamma)^{trunc}, \delta\mathbf{I}\right),$$

where δ is selected to make the acceptance ratio within the range of 0.20 and 0.60 and

$$\bigtriangledown(\gamma) = \frac{\partial}{\partial\gamma}\log(f(\gamma|\mathbf{x}))$$

with $f(\cdot)$ the likelihood function. The Langevin-Hastings Hybrid algorithm is a truncated version of the Langevin-Hastings algorithm obtained by replacing $\bigtriangledown(\gamma)$ in the proposal distribution by

$$\bigtriangledown(\gamma)^{trunc} = -\gamma + \max(H, \frac{\partial}{\partial\gamma}\log(l(\mathbf{x}|\gamma))$$

with $H > 0$ a user-specified parameter (Møller and Waagepetersen, 2004, p. 192). The Langevin-Hastings Hybrid algorithm has an advantage in terms of its convergence performance. The posterior samples drawn according to the Langevin-Hastings Hybrid algorithm converge faster than those drawn according to the Metropolis-Hastings algorithm.

11.2.3. *MCMC Estimation of Model Parameters*

We will now apply both the Gibbs sampler and the Langevin-Hastings Hybrid algorithm to generate a sequence of samples from the joint posterior distribution of the random variables β, σ^2, $\mathbf{a_{sh}}$, $\mathbf{a_{sc}}$ and $\mathbf{P}$ that appear in our model. The first step in the process is to derive a likelihood function for the parameters we are interested in and put priors on each parameter in order to find the posterior distribution. Thus, let $\Theta = (\beta^T, \sigma^2, \mathbf{a_{sh}}^T, \mathbf{a_{sc}}^T, vec(\mathbf{P})^T, \Sigma_{vec(\mathbf{P})})^T$, where $vec(\mathbf{P})$ stacks the columns of $\mathbf{P}^T$ in a single vector form. We then consider the likelihood of our model to be that of a multivariate normal having the form

$$l(\mathbf{x}|\Theta) = \frac{1}{(2\pi)^{N/2}\sigma^{N/2}}\exp\left\{-\frac{1}{2\sigma^2}\sum_{i=1}^{M}\sum_{j=1}^{n_i}(x_{ij} - a_{sh_i} - a_{sc_i}\mathbf{g}_{ij}^T\beta)^2\right\},$$

with $N = \sum_{i=1}^{M} n_i$ the total number of observations.

We choose an improper prior for the coefficient vector β with the prior density $p(\beta)$ proportional to the uniform function. The posterior distribution for β is then

$$p(\beta|\mathbf{x}) \propto \exp\left\{-\frac{1}{2}(\beta - \boldsymbol{\mu}_\beta)^T\Sigma_\beta^{-1}(\beta - \boldsymbol{\mu}_\beta)\right\}$$

with

$$\boldsymbol{\mu}_\beta = \frac{1}{\sigma^2}\Sigma_\beta\sum_{i=1}^{M}\sum_{j=1}^{n_i}(x_{ij}a_{sc_i}\mathbf{g}_{ij}^T - a_{sh_i}a_{sc_i}\mathbf{g}_{ij}^T)^T,$$

$$\Sigma_\beta = \sigma^2\left[\sum_{i=1}^{M}\sum_{j=1}^{n_i}(a_{sc_i}^2\mathbf{g}_{ij}^T\mathbf{g}_{ij})\right]^{-1}.$$

We next choose the prior for σ^2 to be an improper prior proportional to $1/\sigma^2$. The posterior distribution of σ^2 is then found to be that of an inverse gamma with parameters

$$\alpha_{\sigma^2} = \frac{N}{2},$$

and

$$\beta_{\sigma^2} = \frac{1}{2}\sum_{i=1}^{M}\sum_{j=1}^{n_i}(x_{ij} - a_{sh_i} - a_{sc_i}\mathbf{g}_{ij}^T\boldsymbol{\beta})^2 + \frac{vec(\mathbf{P})^T\Sigma_{vec(\mathbf{P})}^{-1}vec(\mathbf{P})}{2}.$$

In the appendix we show that our prior specifications leads to a proper joint distributions for all the model parameters.

We model the shift effects as being independent with a_{sh_i} having a $\mathcal{N}(0,\sigma_{sh}^2)$ distribution for $i = 1,\ldots,M$, where σ_{sh}^2 is a hyper parameter. Then, the posterior distribution of a_{sh_i} follows a normal distribution with mean

$$\frac{\sigma_{sh}^2}{n_i\sigma_{sh}^2 + \sigma^2}\sum_{j=1}^{n_i}(x_{ij} - a_{sc_i}\mathbf{g}_{ij}^T\boldsymbol{\beta})$$

and variance

$$\sigma^2\sigma_{sh}^2/(n_i\sigma_{sh}^2 + \sigma^2).$$

The hyper parameter σ_{sh}^2 is given an inverse Gamma distribution with shape parameter $\alpha_{\sigma_{sh}^2} = 2$ and scale parameter $\beta_{\sigma_{sh}^2} = 0.5$. These choices for the shape and scale parameters make the hyper-prior distribution diffuse. The posterior distribution of σ_{sh}^2 is again an inverse Gamma with shape parameter $M/2 + 2$ and scale parameter $\sum_{i=1}^{M}a_{sh_i}^2/2 + 0.5$.

In our setting, the scale factors a_{sc_i} are all positive. To ensure this property, we will perform the transformation $a_{sc_i} = e^{\tilde{\alpha}_i}$, where $\tilde{\alpha}_i = \sigma_{sc}\alpha_i$ with σ_{sc}^2 being a hyper parameter and α_i being unconstrained random parameters. The hyper parameter σ_{sc}^2 is given an inverse Gamma distribution with shape parameter $\alpha_{\sigma_{sc}^2} = 2$ and scale parameter $\beta_{\sigma_{sc}^2} = 0.5$. The posterior distribution of σ_{sc}^2 is again an inverse Gamma with shape parameter $M/2 + 2$ and scale parameter $\sum_{i=1}^{M}\tilde{\alpha}_i^2/2 + 0.5$.

If we now take $\boldsymbol{\alpha} = (\alpha_1,\ldots,\alpha_M)^T$, the prior distribution of $\boldsymbol{\alpha}$, or $\pi(\boldsymbol{\alpha})$, is chosen to be $\mathcal{N}(\mathbf{0},\mathbf{I})$ with $\mathbf{I}$ being the $M \times M$ identity matrix. Because the posterior distribution of $\boldsymbol{\alpha}$ is not of a form that allows us to easily generate random ordinates, we will apply the Langevin-Hastings Hybrid algorithm to update the posterior samples of $\boldsymbol{\alpha}$. Thus, let

$$\nabla(\boldsymbol{\alpha}) = \frac{\partial}{\partial\boldsymbol{\alpha}}\log[\pi(\boldsymbol{\alpha})l(\mathbf{x}|\boldsymbol{\alpha})]$$

$$= -\boldsymbol{\alpha} + \frac{1}{\sigma^2}\left\{\sum_{j=1}^{n_i}(x_{ij} - a_{sh_i} - e^{\sigma_{sc}\alpha_i}\mathbf{g}_{ij}^T\boldsymbol{\beta})\sigma_{sc}e^{\sigma_{sc}\alpha_i}\mathbf{g}_{ij}^T\boldsymbol{\beta}\right\}_{i=1}^{M}, \quad (11.11)$$

where $\{f_i(\cdot)\}_{i=1}^M$ denotes a vector of size M with ith element $f_i(\cdot)$. The proposal density for $\boldsymbol{\alpha}^*$ is then given by

$$p(\boldsymbol{\alpha}^*|\boldsymbol{\alpha}) = \mathcal{N}\left(\boldsymbol{\alpha} + \frac{\delta_1}{2}\nabla(\boldsymbol{\alpha})^{trunc}, \delta_1 \mathbf{I}\right) \qquad (11.12)$$

The updated Langevin-Hastings Hybrid algorithm for $\boldsymbol{\alpha}$ works as follows.

(1) Generate $\boldsymbol{\alpha}^*$ according to (11.12), where δ_1 is chosen to make the acceptance ratio within the range of 0.20 to 0.60.
(2) Calculate

$$a_1 = \frac{\pi(\boldsymbol{\alpha}^*)}{\pi(\boldsymbol{\alpha})}\frac{l(\mathbf{x}|\boldsymbol{\alpha}^*)}{l(\mathbf{x}|\boldsymbol{\alpha})}\frac{p(\boldsymbol{\alpha}|\boldsymbol{\alpha}^*)}{p(\boldsymbol{\alpha}^*|\boldsymbol{\alpha})}, \qquad (11.13)$$

where

$$\frac{\pi(\boldsymbol{\alpha}^*)}{\pi(\boldsymbol{\alpha})} = \exp\left\{-\frac{\boldsymbol{\alpha}^{*T}\boldsymbol{\alpha}^*}{2} + \frac{\boldsymbol{\alpha}^T\boldsymbol{\alpha}}{2}\right\}, \qquad (11.14)$$

$$\frac{l(\mathbf{x}|\boldsymbol{\alpha}^*)}{l(\mathbf{x}|\boldsymbol{\alpha})}$$
$$= \exp\Big\{-\frac{1}{2\sigma^2}[2(\text{diag}(\mathbf{1}_{n_1}\ldots,\mathbf{1}_{n_M})\mathbf{a}_{sh})^T\text{diag}(\mathbf{g}^T\boldsymbol{\beta})$$
$$\times (e^{\sigma_{sc}\boldsymbol{\alpha}^*} - e^{\sigma_{sc}\boldsymbol{\alpha}}) - 2\mathbf{x}^T\text{diag}(\mathbf{g}^T\boldsymbol{\beta})(e^{\sigma_{sc}\boldsymbol{\alpha}^*} - e^{\sigma_{sc}\boldsymbol{\alpha}})$$
$$+ e^{\sigma_{sc}\boldsymbol{\alpha}^{*T}}(\text{diag}(\mathbf{g}^T\boldsymbol{\beta}))^T\text{diag}(\mathbf{g}^T\boldsymbol{\beta})e^{\sigma_{sc}\boldsymbol{\alpha}^*}$$
$$- e^{\sigma_{sc}\boldsymbol{\alpha}^T}(\text{diag}(\mathbf{g}^T\boldsymbol{\beta}))^T\text{diag}(\mathbf{g}^T\boldsymbol{\beta})e^{\sigma_{sc}\boldsymbol{\alpha}}]\Big\}, \qquad (11.15)$$

and

$$\frac{p(\boldsymbol{\alpha}|\boldsymbol{\alpha}^*)}{p(\boldsymbol{\alpha}^*|\boldsymbol{\alpha})}$$
$$= \exp\Big\{-\frac{1}{2\delta_1}[(\boldsymbol{\alpha} - \boldsymbol{\alpha}^* - \frac{1}{2}\delta_1\nabla(\boldsymbol{\alpha}^*))^T(\boldsymbol{\alpha} - \boldsymbol{\alpha}^* - \frac{1}{2}\delta_1\nabla(\boldsymbol{\alpha}^*))$$
$$- (\boldsymbol{\alpha}^* - \boldsymbol{\alpha} - \frac{1}{2}\delta_1\nabla(\boldsymbol{\alpha}))^T(\boldsymbol{\alpha}^* - \boldsymbol{\alpha} - \frac{1}{2}\delta_1\nabla(\boldsymbol{\alpha}))]\Big\}. \qquad (11.16)$$

(3) For $t = 1, 2, \ldots, n_{iter}$, update $\boldsymbol{\alpha}^{t+1} = \boldsymbol{\alpha}^*$ if a_1 in step 2 is greater than 1. Otherwise, generate a sample u from the uniform distribution on $[0, 1]$. Compare the value of u with that of a_1. If $u < a_1$, then let $\boldsymbol{\alpha}^{t+1} = \boldsymbol{\alpha}^*$. Otherwise, keep the value of $\boldsymbol{\alpha}^t$: that is, $\boldsymbol{\alpha}^{t+1} = \boldsymbol{\alpha}^t$.

We also choose the prior for the random effect $vec(\mathbf{P})$ to be that of a normal density with mean $\mathbf{0}$ and variance covariance matrix $\sigma^2 \Sigma_{vec(\mathbf{P})}$. The hyper-parameter $\Sigma_{vec(\mathbf{P})}$ has an inverse Wishart distribution with degrees of freedom ν and scale matrix Ψ, which is obtained by deleting the first and last columns and the first and last rows of $\tilde{\Psi}$, with

$$\tilde{\Psi} = \lambda_0 \left\{ \int_0^1 \mathbf{U}''(t)\mathbf{U}''(t)^T dt \right\}, \tag{11.17}$$

where $\mathbf{U}''$ is the second derivative of the vector $\mathbf{U}$ in equation (11.4) and λ_0 is some user specified small value. The posterior distribution of $\Sigma_{vec(\mathbf{P})}$ is given by an inverse Wishart distribution with parameters $\nu + (M-1)(D+1) + 1$ and scale matrix $\tilde{\tilde{\Psi}} = (1/\sigma^2)vec(\mathbf{P})vec(\mathbf{P})^T + \Psi$.

The posterior distribution of $vec(\mathbf{P})$ does not have a standard form. As a result we will apply the Langevin-Hasting's algorithm to update $vec(\mathbf{P})$ in the MCMC iteration.

To accomplish this, first write $vec(\mathbf{P}) = \sigma \Sigma_{vec(\mathbf{P})}^{1/2} \boldsymbol{\gamma}$, so that $\boldsymbol{\gamma}$ is a $(M-1)(D+1) \times 1$ vector, which has a prior distribution that is multivariate normal with mean $\mathbf{0}$ and identity variance matrix. Then, take the proposal distribution of $\boldsymbol{\gamma}^*$ to be $p(\boldsymbol{\gamma}^*|\boldsymbol{\gamma}) = \mathcal{N}\left(\boldsymbol{\gamma} + \frac{\delta_2}{2}\nabla(\boldsymbol{\gamma})^{\text{trunc}}, \delta_2\mathbf{I}\right)$, with

$$\nabla(\boldsymbol{\gamma}) = \frac{\partial}{\partial\boldsymbol{\gamma}}\log(f(\mathbf{x}|\boldsymbol{\gamma}))$$

$$= \frac{d}{d\boldsymbol{\gamma}}\log\pi(\boldsymbol{\gamma})\frac{d}{d\boldsymbol{\gamma}}\log l(\mathbf{x}|\boldsymbol{\gamma})$$

$$= -\boldsymbol{\gamma} + \tilde{\nabla}(\boldsymbol{\gamma}), \tag{11.18}$$

where $\pi(\boldsymbol{\gamma})$ and $l(\mathbf{x}|\boldsymbol{\gamma})$ are, respectively, the prior for $\boldsymbol{\gamma}$ and the likelihood of the model with respect to $\boldsymbol{\gamma}$. We write $\frac{d}{d\boldsymbol{\gamma}}\log l(\mathbf{x}|\boldsymbol{\gamma})$ as $\tilde{\nabla}(\boldsymbol{\gamma})$ in equation (11.18) and observe that it can be calculated through the relation

$$\tilde{\nabla}(\boldsymbol{\gamma}) = \sigma \Sigma_{vec(\mathbf{P})}^{1/2} \frac{d}{d(vec(\mathbf{P}))}\log l(\mathbf{x}|vec(\mathbf{P}))$$

$$= \sigma \Sigma_{vec(\mathbf{P})}^{1/2}|J|^{-1}\tilde{\nabla}(vec(\tilde{\Phi})) \tag{11.19}$$

where $vec(\tilde{\Phi})$ stacks the columns of $\tilde{\Phi}$ into a single vector and J is the Jacobian matrix.

Let us take a further look at the Jacobian matrix. By the nature of the Jupp transformation, J is a $(M-1) \times (M-1)$ block tri-diagonal matrix where a particular block J_i has element $(J_i)_{lk}$, $l, k = 1, \ldots, D+1$ in its lth row and kth column with

$$(J_i)_{lk} = \begin{cases} \dfrac{\phi_{il}-\phi_{i(l+2)}}{(\phi_{i(l+2)}-\phi_{i(l+1)})(\phi_{i(l+1)}-\phi_{il})}, & \text{if } l = k, \\ \dfrac{1}{\phi_{i(l+2)}-\phi_{i(l+1)}}, & \text{if } k = l + 1, \\ 0, & \text{if } k > l + 1, \\ (J_i)_{kl}, & \text{if } l > k. \end{cases} \quad (11.20)$$

The quantity $\tilde{\nabla}(vec(\tilde{\Phi}))$ is calculated via the relation

$$\begin{aligned}\tilde{\nabla}(vec(\tilde{\Phi})) &= \frac{\partial}{\partial(vec(\tilde{\Phi}))}(\log(\mathbf{x}|vec(\tilde{\Phi}))) \\ &= \frac{\partial}{\partial(vec(\tilde{\Phi}))}\left[\frac{-\sum_{i=1}^{M}\sum_{j=1}^{n_i}(x_{ij}-a_{sh_i}-a_{sc_i}\mathbf{g}_{ij}^T\boldsymbol{\beta})^2}{2\sigma^2}\right] \\ &= \frac{\sum_{j=1}^{n_i}(x_{ij}-a_{sh_i}-a_{sc_i}\mathbf{g}_{ij}^T\boldsymbol{\beta})a_{sc_i}(\mathbf{g}_{ij}^T)'\boldsymbol{\beta}}{\sigma^2}\end{aligned} \quad (11.21)$$

with $\mathbf{g}_{ij}^T$ a function of $vec(\tilde{\Phi})$ through equation (11.6) and $(\mathbf{g}_{ij}^T)'$ being the partial derivative of $\mathbf{g}_{ij}^T$ with respect to $vec(\tilde{\Phi})$.

The updated Langevin-Hastings Hybrid algorithm for $\boldsymbol{\gamma}$ now works as follows:

(1) Generate $\boldsymbol{\gamma}^*$ according to $\mathcal{N}(\boldsymbol{\gamma}+\frac{\delta_2}{2}\nabla(\boldsymbol{\gamma})^{\text{trunc}}, \delta_2\mathbf{I})$, where δ_2 is chosen to make the acceptance ratio within the range of 0.20 and 0.60.
(2) Calculate

$$a_2 = \frac{\pi(\boldsymbol{\gamma}^*)}{\pi(\boldsymbol{\gamma})}\frac{l(\mathbf{x}|\boldsymbol{\gamma}^*)}{l(\mathbf{x}|\boldsymbol{\gamma})}\frac{p(\boldsymbol{\gamma}|\boldsymbol{\gamma}^*)}{p(\boldsymbol{\gamma}^*|\boldsymbol{\gamma})}, \quad (11.22)$$

where

$$\frac{\pi(\boldsymbol{\gamma}^*)}{\pi(\boldsymbol{\gamma})} = \exp\left\{-\frac{\boldsymbol{\gamma}^{*T}\boldsymbol{\gamma}^*}{2}+\frac{\boldsymbol{\gamma}^T\boldsymbol{\gamma}}{2}\right\}, \quad (11.23)$$

$$\begin{aligned}\frac{l(\mathbf{x}|\boldsymbol{\gamma}^*)}{l(\mathbf{x}|\boldsymbol{\gamma})} = \exp\Big\{&-\frac{1}{2\sigma^2}\{\sum_{i=1}^{M}\sum_{j=1}^{n_i}[-2x_{ij}a_{sc_i}(\mathbf{g}_{ij}^{*T}-\mathbf{g}_{ij}^T)\boldsymbol{\beta} \\ &+2a_{sh_i}a_{sc_i}(\mathbf{g}_{ij}^{*T}-\mathbf{g}_{ij}^T)\boldsymbol{\beta}+a_{sc_i}^2\boldsymbol{\beta}^T(\mathbf{g}_{ij}^*\mathbf{g}_{ij}^{*T}-\mathbf{g}_{ij}\mathbf{g}_{ij}^T)\boldsymbol{\beta}]\}\Big\},\end{aligned} \quad (11.24)$$

and

$$\begin{aligned}\frac{p(\boldsymbol{\gamma}|\boldsymbol{\gamma}^*)}{p(\boldsymbol{\gamma}^*|\boldsymbol{\gamma})} = \exp\Big\{&-\frac{1}{2\delta_2}[(\boldsymbol{\gamma}-\boldsymbol{\gamma}^*-\frac{1}{2}\delta_2\nabla(\boldsymbol{\gamma}^*))^T(\boldsymbol{\gamma}-\boldsymbol{\gamma}^*-\frac{1}{2}\delta_2\nabla(\boldsymbol{\gamma}^*)) \\ &-(\boldsymbol{\gamma}^*-\boldsymbol{\gamma}-\frac{1}{2}\delta_2\nabla(\boldsymbol{\gamma}))^T(\boldsymbol{\gamma}^*-\boldsymbol{\gamma}-\frac{1}{2}\delta_2\nabla(\boldsymbol{\gamma}))]\Big\}.\end{aligned} \quad (11.25)$$

(3) For $t = 1, 2, \ldots, n_{iter}$, update $\gamma^{t+1} = \gamma^*$ if a_2 in step 2 is greater than 1. Otherwise, generate a sample u from the uniform distribution on $[0, 1]$. Compare the value of u with that of a_2. If $u < a_2$, then let $\gamma^{t+1} = \gamma^*$. Otherwise, keep the value of γ^t: that is, $\gamma^{t+1} = \gamma^t$.

Combining all of our estimation steps, we can now draw a sequence of samples for

$$\Theta = (\boldsymbol{\beta}^T, \sigma^2, \mathbf{a_{sh}}^T, \sigma_{sh}^2, \mathbf{a_{sc}}^T, \sigma_{sc}^2, vec(\mathbf{P})^T, \Sigma_{vec(\mathbf{P})})^T.$$

The components of $\boldsymbol{\beta}$, σ^2, $\mathbf{a_{sh}}$, σ_{sh}^2, σ_{sc}^2, $\Sigma_{vec(\mathbf{P})}$ are updated by a Gibbs sampler algorithm, while the component of $\mathbf{a_{sc}}$ and $vec(\mathbf{P})$ are updated by the Langevin-Hastings Hybrid algorithm. The procedure is as follows.

(1) Assign starting values for

$$\Theta^{(0)} = (\boldsymbol{\beta}^{(0)T}, \sigma^{2(0)}, \mathbf{a_{sh}}^{(0)T}, \sigma_{sh}^{2(0)}, \mathbf{a_{sc}}^{(0)T}, \sigma_{sc}^{2(0)}, vec(\mathbf{P})^{(0)T}, \Sigma_{vec(\mathbf{P})}^{(0)})^T. \tag{11.26}$$

(2) Find the corresponding $\Phi^{(0)}$ and $\mathbf{g}_{ij}^{(0)T}$.
(3) For $t = 1, 2, \ldots, n_{iter}$

Step 1 : Generate $\boldsymbol{\beta}^{(t)}$ according to $\mathcal{N}(\boldsymbol{\mu}_\beta, \Sigma_\beta)$.
    Update $\boldsymbol{\beta}$ to $\boldsymbol{\beta}^{(t)}$, then generate $\sigma^{2(t)}$ from a $InvGamma(\alpha_{\sigma^{2(t-1)}}, \beta_{\sigma^{2(t-1)}})$ distribution.
Step 2 : Update $\mathbf{a_{sh}}$ to $\mathbf{a}_{sh}^{(t)}$ and update σ_{sh}^2 to $\sigma_{sh}^{2(t)}$. Then, update $\boldsymbol{\alpha}$ according to the Langevin-Hastings Hybrid algorithm and update σ_{sc}^2 to $\sigma_{sc}^{2(t)}$.
Step 3 : Find $\mathbf{a}_{sc}$ according to $\mathbf{a}_{sc} = e^{\sigma_{sc}}\boldsymbol{\alpha}$.
Step 4 : Update γ according to the Langevin-Hastings Hybrid algorithm. Find $vec(\mathbf{P})$ according to $vec(\mathbf{P}) = \sigma\Sigma_{vec(\mathbf{P})}^{1/2}\gamma$ and update $\mathbf{P}$ accordingly. Calculate Φ by $\Phi = \text{Jupp}^{-1}([\mathbf{0} \quad \mathbf{Z}^T\mathbf{P} \quad \mathbf{0}] + \text{Jupp}(\mathbf{S}))$ and then update $\mathbf{g}$ using the new values in Φ.
Step 5 : Update the hyper prior $\Sigma_{vec(\mathbf{P})}$ to its posterior distribution using the new values in $vec(\mathbf{P})$ and return to Step 1.

(4) Iterate until all posterior distributions converge.

To conclude let us return to the Bayesian curve registration paper by Telesca and Inoue (2008) that was mentioned in the introduction. Similar to our approach, they employ a common shape function and model this as well as their warping functions as linear combinations of B-spline basis functions. However, they use a random walk procedure to draw posterior samples of the ϕ_i which is known to converge slowly if the proposal density is not chosen to be close to the true distribution. As a result, our use of the Langevin-Hastings Hybrid algorithm can be expected to produce faster convergence performance. Other differences between our methodology and that of Telesca/Inoue include how the hyperparameters are modeled. In particular, we allow the variance covariance matrix of $\mathbf{P}$ to be chosen as a hyper-parameter with an inverse Wishart prior distribution to allow for more flexibility in the estimation process. Perhaps the most significant difference between the Telesca/Inoue formulation and ours concerns the way the ϕ_i are modelled. Telesca and Inoue (2008) employ a multivariate normal prior for the ϕ_i which has the consequence of not ensuring monotonicity of the warping functions. By using the Jupp transformation (Jupp, 1978) as we have done here it is possible to avoid such difficulties.

11.3. Empirical Study

In this section, we examine the performance of our registration methodology in a small empirical experiment. In this simulation study we will obtain Bayesian estimators for the parameters in our model and thereby obtain registered curves that can be employed in a cross-sectional sample mean for estimation of the process mean. After that, we will compare the performance of our methodology with results obtained using the Ramsay and Li (1996) approach and with the unregistered cross-sectional mean. At the end of this section, we will discuss the sensitivity of our results to the choice of the prior distributions.

All the computations in this section were carried out in R. Original source code was created for our Bayesian algorithm while the Ramsay and Li (1996) methodology was implemented using the register.fd function from the R fda package.

The design of the simulation is motivated by the empirical work in Gervini and Gasser (2004). In particular, we use a sine function as the prototype process mean in our simulation which has the consequence of creating problems for estimation of peaks and valleys of a curve.

Let the true mean function be

$$\mu(t) = sin(2\pi t), \qquad (11.27)$$

for $t \in [0,1]$. This function has a peak at $\tau_{01} = 0.25$ and a valley at $\tau_{02} = 0.75$, with maximum value 1 and minimum value -1. For each sample curve, we then choose n uniformly spaced input values $t_j = (j-1)/(n-1)$, $j = 1, \ldots, n$, and generate data as follows.

(1) We first generate the location for the peak and valley for each curve randomly. Specifically, for curve i, let τ_{i1}, τ_{i2} be the locations for the peak and valley, respectively. We then take $\tau_{ik} = \tau_{01} + \xi_{ik}$, $k = 1, 2$ with $\xi_{ik} = \min\{\max\{V_{ik}/12, -0.24\}, 0.24\}$ and the V_{ik} being generated independently from a $\mathcal{N}(0, 1)$ distribution. The truncation employed to produce the ξ_{ik} is to ensure that $\tau_{i1} < \tau_{i2}$ and τ_{i1} and τ_{i2} are within $[0, 1]$.

(2) Secondly, we choose the warping function $h_i(t)$ to be piecewise linear functions with $h_i(0) = 0$, $h_i(\tau_{01}) = \tau_{i1}$, $h_i(\tau_{02}) = \tau_{i2}$ and $h_i(1) = 1$.

(3) We next generate the responses as $x_{ij} = a_i + b_i \mu(h_i^{-1}(t_{ij})) + \epsilon_{ij}$, where a_i and b_i are independent and identically distributed as $\mathcal{N}(0, 0.1(1 - 1/\sqrt{2}))$ and $\mathcal{N}(1, 0.1)$ random variables, respectively and the ϵ_{ij} are random errors following a $\mathcal{N}(0, 0.1)$ distribution. These choices of the variances of a_i, b_i and ϵ_{ij} make the error-to-model ratio 0.5.

(4) We generate $M = 10$ sample curves with each curve being sampled at either $n = 20$ or 50 points to produce the actual data. Thirty replications for each combination were generated.

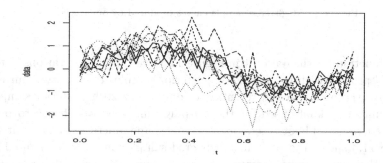

Fig. 11.1.   10 simulated sample curves.

A specific data set with 10 curves each sampled at 50 points is shown in Figure 11.1. The curves in the figure include noise. Figure 11.2 displays smoothed versions of the curves in Figure 11.1. We choose a cubic B-spline basis with 18 knots to smooth the curves. In order to estimate the

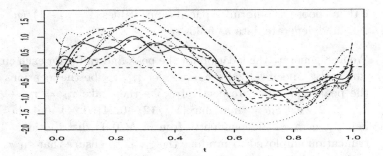

Fig. 11.2.   10 smoothed sample curves.

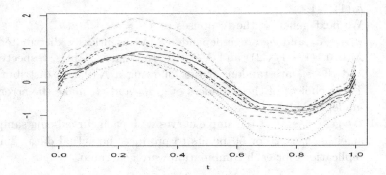

Fig. 11.3.   Registered sample curves produced by our method.

mean function for the data set in Figure 11.1, we ran 6500 iterations of the MCMC process with the first 2500 posterior samples representing burn-in and with thinning through retention of only every 20th posterior sample. The number of knots used for the warping functions was chosen to be 3 and the number of knots used for the global shape function μ was set at 18. Figure 11.3 shows the resulting registered sample curves and Figure 11.4 shows the corresponding registered mean curve together with associated 95% credible intervals. It clearly shows that our registered cross-sectional mean curve has a peak at approximately 0.25, with maximum around 1. In addition, the value of the true mean function falls within the 95% credible intervals. The curve in dotted lines in Figure 11.4 is the cross-sectional sample mean of the smoothed sample curves. Our registered mean can be seen to do a better job of estimating the timing of the valley. Figure 11.5 shows

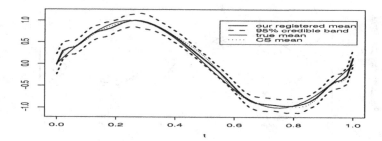

Fig. 11.4. Our registered mean curve (in bold solid lines) with 95% credible intervals (in bold dashed lines), true mean function (in solid lines) and the cross-sectional (CS) sample mean (in dotted lines).

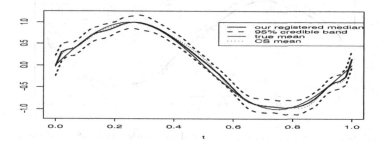

Fig. 11.5. Our registered median curve (in bold solid lines) with 95% credible intervals (in bold dashed lines), true mean function (in solid lines) and the cross-sectional (CS) sample mean (in dotted lines).

the registered posterior median with 95% credible intervals. Comparison with Figure 11.4 suggests that the results are similar to those obtained using the registered posterior mean. The curve in dotted lines in Figure 11.4 is the cross-sectional sample mean of the smoothed sample curves. Our registered mean can be seen to do a better job of estimating the timing of the valley.

Figure 11.5 shows the registered posterior median with 95% credible intervals. Comparison with Figure 11.4 suggests that the results are similar to those obtained using the registered posterior mean. We also compared the performance of our method for the data in Figure 11.1 to results obtained from the Ramsay and Li (1996) continuous monotone registration method. Their method was applied twice in this example. First we chose the cross-sectional mean function as the target function and used their technique with smoothing parameter 0.005 to get an initial set of registered curves. Then

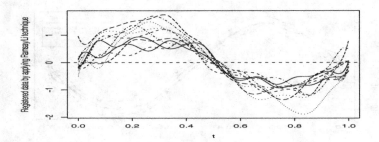

Fig. 11.6.   Registered sample curves produced by the Ramsay-Li method.

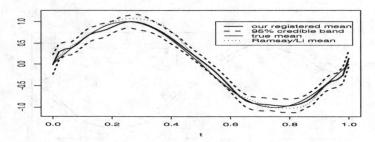

Fig. 11.7.   Our registered mean curve (in bold solid lines) with 95% credible intervals
(in bold dashed lines), true mean function (in solid lines) and the Ramsay-Li registered
cross-sectional mean (in dotted lines).

we computed a new cross-sectional mean using these registered curves and
treated the new cross-sectional mean as the target function for a second
application of their technique.

Figure 11.6 shows the registered sample curves produced by the Ramsay-
Li method while Figures 11.7 and 11.8 show the cross-sectional mean obtained through the Ramsay-Li method in dotted lines. Their approach
appears to be over-estimating the peak and under-estimating the valley of
the true regression curve. There is also a bias present in estimating the
timing of the valley that is visually evident in the plot. One could choose
to apply the above process for a third time or, more generally, attempt to
iterate to some type of convergence. But, in practice our experience has
been that there is not much improvement to be gained by doing this. For
this particular data set, we obtained the posterior mean and median of
the error variance σ^2 to be 0.2 with corresponding 95% credible interval
$(0.17, 0.25)$. This over-estimates the true value of 0.1. We also investigated

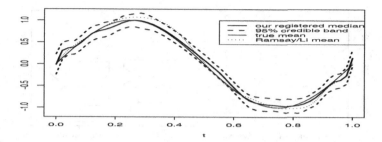

Fig. 11.8. Our registered median curve (in bold solid lines) with 95% credible intervals (in bold dashed lines), true mean function (in solid lines) and the Ramsay-Li registered cross-sectional mean (in dotted lines).

the hyper parameters σ_{sh}^2 and σ_{sc}^2 and found that their true values fall in their associated 95% credible intervals. We used root average squared error ($RASE$) as a measure of performance for our mean function estimator where

$$RASE(\hat{\mu}) = \left[\sum_{j=1}^{n} \{\hat{\mu}(t_j) - \mu(t_j)\}^2 / n \right]^{1/2} . \tag{11.28}$$

Figure 11.9 summarizes the $RASE$ produced by four estimated mean curves: our posterior mean, our posterior median, the sample cross-sectional mean and the registered cross-sectional mean obtained through the Ramsay-Li method for sampling at $n = 20$ points per curve. The median of $RASE$ for our method is smaller than the other two.

Figure 11.10 provides the same information for $n = 50$ sample points per curve.

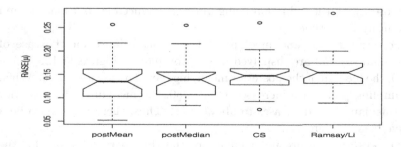

Fig. 11.9. Simulated root average squared errors of $\hat{\mu}$ for $n = 20$.

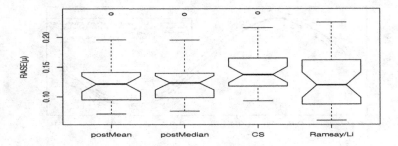

Fig. 11.10.   Simulated root average squared errors of $\hat{\mu}$ for $n = 50$.

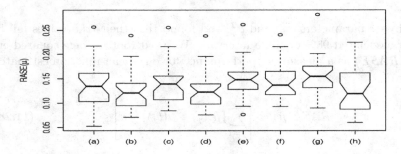

Fig. 11.11.   Simulated root average squared errors of $\hat{\mu}$: (a) Our posterior mean for $n = 20$, (b) our posterior mean for $n = 50$, (c) our posterior median for $n = 20$, (d) our posterior median for $n = 50$, (e) cross-sectional mean for $n = 20$, (f) cross-sectional mean for $n = 50$, (g) Ramsay-Li cross-sectional mean for $n = 20$, (h) Ramsay-Li cross-sectional mean for $n = 50$.

We also compare the combined *RASE* results for the case of $n = 20$ and $n = 50$ in Figure 11.11. For all the four methods, *RASE* decreases as n increases, with the reduction being most pronounced for our method and the Ramsay-Li method.

The posterior means and medians of the error variance for the case of $n = 20$ and $n = 50$ are displayed in the box plot in Figure 11.12. When $n = 20$, the median of the posterior mean/median of σ^2 is about 0.16. When the sampling points are increased to $n = 50$, the median of the posterior mean/median of σ^2 decreases to about 0.14. Thus, there appears to be a positive bias in estimation of σ^2.

The posterior means and medians of the shift parameter $\mathbf{a}_{sh}$ for the $n = 20$ and $n = 50$ cases are shown in Figure 11.13.

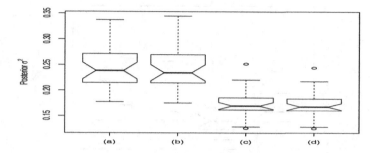

Fig. 11.12. Box plots of estimated σ^2. (a) Posterior means of σ^2 for $n = 20$, (b) posterior medians of σ^2 for $n = 20$, (c) posterior means of σ^2 for $n = 50$, and (d) posterior medians of σ^2 for $n = 50$. The true value of σ^2 is 0.1.

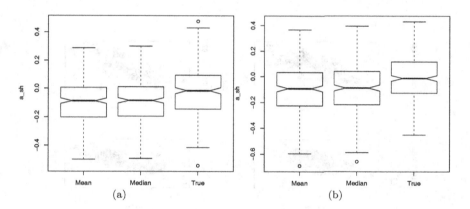

Fig. 11.13.   Box plots of (a) a_{sh} for $n = 20$ and (b) $n = 50$.

The median of the posterior means of $\mathbf{a}_{sh}$ is lower than 0 and the variability is about the same as that of the true values of the shift parameter, as indicated by the notches in Figure 11.13. We looked at the 95% credible intervals for $\mathbf{a}_{sh}$ and found that about 98% of them contain the true values. The posterior mean and median for the hyper parameter σ^2_{sh} are shown in Figure 11.14.

About 90% of the 95% credible intervals of σ^2_{sh} contain the true value 0.03. We plotted the true shift parameter values versus the estimated values in Figure 11.15, which shows linear trends with slopes around 0.6 for $n = 20$ and around 0.8 for $n = 50$. Both the slopes and correlation coefficients increase as the number of sampling points increases. Estimation results

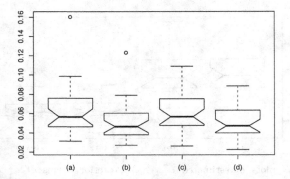

Fig. 11.14. Box plots of estimated σ^2_{sh}. (a) Posterior means of σ^2_{sh} for $n = 20$, (b) posterior medians of σ^2_{sh} for $n = 20$, (c) posterior means of σ^2_{sh} for $n = 50$, and (d) posterior medians of σ^2_{sh} for $n = 50$. The true value of σ^2_{sh} is 0.03.

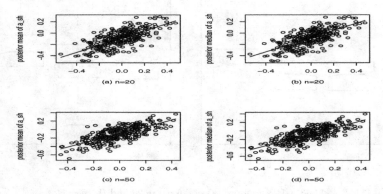

Fig. 11.15. Sample curve shifts versus estimated values. (a) Slope 0.636679 and correlation coefficient 0.6994137, (b) slope 0.628929 and correlation coefficient 0.7012147, (c) slope 0.799683 and correlation coefficient 0.738109 and (d) slope 0.795091 and correlation coefficient 0.7416735.

for the $\mathbf{a}_{sc}$ are summarized by the box plots of the posterior mean/median estimators in Figure 11.16.

The median of the posterior mean/median of $\mathbf{a}_{sc}$ is about 1, which is the true mean for the slopes that were used to generate the data. The box plot of the hyper parameter $\sigma_{sc}{}^2$ is shown in Figure 11.17. The posterior means/medians of the hyper parameter $\sigma_{sc}{}^2$ become closer to the true value 0.1 as the number of sampling points increase. In addition, all the 95% credible intervals include the true value 0.1.

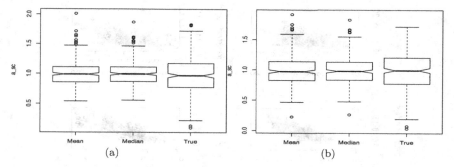

Fig. 11.16.   Box plots of a_{sc} for (a) $n = 20$ and (b) $n = 50$.

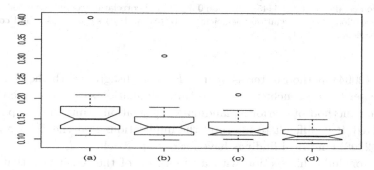

Fig. 11.17.   Box plots of estimated σ_{sc}^2.   (a) Posterior means of σ_{sc}^2 for $n = 20$, (b) posterior medians of σ_{sc}^2 for $n = 20$, (c) posterior means of σ_{sc}^2 for $n = 50$, and (d) posterior medians of σ_{sc}^2 for $n = 50$. The true value of σ_{sc}^2 is 0.1.

Figure 11.18 shows the scatter plots of the scale parameter values for the data and the corresponding estimated values. Again, both the slopes and the correlation coefficients increase as n increases.

Empirical work in Telesca and Inoue (2008) compares their approach with landmark registration and the Gervini and Gasser (2004) self-modeling warping function method under a simulation framework that is virtually identical to the one employed in this section. In combination with our comparisons with the Ramsay and Li (1996) approach, one could view the combined implication as suggesting superior performance may be obtained using a Bayesian approach (either ours or that of Telesca and Inoue (2008)) than from two of the more popular methods for curve registration. However, we are puzzled by the Telesca and Inoue (2008) result that reports superior empirical performance for landmark registration over the Gervini and

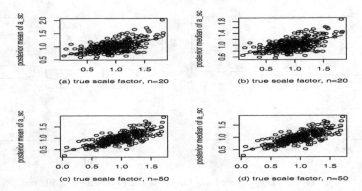

Fig. 11.18. Sample curve scale factors versus estimated values. (a) Slope 0.39190 and correlation coefficient 0.5694445, (b) slope 0.36656 and correlation coefficient 0.5669515, (c) slope 0.62638 and correlation coefficient 0.7475686 and (d) slope 0.58813 and correlation coefficient 0.7449118.

Gasser (2004) method in terms of $RASE$. This disagrees with the Gervini and Gasser (2004) conclusion for this same simulation scenario where they find their method outperforms landmark registration. Until this disparity is resolved, it is difficult to interpret the implications that the Telesca and Inoue (2008) empirical findings have for our approach.

We conclude this section with a discussion of the sensitivity that we observed from the choices of the prior distributions in our empirical work. First, we note that we have tried both proper and improper prior for σ^2 and β in our particular simulated data sets and found little difference in the estimation of the posterior mean and median of the parameters σ^2, β, a_{sh}, a_{sc}, σ^2_{sh} and σ^2_{sc}.

We tried proper prior distributions for σ^2 that were inverse Gamma with shape parameter 2 and scale parameters 1, 0.5 or 0.1. Proper prior distributions for β that were multivariate normal with identity variance matrix were used with mean vectors of **0** or **1**. These choices did not affect the performance of the registered curves and the registered mean curves. For the specific data set in Figure 11.5, all the proper priors and the improper prior for σ^2 and β give results with $RASE$ values all around 0.08.

For the hyper parameters σ^2_{sh} and σ^2_{sc}, we gave them hyper prior inverse gamma distribution with scale parameters of 1.5, 1 or 0.5 so that the resulting prior mean is 1.5, 1 or 0.5. These choices of the prior distribution do not influence the performance of the registered curves and the mean curve. The posterior mean and median of σ_{sh}^2 and σ_{sc}^2, however, increase as their prior mean increases. But the true values still fall in the 95% credible intervals.

11.4. Summary and Conclusions

Functional data analysis has been an active research area in recent years. As a result the curve registration problem in functional data has received much attention in the literature. In the case of time related data, a registered process mean function provides a very useful tool for understanding the timings and values of certain features that are of interest in a functional data analysis context. Most existing registration methods rely on the estimation of warping functions. One challenge here is that warping functions need to be modeled as monotone increasing function. To solve this problem, we employed monotone spline functions with coefficients produced from the Jupp transformation (as outlined in Jupp, 1978) whose inverse has the monotone increasing property.

In order to estimate the mean function, we framed the registration problem in a shape invariant model setting. This approach has the consequence that registered sample curves preserve the same shape but are stretched, shrunk or shifted in some way. We then developed a method to estimate the shape and amplitude from a Bayesian perspective. Our formulation treated the cross-sectional mean curves and warping functions as random realizations of certain distributions. We also treat the shifts and the amplitudes as random covariates in the model. Those parameters are estimated by taking samples through an MCMC algorithm. In the MCMC process, we incorporated the Gibbs sampler and the Langevin-Hastings Hybrid algorithms to obtain a fast converging path for the posterior samples. Limited empirical comparisons with the continuous monotone registration method of Ramsay and Li (1996) suggests that our method is competitive with others that are commonly used in this area while having the advantage of providing the ability to obtain credible intervals for the model parameters and process mean function.

One extension of our registration method concerns the way of choosing the number and locations of knots of the B-spline basis functions that are used to model both the process mean and warping functions. The current approach is ad hoc and does not allow for data fitting adaptation in terms of knot locations. One possible way to allow for knot location flexibility is to choose a large number of potential knots from which a subset of knots can be selected. The problem then turns into a variable selection problem that could be addressed using Bayesian variable selection techniques as in Smith and Kohn (1996). We intend to investigate this approach in future work.

Another aspect of our model we want to further examine is the choice of λ_0 in the hyper-prior $\Sigma_{vec(\mathbf{P})}$. In particular, we want to explore the ufe of cross-validation methods or (Bayesian) generalized maximum likelihood to adaptively estimate this parameter. It is also of interest to further examine choice of the prior distribution for $\boldsymbol{\beta}$. Specifically, we would like to assess the effect of choosing a prior distribution such as the one we chose for the hyper-prior $\Sigma_{vec(\mathbf{P})}$. An adaptive choice of the "smoothing parameter" could then be implemented here as well.

Acknowledgement

Majumdar's and Eubank's research was supported by grants from NSF. The authors would like to expresstheir their appreciation to Lyndia Brumback and Mary Lindstrom for making their code available to us.

11.5. Appendix

Since we have introduced the improper prior $\pi(\boldsymbol{\beta}, \sigma^2) = 1/\sigma^2$ into the model defined in Section 11.2, we would like to show that the joint distribution is proper. That is, we would like to prove that

$$\int p(\boldsymbol{\beta}, \sigma^2, \boldsymbol{\theta}; \mathbf{x}) d\boldsymbol{\beta} d\sigma^2 d\boldsymbol{\theta} d\mathbf{x} = 1, \qquad (11.29)$$

where $\boldsymbol{\theta}$ is the vector of all the other parameters used in the model, other than $(\boldsymbol{\beta}, \sigma^2)$. To undertake this proof, we state two important results from Gelman et al. (2004, p. 356).

Result 1: Suppose that $\mathbf{y}$ is a $N \times 1$ vector and $\boldsymbol{\beta}$ is a $(Q+4) \times 1$ vector such that

$$\mathbf{y}|\boldsymbol{\beta}, \sigma^2, \mathbf{X} \sim \mathcal{N}(\mathbf{X}\boldsymbol{\beta}, \sigma^2 \mathbf{I}_N)$$

with $\mathbf{I}_N$ being the identity matrix of dimension N and the prior distribution is given by

$$p(\boldsymbol{\beta}, \sigma^2|\mathbf{X}) = \frac{1}{\sigma^2}. \qquad (11.30)$$

Then, it follows that

$$\beta | \sigma^2, \mathbf{y}, \mathbf{X} \sim \mathcal{N}(\beta, \sigma^2 V_\beta)$$

and

$$\sigma^2 | \mathbf{y}, \mathbf{X} \sim \text{Inv-}\chi^2 \left(N - (Q+4), \frac{1}{N-(Q+4)} (\mathbf{y} - \mathbf{X}\hat{\beta})^T (\mathbf{y} - \mathbf{X}\hat{\beta}) \right),$$

(11.31)

where $\hat{\beta} = (\mathbf{X}^T\mathbf{X})^{-1}\mathbf{X}^T\mathbf{y}$, $V_\beta = (\mathbf{X}^T\mathbf{X})^{-1}$ and Inv-χ^2 is the inverse χ^2 distribution.

From result 1, it is clear that the joint posterior distribution of (β, σ^2) given $(\mathbf{y}, \mathbf{X})$ is valid. Hence,

$$p(\beta, \sigma^2, \mathbf{y}, \mathbf{X}) = \frac{l(\mathbf{y}|\beta, \sigma^2, \mathbf{X})}{\sigma^2}$$

(11.32)

defines a valid joint density for $(\beta, \sigma^2, \mathbf{y})$ with $l(\mathbf{y}|\beta, \sigma^2, \mathbf{X})$ being the likelihood. This leads to our second result:

Result 2: $\int \int \int p(\beta, \sigma^2, \mathbf{y}|\mathbf{X}) d\beta d\sigma^2 d\mathbf{y} = 1$: i.e.,

$$\int \int \int \frac{1}{(2\pi\sigma^2)^{N/2}\sigma^2} \exp \left\{ -\frac{(\mathbf{y} - \mathbf{X}\beta)^T(\mathbf{y} - \mathbf{X}\beta)}{2\sigma^2} \right\} d\beta d\sigma^2 d\mathbf{y} = 1.$$

(11.33)

Next, we turn to our model (11.1). Let us define

$$\mathbf{y}(\mathbf{x}, \mathbf{a}_{sh}) = \mathbf{x} - \text{diag}(\mathbf{1}_{n_1}, \dots, \mathbf{1}_{n_M})\mathbf{a}_{sh}$$

(11.34)

and

$$\mathbf{X}(\Phi, \mathbf{a}_{sc}) = \text{diag}\{\mathbf{B}[\mathbf{U}(\mathbf{t}_1)\phi_1]\beta, \dots, \mathbf{B}[\mathbf{U}(\mathbf{t}_M)\phi_M]\}\mathbf{a}_{sc}.$$

(11.35)

For simplicity, we can denote $\mathbf{y} = \mathbf{y}(\mathbf{x}, \mathbf{a}_{sh})$ and $\mathbf{X} = \mathbf{X}(\Phi, \mathbf{a}_{sc})$. Then, we observe that the likelihood of the model is given by

$$l(\beta, \sigma^2, \theta; \mathbf{x}) = \frac{1}{(2\pi\sigma^2)^{N/2}} \exp \left\{ -\frac{(\mathbf{y} - \mathbf{X}\beta)^T(\mathbf{y} - \mathbf{X}\beta)}{2\sigma^2} \right\}.$$

(11.36)

Hence it follows from Result 2 that

$$\int \int \int l(\beta, \sigma^2, \theta; \mathbf{x})/\sigma^2 d\beta d\sigma^2 d\mathbf{y} = 1.$$

(11.37)

Note that the prior distribution for θ is proper in that

$$\int \pi(\theta) d\theta = 1.$$

(11.38)

So we have

$$\int \int \int \int \pi(\boldsymbol{\theta})l(\boldsymbol{\beta}, \sigma^2, \boldsymbol{\theta}; \mathbf{x})/\sigma^2 d\boldsymbol{\beta} d\sigma^2 d\mathbf{y} d\boldsymbol{\theta} = 1, \qquad (11.39)$$

which proves that

$$p(\boldsymbol{\beta}, \sigma^2, \boldsymbol{\theta}, \mathbf{x}) = \pi(\boldsymbol{\theta})l(\boldsymbol{\beta}, \sigma^2, \boldsymbol{\theta}; \mathbf{x})/\sigma^2$$

is a valid density.

References

1. Alshabani, A. K. S., Dryden, I. L., Litton, C. D. and Richardson, J. (2007). *Bayesian analysis of human movement curves, Applied Statistics*, Vol. 56, pp. 415–428.
2. de Boor, C. (1978), *A Practical Guide to Splines*. New York: Springer-Verlag.
3. Brumback, L. and Lindstrom, M. (2004). Self modeling with flexible, random time transformations. *Biometrics*, **60**, 461–470.
4. Gasser, T. and Kneip, A. (1995). *Searching for structure in curve samples. J. Amer. Statist. Assoc.* **90**, 1179–1188.
5. Gelman, A., Carlin, J. B., Stern, H. S. and Rubin, D. B. (2004). *Bayesian Data Analysis*. Boca Raton: Chapman & Hall/CRC.
6. Gervini, D. and Gasser, T. (2004). Self-modeling warping functions. *J. Roy Statist. Soc.* B **66**, 959–971.
7. Jupp, D. (1978). Approximation to data by splines with free knots. *SIAM Journal on Numerical Analysis* **15**, 328–343.
8. Lawton, W., Sylvestre, E. and Maggio, M. (1972). *Self modeling nonlinear regression. J. Amer. Statist. Assoc.* **14**, 513–532.
9. McKeague, I. (2005). A statistical model for signature verification. *J. Amer. Statist. Assoc.* **100**, 231–241.
10. Møller, J. and Waagepetersen, R. P. (2004). *Statistical Inference and Simulation for Spatial Point Processes*, Boca Raton: Chapman & Hall/CRC.
11. Ramsay, J. and Li, X. (1996). *Curve registration, Journal of the Royal Statistical Society* **60**, 351–363.
12. Ramsay, J. and Silverman, B. (1997). *Functional Data Analysis*. New York: Springer.
13. Ramsay, J. and Silverman, B. (2002). *Applied Functional Data Analysis: Methods and Case Studies*. New York: Springer.
14. Sakoe, H. and Chiba, S. (1978). *Dynamic programming algorithm optimization for spoken word recognition, IEEE Transactions on Signal Processing*, **26**, 43–49.
15. Smith, M. and Kohn, R. (1996). *Nonparametric regression using Bayesian variable selection, Journal of Econometrics*, **75**, 317–343.
16. Telesca, D. and Inoue, L. Y. T. (2008). *Bayesian Hierarchical curve registration, J. Amer. Statist. Assoc.* **103**, 328–339.

PART 4
NON-PARAMETRIC & PROBABILITY

Chapter 12

Nonparametric Estimation in a One-Way Error Component Model: A Monte Carlo Analysis

Daniel J. Henderson[1] and Aman Ullah[2]

[1]*State University of New York, USA*
[2]*University of California, Riverside, USA*

This paper considers the problem of nonparametric kernel estimation in panel data models. We examine the finite sample performance of several estimators for estimating a one-way random effects error component model. Monte Carlo experiments show that the pooled estimator that ignores the dependence structure in the model performs well in each trial, but it is not the most efficient estimator since it is generally outperformed in the mean squared sense by the nonparametric feasible generalized least squares estimator, the two-step estimator and an iterative estimator. Although the asymptotic bias and variance of many of the estimators converge at the same rate, the iterative and Two-Step estimators, which have asymptotic improvements, outperform the pooled estimator, but they are not always the best estimators in the simulations. Finally, we employ the estimators in an empirical example focusing on the public capital productivity puzzle to showcase their performance in a real data setting.

Keywords: Nonparametric kernel, Panel data, Productivity, Public capital.

Contents

12.1. Introduction

Recently nonparametric modeling and estimation has attracted much attention among statisticians and econometricians in panel data models. The most popular model studied is the nonparametric one-way random effects error component model. In addition to relaxing the restrictive parametric assumptions on the functional form of the model, effort has been made to incorporate the dependence structure in the data within the model. In this setup, several estimators of the nonparametric one-way error component have been proposed.

The recently developed nonparametric estimators consists of estimators ignoring the dependence structure in the errors and estimators which incorporate the dependence structure, see Henderson and Ullah (2005), Li and Ullah (1998), Lin and Carroll (2000, 2001, 2006), Ruckstuhl, Welsh and Carroll (2000), Ullah and Roy (1998), and Wang (2003), among others. The puzzling feature about many of these estimators is that the pooled estimator, which ignores the dependence structure, often converge at the same rate as estimators which do take the correlation structure into account. Ruckstuhl, Welsh and Carroll (0) argue that "the simple pooled estimator which ignores the dependence structure performs well asymptotically. Intuitively this is because dependence is a global property of the error structure which is not important to methods which act locally in the covariate space." To combat this problem two estimators have been derived which achieve asymptotic improvements over the pooled estimator. However, except for comparisons with the pooled estimator, little work has been done to compare any of these estimators against one another in a finite sample setting. In fact, little, if anything, is known about the performance of each of these estimators against one another. Although the estimators found in Ruckstuhl, Welsh and Carroll (2000) and Wang (2003) have asymptotic improvements over the pooled estimator in terms of the variance of the estimators, it is unclear which estimator is preferable, if either, in a finite setting.

In this paper we examine the finite sample performance of several nonparametric techniques for estimating a random effects one-way error component model. First, we employ Monte Carlo simulations to compare the finite sample performance of estimators which do or do not take the inherent dependence structure into account. Our results show that the pooled estimator performs very well. However, it is generally not the most efficient estimator in our trials, since a feasible generalized least squares estimator incorporating the dependence structure is found to outperform it in a mean squared sense. Further, the iterative and Two-Step estimators, each of which have or can have smaller asymptotic variance than the pooled estimator, outperform most models when estimating the unknown function. In a direct comparison to one another, there does not exist a uniform dominance. The iterative estimator estimates the unknown function very well, but is poor at estimating the varying slope parameter when the technology is nonlinear. At the same time, the Two-Step estimator estimates the varying slope parameter very well, but is not as efficient at estimating the unknown function as the iterative estimator.

Second, we employ each of the estimators to U.S. state level data to examine the public capital productivity puzzle. The procedures are used to estimate the returns to public capital, private capital, and employment in gross state product from a panel of 48 states over 17 years. In contrast to previous parametric studies and consistent with past nonparametric studies, we find that, in general, the return to public capital is positive and significant. Further, the estimators which suggest that the return to public capital is insignificant at the median are primarily the estimators which perform poorly in the Monte Carlo study.

The remainder of the paper is organized as follows: section 2 presents the nonparametric estimators whereas the third section defines the Monte Carlo setup and summarizes the results of the experiment. Section 4 gives the empirical study and the fifth section concludes.

12.2. Methodology

Here we consider the nonparametric random effects panel data model

$$y_{it} = m(x_{it}) + \varepsilon_{it}, \qquad (12.1)$$

where $i = 1, 2, ..., N$, $t = 1, 2, ..., T$, y_{it} is the endogenous variable, x_{it} is a vector of k exogenous variables and $m(\cdot)$ is an unknown smooth function. We consider the case that the error ε_{it} follows a one-way error component specification

$$\varepsilon_{it} = u_i + v_{it}, \qquad (12.2)$$

where u_i is *i.i.d.* $(0, \sigma_u^2)$, v_{it} is *i.i.d.* $(0, \sigma_v^2)$ and u_i and v_{jt} are uncorrelated for all i and j, $j = 1, 2, \ldots, N$.

If $\varepsilon_i = [\varepsilon_{i1}, \varepsilon_{i2}, \ldots, \varepsilon_{iT}]'$ is a $T \times 1$ vector, then $V \equiv E(\varepsilon_i \varepsilon_i')$, takes the form

$$V = \sigma_v^2 \mathbf{I}_T + \sigma_u^2 \mathbf{i}_T \mathbf{i}_T', \tag{12.3}$$

where $\mathbf{I}_T$ is an identity matrix of dimension T and $\mathbf{i}_T$ is a $T \times 1$ column vector of ones. Since the observations are independent over i and j, the covariance matrix for the full $NT \times 1$ disturbance vector ε, $\Omega = E(\varepsilon \varepsilon')$ is

$$\Omega = \mathbf{I}_N \otimes V. \tag{12.4}$$

We are interested in estimating the unknown function $m(x)$ at a point x and the slope of $m(x)$, $\beta(x) = \nabla m(x)$, where ∇ is the gradient vector of $m(x)$. The parameter $\beta(x)$ is interpreted as a varying coefficient. We consider the usual panel data situation of large N and finite T.

12.2.1. *Ignoring the error structure*

12.2.1.1. *Pooled estimator*

The simpliest consistent method to estimate the parameters of the model is simply to ignore the correlation structure. Nonparametric kernel estimation of $m(x)$ and $\beta(x)$ can be obtained by using local-linear least-squares (LLLS) estimation. This is obtained by minimizing the local least squares of errors

$$\sum_{i=1}^{N} \sum_{t=1}^{T} (y_{it} - X_{it}\delta(x))^2 K\left(\frac{x_{it} - x}{h}\right) = (y - X\delta(x))' K(x)(y - X\delta(x)) \tag{12.5}$$

with respect to $m(x)$ and $\beta(x)$, where y is a $NT \times 1$ vector, X is a $NT \times (k+1)$ matrix generated by $X_{it} = (1, (x_{it} - x))$, $\delta(x) = (m(x), \beta(x))'$ is a $(k+1) \times 1$ vector, $K(x)$ is an $NT \times NT$ diagonal matrix of kernel functions $K(\frac{x_{it} - x}{h})$ and h is the bandwidth (smoothing) parameter. The estimator obtained is

$$\widehat{\delta}(x) = (X'K(x)X)^{-1} X'K(x)y \tag{12.6}$$

and is called the LLLS estimator (see Fan and Gijbels, 1992; Li and Racine, 2006 or Pagan and Ullah, 1999 for details on this estimator and the choice of h) or the 'working independence' estimator (Lin and Caroll, 2000).

12.2.1.2. *Component estimator*

The LLLS estimator simply fits a single nonparametric regression model through all the data. An alternative approach, which Ruckstuhl, Welsh and Carroll (2000, page 54) call the component estimator, involves fitting separate nonparametric models relating the tth component (time period) of y to the tth component (time period) of x and then combines these estimators to produce an overall estimator of the common regression function. The component estimator of $\delta(x)$ is defined as the weighted average of the component estimators given by

$$\widehat{\delta}_C(x) = \sum_{t=1}^{T} c_t \, \widehat{\delta}_t(x), \qquad (12.7)$$

where $\sum_{t=1}^{T} c_t = 1$ and

$$c_t = f_t(x) \left\{ \sum_{k=1}^{T} f_k(x) \right\}^{-1} \quad \text{for } t = 1, 2, ..., T,$$

$\widehat{\delta}_t(x)$ is the local linear kernel regression estimator of the y_{it} on the x_{it} with bandwidth h_t and f_l is the marginal density of x_{il}. Ruckstuhl, Welsh and Carroll (2000, page 63) suggest using the common optimal bandwidth for the component estimator, which minimizes its mean squared error, defined as

$$h_{RWC} = \left(\gamma(0) \left(\sigma_u^2 + \sigma_\varepsilon^2 \right) \left[\left\{ m^{(2)}(x) \right\}^2 N \sum_{t=1}^{T} f_t(x) \right]^{-1} \right)^{\frac{1}{5}}, \qquad (12.8)$$

where $\gamma(0) = \int K^2(z)dz$ and $m^{(2)}(x)$ is the second derivative of $m(x)$. Here we note that Ruckstuhl, Welsh and Carroll (2000) only define the result for $m(x)$ and not $\beta(x)$. Ruckstuhl, Welsh and Carroll (2000) further note that the LLLS and component estimators are asymptotically equivalent. However, LLLS estimation has the advantage of simplicity. It is obvious that the relative performance of the estimators will depend on the sample size and on the data generating process. In practice, one way want to apply the nonparametric poolability test proposed by Baltagi, Hidalgo and Li (1996).

12.2.2. *Incoporating the error structure*

12.2.2.1. *Two-step estimator*

Both of these estimators, however, ignore the information contained in the disturbance vector covariance matrix Ω. In view of this, Ruckstuhl, Welsh and Carroll (2000) develop the Two-Step estimator. This estimator attempts to make use of the known variance structure to achieve an asymptotic improvement over the pooled LLLS estimator. Specifically, the estimator is obtained by first defining

$$z = \tau \Omega^{-\frac{1}{2}} y + \left(\mathbf{I}_{NT} - \tau \Omega^{-\frac{1}{2}} \right) \widehat{m}(x), \qquad (12.9)$$

where τ is a constant and $\widehat{m}(x)$ is the first stage estimator of $\widehat{\delta}(x)$ in (12.6). Estimation of $\delta_{TS}(x) = (m_{TS}(x), \beta_{TS}(x))'$ can now be obtained by running LLLS on the following equation

$$z_{it} = m_{TS}(x_{it}) + v_{it}. \qquad (12.10)$$

Further, the value for τ can be obtained from

$$\tau^2 = \sigma_v^2 \left[1 - \left\{ 1 - (1 - d_T)^{\frac{1}{2}} \right\} / T \right]^{-2}, \qquad (12.11)$$

where $d_T = T\sigma_u^2 / \left(\sigma_v^2 + T\sigma_u^2 \right)$. Ruckstuhl, Welsh and Carroll (2000) show that the order of magnitude of the variance of the Two-Step estimator can be smaller than the pooled LLLS estimator. They further show that the asymptotic bias can also be smaller than that of the pooled LLLS estimator. For example, when $m(\cdot)$ is a quadratic function, the bias decreases monotonically to zero as the ratio σ_u^2/σ_v^2 increases.[a]

12.2.2.2. *Local-linear weighted least-squares*

Henderson and Ullah (2005) also attempt to model the information contained in the distrubance vector covariance matrix. They introduce an estimator, Local-Linear Weighted Least-Squares (LLWLS), by minimizing

$$(y - X\delta(x))'W(x)(y - X\delta(x)) \qquad (12.12)$$

---

[a]Ruckstuhl, Welsh and Carroll (2000) alternatively suggest defining τ to be equal to σ_v^{-1}, which when employed in the estimation of (12.10) has smaller asymptotic variance than estimation performed with (12.11). Although defining τ in this manner works well, it does not perform as well as (12.11) in small samples. Further, in our empirical example we found it to give weaker estimates then when using (12.11). The results for both the Monte Carlo and the empirical exercise using the alternative τ can be obtained from the authors upon request.

with respect to $\delta(x)$, where $W(x)$ is a kernel based weight matrix. This provides the kernel estimating equations for $\delta(x)$ as $0 = X'W(x)(y - X\delta(x))$, which gives

$$d(x) = (X'W(x)X)^{-1} X'W(x)y. \qquad (12.13)$$

They consider the following cases of (12.13),

$$d_r(x) = (X'W_r(x)X)^{-1} X'W_r(x)y, \qquad (12.14)$$

where $d_r(x) = (m_r(x), \beta_r(x))'$, and for $r = 1, 2, 3$, $W_1(x) = \sqrt{K(x)}\Omega^{-1}\sqrt{K(x)}$, $W_2(x) = \Omega^{-1}K(x)$ and $W_3(x) = \Omega^{-\frac{1}{2}}K(x)\Omega^{-\frac{1}{2}}$. The estimators $d_1(x)$ and $d_2(x)$ are as given in Lin and Carroll (2000), and $d_3(x)$ is as given in Ullah and Roy (1998). When the matrix V, and hence Ω, is a diagonal matrix, then $W_1(x) = W_2(x) = W_3(x)$, and hence $d_1(x) = d_2(x) = d_3(x)$. Further, in the special case when $\Omega = \mathbf{I}_{NT}$, $d_1(x) = d_2(x) = d_3(x) = \widehat{\delta}(x)$ in (12.6). In general, however, $d_1(x)$, $d_2(x)$, and $d_3(x)$ are often different. Ruckstuhl, Welsh and Carroll (2000) and Lin and Carroll (2000) provide the asymptotic bias and variance of $d_1(x)$ and $d_2(x)$ for the case of a single regressor. These results provide the consistency of these estimators. However, unlike the Two-Step estimator, Henderson and Ullah (2005) note that the rates of convergence of $d_1(x)$, $d_2(x)$ and $d_3(x)$ are the same as the pooled LLLS estimator for any number of regressors.

12.2.2.3. *Feasible estimators*

Two-Step estimation in (12.10) and the LLWLS estimator in (12.13), however, depend upon the unknown parameters σ_u^2 and σ_v^2. Henderson and Ullah (2005) use the spectral decomposition of Ω to obtain consistent estimators of the variance components as

$$\widehat{\sigma}_1^2 = T \sum_{i=1}^{N} \widehat{\overline{\varepsilon}}_i^2 / N, \qquad \widehat{\sigma}_v^2 = \frac{1}{N(T-1)} \sum_{i=1}^{N} \sum_{t=1}^{T} \left(\widehat{\varepsilon}_{it} - \widehat{\overline{\varepsilon}}_i \right)^2, \qquad (12.15)$$

where $\sigma_1^2 \equiv T\sigma_u^2 + \sigma_v^2$, $\widehat{\overline{\varepsilon}}_i = \frac{1}{T} \sum_{t=1}^{T} \widehat{\varepsilon}_{it}$ and $\widehat{\varepsilon}_{it} = y_{it} - \widehat{m}(x_{it})$ is the LLLS residual based on the first stage estimator of $\widehat{\delta}(x)$ in (12.6). Further, the estimate of σ_u^2 is obtained as $\widehat{\sigma}_u^2 = \left(\widehat{\sigma}_1^2 - \widehat{\sigma}_v^2 \right) / T$.

Ruckstuhl, Welsh and Carroll (2000, page 59) provide an alternative approach to estimating the variance components. They estimate the variance-covariance matrix by pretending that the residuals have mean zero and that the covariance matrix is the same as if $m(\cdot)$ were known. Specifically, the consistent estimators of the elements of $\widetilde{\Omega}$ are obtained as

$$\widetilde{\sigma}_1^2 = \frac{T}{N} \sum_{i=1}^{N} \left(\overline{y}_i - \widehat{\overline{m}}_i \right)^2, \tag{12.16}$$

where $\overline{y}_i = \frac{1}{T} \sum_{t=1}^{T} y_{it}$, $\widehat{\overline{m}}_i = \frac{1}{T} \sum_{t=1}^{T} \widehat{m}(x_{it})$, and

$$\widetilde{\sigma}_v^2 = \frac{1}{N(T-1)} \sum_{i=1}^{N} \sum_{t=1}^{T} \left(y_{it} - \widehat{m}(x_{it}) - \left(\overline{y}_i - \widehat{\overline{m}}_i \right) \right)^2, \tag{12.17}$$

when $\widetilde{\sigma}_1^2 > \widetilde{\sigma}_v^2$ and

$$\widetilde{\sigma}_1^2 = \widetilde{\sigma}_v^2 = \frac{1}{NT} \sum_{i=1}^{N} \sum_{t=1}^{T} (y_{it} - \widehat{m}(x_{it}))^2, \tag{12.18}$$

otherwise. Similarly, the estimate of σ_u^2 is obtained as $\widetilde{\sigma}_u^2 = \left(\widetilde{\sigma}_1^2 - \widetilde{\sigma}_v^2 \right) / T$.

Substituting estimates of σ_u^2 and σ_v^2 from (12.15), (12.16) and (12.17), or (12.18) into (12.10) and (12.14) give the feasible Two-Step and the Local-Linear Feasible Weighted Least-Squares (LLFWLS) or Nonparametric Feasible Weighted Least Squares (NPFWLS) estimators as

$$\widehat{\delta}_{TS}(x) = (X'K(x)X)^{-1}X'K(x)\widehat{z}, \tag{12.19}$$

$$\widetilde{\delta}_{TS}(x) = (X'K(x)X)^{-1}X'K(x)\widetilde{z}, \tag{12.20}$$

$$\widehat{\delta}_r(x) = (X'\widehat{W}_r(x)X)^{-1}X\widehat{W}_r(x)y, \tag{12.21}$$

and

$$\widetilde{\delta}_r(x) = (X'\widetilde{W}_r(x)X)^{-1}X\widetilde{W}_r(x)y, \tag{12.22}$$

where $\widehat{z}$, $\widetilde{z}$, $\widehat{W}_r(x)$ and $\widetilde{W}_r(x)$ are the same as above, with Ω replaced by the consistent estimators $\widehat{\Omega}$ and $\widetilde{\Omega}$, respectively. Further, following Li and Ullah (1998), we can show that the consistency of $\widehat{\delta}_{TS}(x)$, $\widetilde{\delta}_{TS}(x)$, $\widehat{\delta}_r(x)$ and $\widetilde{\delta}_r(x)$ follow from the consistency of $\delta_{TS}(x)$ and $d_r(x)$ for known Ω. Finally, for the remainder of the paper we will refer to $\delta_1(x)$ as the nonparametric generalized least squares (NPGLS) estimator, $\delta_2(x)$ as the Lin and Carroll estimator and $\delta_3(x)$ as the Ullah and Roy estimator.

12.2.3. *Wang's iterative estimator*

The counter-intuitive result above is that ignoring the within-cluster correlation leads to the same rate of convergence for the estimators. This occurs because a critical assumption in nonparametric methods is that the bandwidth shrinks towards zero as the number of cross-sectional units grows. Thus, asymptotically, there is effectively one data point per cluster that contributes to the estimate of the unknown function for a specific value of x. Since we assume that data from different clusters are independent, the estimator with the smallest variance under the given kernel weighted structure can be obtained by ignoring the off-diagonal elements.

In response to this odd result, Wang (2003) develops an iterative procedure which eliminates biases and reduces the variation simultaneously. The basic idea behind her estimator is that once a data point within a cluster has a value within a bandwidth of the x value, and is used to estimate the unknown function, all points in that cluster are used. For data points which lie outside the bandwidth, the contributions of the other points in the local estimate are through their residuals. Formally, Wang (2003) suggests estimating the unknown function and its derivative by solving the first order condition

$$
0 = \sum_{i=1}^{N} \sum_{t=1}^{T} K\left(\frac{x_{it} - x}{h}\right) \left(\frac{1}{\frac{x_{it}-x}{h}}\right) \mathcal{L}_{i,tm} \left(\begin{array}{c} y_i, \widehat{m}\left(x_{i1}\right), \ldots, \widehat{m}\left(x\right) \\ + \left(\frac{x_{it}-x}{h}\right)\widehat{\beta}\left(x\right), \ldots, \widehat{m}\left(x_{iT}\right) \end{array} \right),
$$

$$(12.23)$$

where $y_i = (y_{i1}, y_{i2}, \ldots, y_{iT})$, $\mathcal{L}_i$ is the criterion function for individual i and $\mathcal{L}_{i,tm}$ is the partial derivative of $\mathcal{L}_i$ with respect to $m\left(x_{it}\right)$. This first order condition suggests an iterative estimation procedure. Define $\widehat{m}_{[l-1]}\left(x\right)$ as the current estimate of the unknown function at the $[l-1]$th step. Define $\widehat{\beta}_{[l-1]}\left(x\right)$ similarly. Then $\widehat{m}_{[l]}\left(x\right)$ denotes the next stage estimator of the unknown function where $\widehat{m}_{[l]}\left(x\right)$ and $\widehat{\beta}_{[l]}\left(x\right)$ are the solutions to

$$
0 = \sum_{i=1}^{N} \sum_{t=1}^{T} K\left(\frac{x_{it} - x}{h}\right) \left(\frac{1}{\frac{x_{it}-x}{h}}\right) \mathcal{L}_{i,tm} \left(\begin{array}{c} y_i, \widehat{m}_{[l-1]}\left(x_{i1}\right), \ldots, \widehat{m}_{[l]}\left(x\right) \\ + \left(\frac{x_{it}-x}{h}\right)\widehat{\beta}_{[l]}\left(x\right), \ldots, \widehat{m}_{[l-1]}\left(x_{iT}\right) \end{array} \right).
$$

$$(12.24)$$

This leads to the estimator of the unknown function and its derivative at the lth step as

$$\widehat{\delta}_W(x) = \begin{pmatrix} \widehat{m}_{[l]}(x) \\ \widehat{\beta}_{[l]}(x) \end{pmatrix} = D_1(x)^{-1} D_2(x) \qquad (12.25)$$

$$= \left[\sum_{i=1}^{N} \sum_{t=1}^{T} K\left(\frac{x_{it} - x}{h}\right) \sigma^{tt} \left(\frac{1}{\frac{x_{it}-x}{h}}\right) \left(1 \ \frac{x_{it}-x}{h}\right) \right]^{-1}$$

$$\times \left[\sum_{i=1}^{N} \sum_{t=1}^{T} K\left(\frac{x_{it} - x}{h}\right) \sigma^{tt} \left(\frac{1}{\frac{x_{it}-x}{h}}\right) \right]$$

$$\times \left[\sigma^{tt} y_{it} + \sum_{s \neq t}^{T} \sigma^{st} \left(y_{is} - \widehat{m}_{[l-1]}(x_{is})\right) \right],$$

where σ^{st} is the (t,s)th element of V^{-1}. For a feasible estimator, the estimates for σ^{st} can be obtained by using the residuals from the pooled estimator. It should be noted that Wang (2003) argues that the once-iterated estimator has the same asymptotic behavior as the fully iterated estimator. Her Monte Carlo exercise shows that it performs well for the case of a single regressor, but there is no discussion for the multiple regression case. Here we choose to use the fully iterated model and note that the estimator generally converges in 3 to 4 iterations when $k = 1$. Another point worth noting is that although the estimator looks complicated, it is closely related to the LLLS estimator. The method simply takes the pooled estimator and adds weighted residuals. The only question that remains is how one obtains an initial estimate of the unknown function. One choice is to use the 'working independence' or pooled estimator because it is a consistent estimator and convergence usually can be obtained in a few steps. A second method uses a polynomial regression. We choose this consistent method and note that the results do not significantly differ for this paper.

12.3. Monte Carlo

This section uses Monte Carlo simulations to examine the finite sample performance of the panel data estimators. Following the methodology of Baltagi, Chang and Li (1992), the following data generating process is used:

$$y_{it} = \alpha + x_{it}\beta + x_{it}^2 \gamma + u_i + v_{it},$$

where x_{it} is generated by the method of Nerlove (1971).[b] The value of α is chosen to be 5, β is chosen to be 0.5 and γ takes the values of 0 (linear technology) and 2 (quadratic technology). The distribution of u_i and v_{it} are generated separately as *i.i.d.* Normal. The total variance $\sigma_v^2 + \sigma_u^2 = 20$ and $\rho = \sigma_u^2/(\sigma_u^2 + \sigma_v^2)$ is varied to be 0.1, 0.4, and 0.8. For comparison, we compute the following estimators of δ:

(I)   Parametric (linear) feasible GLS (FGLS) estimator

$$\widehat{\delta} = (X'\widehat{\Omega}^{-1}X)^{-1}X'\widehat{\Omega}^{-1}y.$$

(II)   LLLS (pooled) estimator

$$\widehat{\delta}(x) = (X'K(x)X)^{-1}X'K(x)y.$$

(III)   Component estimator

$$\widehat{\delta}_C(x) = \sum_{t=1}^{T} c_t\, \widehat{\delta}_t(x).$$

(IV)   Feasible Two-Step estimator

$$\widehat{\delta}_{TS}(x) = (X'K(x)X)^{-1}X'K(x)\widehat{z}$$

(V)   Feasible NPGLS (NPFGLS) estimator

$$\widehat{\delta}_1(x) = (X'\sqrt{K(x)}\widehat{\Omega}^{-1}\sqrt{K(x)}X)^{-1}X'\sqrt{K(x)}\widehat{\Omega}^{-1}\sqrt{K(x)}y.$$

(VI)   Feasible Lin and Carroll estimator

$$\widehat{\delta}_2(x) = (X'\widehat{\Omega}^{-1}K(x)X)^{-1}X'\widehat{\Omega}^{-1}K(x)y.$$

(VII)   Feasible Ullah and Roy[c] estimator

$$\widehat{\delta}_3(x) = (X'\widehat{\Omega}^{-\frac{1}{2}}K(x)\widehat{\Omega}^{-\frac{1}{2}}X)^{-1}X'\widehat{\Omega}^{-\frac{1}{2}}K(x)\widehat{\Omega}^{-\frac{1}{2}}y.$$

(VIII)   Wang estimator

$$\widehat{\delta}_W(x) = D_1(x)^{-1}D_2(x).$$

In addition, we also estimate each of the four feasible nonparametric estimators using the estimated omega matrix described in Ruckstuhl, Welsh and Carroll (2000).

(IX)   Feasible Two-Step estimator

$$\widetilde{\delta}_{TS}(x) = (X'K(x)X)^{-1}X'K(x)\widetilde{z}.$$

---

[b]The x_{it} are generated as follows: $x_{it} = 0.1t + 0.5x_{it-1} + w_{it}$, where $x_{i0} = 10 + 5w_{i0}$ and $w_{it} \sim U[-\frac{1}{2}, \frac{1}{2}]$.
[c]This feasible estimator uses a consistent estimator of the omega matrix which is different from that given in Ullah and Roy (1998) and also in Li and Ullah (1998).

(X) NPFGLS estimator

$$\widetilde{\delta}_1(x) = (X'\sqrt{K(x)}\widetilde{\Omega}^{-1}\sqrt{K(x)}X)^{-1}X'\sqrt{K(x)}\widetilde{\Omega}^{-1}\sqrt{K(x)}y.$$

(XI) Feasible Lin and Carroll estimator

$$\widetilde{\delta}_2(x) = (X'\widetilde{\Omega}^{-1}K(x)X)^{-1}X'\widetilde{\Omega}^{-1}K(x)y.$$

(XII) Feasible Ullah and Roy estimator

$$\widetilde{\delta}_3(x) = (X'\widetilde{\Omega}^{-\frac{1}{2}}K(x)\widetilde{\Omega}^{-\frac{1}{2}}X)^{-1}X'\widetilde{\Omega}^{-\frac{1}{2}}K(x)\widetilde{\Omega}^{-\frac{1}{2}}y.$$

12.3.1. *Results*

The parametric estimator is expected to perform best when the parametric model is correctly specified. However, when the parametric model is incorrectly specified, it is expected to lead to inconsistent estimation of $m(\cdot)$ and $\beta(\cdot)$. Further, although the asymptotic bias and variance of most of the nonparametric estimators converge at the same rate, the finite sample performance of the estimators against one another is unknown. Our expectation amongst the nonparametric estimators is that the Two-Step and Wang estimators will generally outperform the other nonparametric estimators.

Reported in Tables 12.1 and 12.2 are the average estimated mean squared errors (MSE) for each estimator. These are computed via $\overline{MSE}(\widehat{m}) = M^{-1}\sum_{j=1}^{M} MSE_j(\widehat{m})$, where $MSE_j(\widehat{m}) = \frac{1}{NT}\sum_{i=1}^{N}\sum_{t=1}^{T}(\widehat{m}(x_{it}) - m^*(x_{it}))^2$ is the MSE of $\widehat{m}$ at the jth replication. Further, M $(j = 1, 2, ..., M)$ is the number of replications, and $\widehat{m}(x)$ is the estimated value of $m^*(x) = \alpha + x\beta + x^2\gamma$, evaluated at each x in a given replication. Similarly, for the varying coefficient parameter, $\overline{MSE}(\widehat{\beta}) = M^{-1}\sum_{j=1}^{M} MSE_j(\widehat{\beta})$, where $MSE_j(\widehat{\beta}) = \frac{1}{NT}\sum_{i=1}^{N}\sum_{t=1}^{T}(\widehat{\beta}(x_{it}) - \beta^*(x_{it}))^2$ is the MSE of $\widehat{\beta}$ at the jth replication. Again, M $(j = 1, 2, ..., M)$ is the number of replications, and $\widehat{\beta}(x)$ is the estimated value of $\beta^*(x) = \beta + 2x\gamma$, evaluated at each x. $M = 499$ is used in all simulations. T is varied to be 5 and 10, while N takes the values 25 and 50. The smallest MSE for each case (for a given N, T, ρ and γ) is shown as a boldface number.

12.3.1.1. *Linear data generating process*

Table 1 reports the results for $\gamma = 0$ (linear technology). As expected, the correctly specified linear parametric model outperforms each of the nonparametric estimators in terms of MSE. The average MSE for the unknown function is generally twice as small as the best nonparametric estimator and several times smaller for estimation of the slope parameter. This latter result is as expected because the local-linear estimator for the unknown function convergences at a faster rate than the varying coefficient parameter. We do note, however, that the performance of the nonparametric estimators improve as the number of cross-sections increases.

Table 12.1. Monte Carlo Results (Average MSE) – Linear Technology ($\gamma = 0$)

	m			β		
	$\rho = 0.1$	$\rho = 0.4$	$\rho = 0.8$	$\rho = 0.1$	$\rho = 0.4$	$\rho = 0.8$
			N = 25, T = 5			
FGLS	0.433	0.513	0.734	0.118	0.057	0.024
LLLS	0.899	0.972	1.223	3.612	3.617	3.283
Components	0.848	0.905	1.181	7.278	6.798	6.099
Two-Step	0.886	0.852	0.876	3.575	2.741	0.98
NPFGLS	0.897	0.955	1.19	3.599	3.139	1.864
Lin and Carroll	0.956	1.415	2.267	5.63	21.617	44.22
Ullah and Roy	1.862	14.275	36.32	3.317	2.398	0.65
Wang	0.844	0.837	0.854	1.04	0.702	0.293
Two-Step RWC	0.893	1.038	1.258	4.196	4.379	2.383
NPFGLS RWC	0.898	0.962	1.206	3.604	3.463	3.051
Lin and Carroll RWC	0.9	0.992	1.256	3.731	4.663	5.054
Ullah and Roy RWC	0.953	1.153	1.417	3.554	3.547	3.116
			N = 50, T = 5			
FGLS	0.173	0.272	0.379	0.038	0.03	0.011
LLLS	0.457	0.529	0.676	2.132	1.992	2.242
Components	0.4	0.486	0.621	2.989	2.945	2.99
Two-Step	0.452	0.465	0.456	2.111	1.573	0.654
NPFGLS	0.456	0.514	0.67	2.136	1.741	1.406
Lin and Carroll	0.47	0.813	1.49	2.895	14.303	36.242
Ullah and Roy	0.897	7.85	31.611	1.842	1.288	0.389
Wang	0.436	0.418	0.495	0.483	0.424	0.183
Two-Step RWC	0.49	0.561	0.595	2.142	2.107	1.692
NPFGLS RWC	0.457	0.526	0.671	2.124	1.917	2.1
Lin and Carroll RWC	0.458	0.535	0.68	2.171	2.274	2.446
Ullah and Roy RWC	0.465	0.553	0.695	2.1	1.951	2.052

The discussion of the performance of the nonparametric estimators relative to one another is not as straightforward.[d] First, we note that several estimators outperform the pooled estimator. The components estimator, which also ignores the correlation structure, generally has a smaller MSE, for both the unknown function and its derivative. Additionally,

---

[d]Except for the component estimator whose bandwidth is shown in (12.8), we use the Silverman (1986) rule of thumb bandwidth in our Monte Carlo simulations.

Table 12.1. (Continued) Monte Carlo Results (Average MSE) – Linear Technology ($\gamma = 0$)

	m			β		
	$\rho = 0.1$	$\rho = 0.4$	$\rho = 0.8$	$\rho = 0.1$	$\rho = 0.4$	$\rho = 0.8$
			N = 25, T = 10			
FGLS	0.204	0.39	0.619	0.081	0.055	0.02
LLLS	0.555	0.784	1.01	5.833	6.503	5.593
Components	0.479	0.688	0.922	9.322	13.246	8.92
Two-Step	0.542	0.664	0.714	5.65	4.55	1.49
NPFGLS	0.554	0.755	1.006	5.605	5.243	3.681
Lin and Carroll	0.629	1.357	2.29	16.324	79.013	156.762
Ullah and Roy	5.485	73.536	183.369	5.188	3.5	1.058
Wang	0.614	0.669	0.889	0.813	0.589	0.262
Two-Step RWC	0.564	0.699	1.139	7.479	4.852	4.991
NPFGLS RWC	0.555	0.783	1.007	5.808	6.37	5.408
Lin and Carroll RWC	0.56	0.791	1.036	6.565	7.134	9.324
Ullah and Roy RWC	0.761	1.407	1.231	5.769	6.344	5.788
			N = 50, T = 10			
FGLS	0.123	0.219	0.339	0.042	0.028	0.009
LLLS	0.345	0.443	0.583	3.733	4.118	4.026
Components	0.284	0.38	0.521	4.746	8.592	5.091
Two-Step	0.335	0.374	0.395	3.554	2.874	1.003
NPFGLS	0.343	0.432	0.586	3.583	3.486	2.556
Lin and Carroll	0.416	0.913	1.628	15.456	74.639	153.669
Ullah and Roy	4.255	42.299	133.789	3.578	2.16	0.625
Wang	0.382	0.361	0.362	0.463	0.451	0.341
Two-Step RWC	0.341	0.447	0.464	3.61	3.633	2.172
NPFGLS RWC	0.344	0.442	0.58	3.71	3.978	3.838
Lin and Carroll RWC	0.346	0.451	0.6	4.064	5.766	7.227
Ullah and Roy RWC	0.418	0.572	0.896	3.753	4.256	4.125

the NPFGLS, Two-Step and Wang (2003) estimators all have smaller average MSE in each scenario. In comparison with one another, we see that the component estimator performs well in terms of estimating $m(\cdot)$ when $\rho = 0.1$. That is, the Ruckstuhl, Welsh and Carroll (2000) asymptotic results go through for small samples. However, as expected, it is often outperformed by the Two-Step estimator. Further, the Wang (2003) estimator gives the smallest value in seven out of the twelve cases. This first set of results shows us that it is not obvious which nonparametric estimator is preferable here.

We now turn to the estimation of the slope parameter. Our attempts at estimating $\beta(\cdot)$ with the component estimator (which was not defined in Ruckstuhl, Welsh and Carroll, 2000) proved to be poor. One possible explanation for this is that the component estimator is designed to minimize the MSE of $m(\cdot)$ and not necessarily $\beta(\cdot)$ (see Ruckstuhl, Welsh and Car-

Table 12.2.   Monte Carlo Results (Average MSE) – Linear Technology ($\gamma = 2$)

	m			β		
	$\rho = 0.1$	$\rho = 0.4$	$\rho = 0.8$	$\rho = 0.1$	$\rho = 0.4$	$\rho = 0.8$
			N = 25, T = 5			
FGLS	11.086	11.216	11.487	62.623	61.385	61.96
LLLS	0.816	0.967	1.215	4.366	4.973	3.85
Components	1.346	1.521	1.726	126.083	10.032	7.427
Two-Step	0.811	0.852	0.923	4.311	4.028	1.948
NPFGLS	0.812	0.932	1.175	4.325	4.1	2.625
Lin and Carroll	0.859	1.395	2.31	6.1	21.116	48.132
Ullah and Roy	6.93	90.185	661.049	6.118	19.069	34.699
Wang	0.807	0.799	0.738	10.14	9.622	9.243
Two-Step RWC	0.938	0.964	1.063	4.314	4.032	4.371
NPFGLS RWC	0.814	0.961	1.206	4.373	4.882	3.737
Lin and Carroll RWC	0.82	0.982	1.24	4.51	5.542	4.99
Ullah and Roy RWC	1.552	2.033	3.529	4.606	5.204	4.348
			N = 50, T = 5			
FGLS	11.207	11.1	11.227	62.811	62.038	61.895
LLLS	0.51	0.551	0.671	3.111	2.987	2.608
Components	0.948	1.037	1.199	5.21	5.224	4.675
Two-Step	0.509	0.482	0.483	3.052	2.465	1.339
NPFGLS	0.505	0.527	0.633	3.027	2.661	1.938
Lin and Carroll	0.535	0.76	1.357	4.126	12.662	31.262
Ullah and Roy	5.873	90.683	712.723	5.107	21.023	37.162
Wang	0.497	0.47	0.382	11.733	11.491	11.463
Two-Step RWC	0.512	0.506	0.544	3.095	2.748	2.628
NPFGLS RWC	0.51	0.549	0.665	3.102	2.961	2.473
Lin and Carroll RWC	0.511	0.552	0.678	3.189	3.062	2.927
Ullah and Roy RWC	0.958	1.203	1.976	3.266	3.201	3.016

roll, 2000, page 63). The Ullah and Roy estimator performs well on average amongst the nonparametric estimators. Unfortunately, it performs poorly in terms of estimating $m(\cdot)$ in our exercise when using $\widehat{\Omega}$. A similar result holds for the Lin and Carroll estimator. It performs poorly in terms of estimating $\beta(\cdot)$ when using $\widehat{\Omega}$. Estimation of both $m(\cdot)$ and $\beta(\cdot)$ improves for the Lin and Carroll estimator and for $m(\cdot)$ (but not for $\beta(\cdot)$) with the Ullah and Roy estimator when employing $\widetilde{\Omega}$. However they are still less efficient than several of the other estimation techniques. Two-Step, NPFGLS (whose average MSE are smaller with $\widehat{\Omega}$) and LLLS are more consistent in terms of their performance relative to the other nonparametric estimators. Although they are not consistently the top nonparametric performers in terms of $m(\cdot)$ or $\beta(\cdot)$, the Two-Step and NPFGLS estimators perform well and better than LLLS in most situations in the table. That being said, the top performer in terms of estimating the varying coefficient parameter is

Table 12.2. (Continued) Monte Carlo Results (Average MSE) – Linear Technology ($\gamma = 2$)

	m			β		
	$\rho = 0.1$	$\rho = 0.4$	$\rho = 0.8$	$\rho = 0.1$	$\rho = 0.4$	$\rho = 0.8$
N = 25, T = 10						
FGLS	5.933	6.139	6.441	64.587	64.073	64.084
LLLS	0.648	0.802	1.094	7.323	7.21	6.613
Components	0.778	0.924	1.225	13.682	13.708	56.857
Two-Step	0.641	0.681	0.806	7.154	5.231	2.118
NPFGLS	0.649	0.788	1.095	7.219	6.206	4.294
Lin and Carroll	0.741	1.486	2.432	19.108	104.74	178.604
Ullah and Roy	19.062	242.553	1456.59	12.958	28.891	41.802
Wang	0.604	0.715	0.711	8.543	8.295	8.131
Two-Step RWC	0.654	0.745	1.005	7.32	5.641	4.637
NPFGLS RWC	0.649	0.799	1.09	7.243	6.964	6.444
Lin and Carroll RWC	0.651	0.817	1.124	7.543	9.759	11.742
Ullah and Roy RWC	1.378	1.49	2.388	7.264	7.39	7.271
N = 50, T = 10						
FGLS	5.951	6.027	6.284	63.824	63.584	64.053
LLLS	0.313	0.442	0.614	4.107	4.77	4.522
Components	0.426	0.558	0.723	5.981	6.652	6.818
Two-Step	0.305	0.376	0.439	4.011	3.531	1.433
NPFGLS	0.313	0.434	0.61	4.06	3.857	2.889
Lin and Carroll	0.367	0.847	1.808	12.991	62.793	163.742
Ullah and Roy	15.351	193.05	1402.883	10.39	34.367	46.022
Wang	0.289	0.36	0.359	9.214	8.988	8.967
Two-Step RWC	0.312	0.392	0.554	4.225	4.489	2.956
NPFGLS RWC	0.313	0.44	0.612	4.091	4.613	4.277
Lin and Carroll RWC	0.315	0.449	0.633	4.5	6.079	7.625
Ullah and Roy RWC	0.653	1.055	1.79	4.381	5.316	5.39

the iterative estimator. The average MSE for the Wang estimator is far smaller than the other nonparametric estimators in each scenario. Given its relatively strong performance in the estimation of $m(\cdot)$ and its dominance in the estimation of the slope parameter, the linear technology points to the iterative estimator as the prefered nonparametric estimation procedure. However, it is premature to declare it the winner as we have yet to see its performance with a nonlinear technology.

12.3.1.2. *Nonlinear data generating process*

The Monte Carlo results for the quadratic technology ($\gamma = 2$) are given in the second table. The now misspecified (linear) parametric estimator provides biased estimates. The estimator has a large MSE which does not decrease as the number of cross-sections grow. Again, the LLLS estimates perform well, but less efficient generally than several other estimators. The results for the nonlinear technology show quite well for both the Two-Step and NPFGLS estimators. The Two-Step estimator outperforms a majority of the other models. However, in two cases, the NPFGLS estimator, which does not give an asymptotic improvement over the pooled estimator, outperforms the Two-Step estimator, which can give an asymptotic impact. In each of these two cases ρ is relatively small. As is the case with the linear technology, the performance of the Two-Step estimator improves greatly when ρ increases. Again, this estimator is not dominant. We see a strong showing again for the Wang (2003) estimator. However, with the quadratic technology it is the unknown function for which the iterative estimator performs best. In all but one case, the average MSE of the Wang estimator is below that of the other nonparametric estimators. That being said, the difference is not nearly as large as that for the slope parameter in the linear data generating process case.

The results for the estimation of the slope parameter are surprising. Here the iterative estimator is generally worse than the other nonparametric estimators. This is unexpected given the strong performance of it in the linear technology. However, it again should be noted that the iterative procedure focuses on the estimation of the unknown function and not the slope parameter. As with the components estimator, there is no reason to necessarily expect it to outperform the other estimators. Wang (2003) only proves that the variance of the iterative estimator of the unknown function is smaller than that of the pooled estimator A possible explanation for the difference between the two technologies is that the linear technology is relatively simple and needs fewer iterations than the quadratic data generating process.

The Two-Step estimator of the varying coefficient parameter outperforms each of the other methods except the NPFGLS estimator for a single case. The Lin and Carroll estimator again performs poorly in terms of $\beta(\cdot)$ and now the Ullah and Roy estimator performs poorly in terms of both $m(\cdot)$ and $\beta(\cdot)$. Even more so than in the linear case, estimation of both estimators improves greatly when employing $\widetilde{\Omega}$ instead of $\widehat{\Omega}$. Again, the component estimator performs well in terms of the estimation of $m(\cdot)$, but not as well in the estimation of $\beta(\cdot)$.

Finally, we estimated each of the feasible estimators assuming that the omega matrix was known. The conclusions of this experiment are not significantly different from the estimated omega matrix calculations for the feasible estimators and are not reported here for sake of brevity but are available from the authors upon request.

12.3.2. Discussion

Based on the above results, we note that although the results differed between the two data generating processes, we were able to learn a great deal about the finite sample performance of the estimators. In summary, our principal conclusions are as follows: (1) As expected, when correctly specified, the parametric estimator outperforms each of the nonparametric estimators. (2) With the linear data generating process, there is no clear cut winner amongst the nonparametric estimators. Although the iterative estimator performed best in terms of estimating the slope parameter, it was not uniformly dominant in terms of the estimation of the unknown function. Although less efficient, the Two-Step estimator, the NPFGLS and to a lesser extent the LLLS estimator consistently performed well in terms of the estimation of $m(\cdot)$ and $\beta(\cdot)$. (3) When the technology became nonlinear, the Two-Step estimator performed well. It was outperformed in two cases by the NPFGLS (when $\rho = 0.1$). LLLS also performed well and occasionally gave a lower MSE than the NPFGLS estimator. The Wang (2003) estimator performed best in terms of estimating the unknown function, but surprisingly performed poorly in estimating the varying coefficient paramter. (4) Finally, the Henderson and Ullah (2005) omega matrix employed in both the estimation of the Two-Step and the NPFGLS estimators worked well, but caused both the Lin and Carroll, and Ullah and Roy estimators to give less efficient results in our exercise. The performance of the latter two estimators improves when employing the Ruckstuhl, Welsh and Carroll (2000) estimator of the variance-covariance matrix.

12.4. Empirical Application

The Monte Carlo results in the previous section compare the finite sample performance of several nonparametric panel data estimators. In this section we apply the aforementioned estimation procedures to the well known public capital productivity puzzle debate. Although numerous authors have examined this puzzle, we will compare our results to the more recent study by Baltagi and Pinnoi (1995). In their paper they consider the following production function

$$y_{it} = \alpha + \beta_1 KG_{it} + \beta_2 KP_{it} + \beta_3 L_{it} + \beta_4 unem_{it} + \varepsilon_{it}, \qquad (12.26)$$

where y_{it} denotes the gross state product of state i ($i = 1, ..., 48$) in period t ($t = 1970, ..., 1986$), public capital (KG) aggregates highways and streets

(KH), water and sewer facilities (KW), and other public buildings and structures (KO), KP is the Bureau of Economic Analysis' private capital stock estimates, and labor (L) is employment in non-agricultural payrolls. Details on these variables can be found in Munnell (1990) as well as Baltagi and Pinnoi (1995). Following Baltagi and Pinnoi (1995) we use the unemployment rate $(unem)$ to control for business cycle effects.

12.4.1. Parametric results

The results based on the Cobb-Douglas production function (linear in logs) are the same as in Baltagi (2001, page 25) and are reported in Table 12.3. The coefficients on both labor and private capital are found to be positive and statistically significant. On the other hand, the coefficient on public capital is quite small and statistically insignificant. These results have caused some to suggest that public capital is unproductive.

Such a large difference in the returns between public and private capital is difficult to explain. Using the estimated elasticities, the marginal products can also be evaluated. For example, the average estimates for KP and KG are found to be 0.263 and 0.042, respectively. If one views this as a problem of allocation of funds between private and public capital, returns from a dollar from public and private investment should be the same. Although an optimizing model would suggest equal returns to private and goverment capital, finding a political process to allocate goverment capital in such an optimal manner is difficult. Given this difficulty, it may be suggested that public capital would have a lower, but positive return. However, much of the recent literature does not find this phenomenon. Since the estimated marginal product of KG is much less than that of KP, there must be some explanation for such a massive overinvestment in public projects. The question is whether such a big difference in the returns can be explained.

The ammunition for this debate has primarily been arrived at by using the results from a Cobb-Douglas production function. However, assuming a particular production function assumes a particular form for the underlying production function, which may or may not be correct. Further, by construction, the elasticities of this model are exactly the same across all states and over all years. Thus, it seems natural to ask whether the results from the Cobb-Douglas model can be trusted. In fact, if the true model is nonlinear and one ignores it, the resulting estimates of returns to inputs are likely to be inconsistent.

12.4.2. *Nonparametric results*

The results for the nonparametric models are also reported in Table 12.3. It should be noted that we generally find similar results as in the Cobb-Douglas case (by using nonparametric regression that captures nonlinearity in the functional form) in terms of private capital, labor and unemployment. However, our results are significantly different in terms of the returns to public capital. Specifically, in a majority of the cases, we find evidence of a significant positive return to public capital. The elasticities here are similar to those found in Henderson and Kumbhakar (2006) and Henderson and Millimet (2005) who first showed the return to public capital to be positive and significant when employing nonparametric techniques.

Somewhat striking are the results for the Lin and Carroll $(\widehat{\delta}_2(x))$ and the Ullah and Roy $(\widehat{\delta}_3(x))$ estimators which use the Henderson and Ullah (2005) omega matrix $(\widehat{\Omega})$. In each case the median estimated elasticity for public capital is small and not significantly different from zero. However, when employing the Ruckstuhl, Welsh and Carroll (2000) estimator of Ω, each of these estimates becomes positive and significant (here they actually give results nearly identical to one another and to that of the LLLS estimator). These results should not be surprising. The Monte Carlo results in Table 2 show that the performance of the Lin and Carroll, and Ullah and Roy estimators when using the Henderson and Ullah (2005) omega matrix are less efficient in small samples and that they can be improved by employing the Ruckstuhl, Welsh and Carroll (2000) omega matrix.

The median results for the Wang (2003) estimator are perhaps more peculiar. Here the coefficient estimates for public capital and unemployment are similar to the FGLS estimator. However, the estimate for both physical capital and labor are far smaller than any of the other models suggest. Given that we can reject the linear model using the Li and Wang (1998) test (p-value $= 0.000$) and given the results from the previous section, it seems plausible that the results provided by the iterative estimator are less than desireable. Although the estimator performs very well in terms of prediction of the dependent variable ($R^2 = 0.993$), it gives parameter estimates at the median which are not economically intuitive. Given these peculiar results, we performed additional robustness checks. The iterative procedure required twelve iterations. We also examined the parameter estimates with a single iteration. The results did not significantly differ. We further tried estimation by using alternative estimators of σ^{st}. The conclusions of the study again did not change.

The results using the Two-Step estimator are equally interesting. When employing the Henderson and Ullah (2005) omega matrix, the median estimate corresponding to public capital (0.136) is slightly less than that of the NPFGLS and LLLS estimates. However, when employing the Ruckstuhl, Welsh and Carroll (2000) omega matrix, the return to public capital (0.153) is closer to that of the other nonparmetric estimators. As stated by Ruckstuhl, Welsh and Carroll (2000) and as shown in the Monte Carlo section of this paper, the estimates of the Two-Step estimator are greatly improved when the ratio σ_u^2/σ_v^2 increases. The Henderson and Ullah (2005) estimated $\rho \left(= \sigma_u^2/(\sigma_u^2 + \sigma_v^2)\right)$ for this particular data set is 0.621, which is between the 0.4 and 0.8 values in the Monte Carlo experiment which show strong performance of the Two-Step estimator. Thus, there is some reason to believe that the other estimators may be slightly overstating the elasticity of public capital in this data set. However, we must take the results from this estimator with a grain of salt as the standard errors of the median parameter estimates in this case are quite large.

When examining the point estimates, we find that for most states the estimated return to public capital is positive a majority of the time when looking at the LLLS, Two-Step or NPFGLS estimates. However, several states still give negative values for a majority of the time periods. For example, when employing the NPFGLS estimator we find that some of these states (New Mexico, North Dakota and South Dakota) give relatively small negative returns (≈ -0.05) while others (Wyoming) give larger negative returns (≈ -0.15) on average (we should further note that although Idaho, Iowa and Montana have positive returns on average, they also posses several negative values). It is interesting to note that this group of states are all plains states. One possible explanation for these negative returns to (or over-investment in) public capital is that they each have large relative investments in highways. These states have major highways running through them (designed to transport goods through their respective states) while at the same time their gross state products are relatively small.

12.4.3. *Discussion*

Here we showed that the popular linear parametric specification of the production function was not supported by the state-level panel data that are used to estimate returns on public and private capital. We argue that the linear technology was unable to capture the nonlinearitiy in the functional relationship underlying the production technology. Consequently, the parametric models are likely to give incorrect estimates of returns to inputs.

To avoid model misspecification we tried several nonparametric estimators for a one-way error component model. We find that the return to public capital is positive and significant for most of these estimators. For the estimators for which public capital was insignificant at the median, we primarily found evidence in the previous section that they did not perform as well in finite samples. This gives us some evidence that the results found in Henderson and Kumbhakar (2006) and Henderson and Millimet (2005) are robust to alternative nonparametric procedures. Thus, we feel that model misspecification is what caused the insignificant coefficients on public capital in parameteric models. This caused former researchers who employed the Cobb-Douglas model to believe that public capital was unproductive. Although this example does not necessarily prove the opposite, it does show that the previous research is not sufficient to condemn the idea that public capital is productive.

12.5. Concluding Remarks

In this paper we considered the problem of estimating a nonparametric panel data model with errors that exhibit a one-way error component structure. Specifically, we examined the finite sample performance of several nonparametric kernel estimators for estimating a random effects model. When the data generating process used in the Monte Carlo exercise was linear, there was no consensus on the best nonparametric estimator. The pooled estimator, which ignores the underlying structure of the data performed well, but was often dominated by several other estimators. At the same time, the Wang estimator, which has an asymptotic improvement over the pooled estimator, performed best in terms of the estimation of the varying coefficient parameter.

When the technology used became nonlinear, the Wang estimator performed best in nearly each trial when estimating the unknown function. However, this dominance was coupled with a weak performance in terms of estimating the slope parameter. Interestingly, the LLLS estimator which ignores the dependence structure in the model and the information contained in the disturbance vector covariance matrix also performed well and occasionaly gave a lower MSE than some of the nonparametric estimators which incorporate the underlying structure of the data. That being said, the Two-Step estimator performed well in terms of estimation of the un-

Table 12.3. Empirical Example – Median Estimated
Elasticities

	β(KG)	β(KP)	β(L)	β(unem)
FGLS	0.004	0.311	0.73	-0.006
	0.023	0.02	0.025	0.001
LLLS	0.155	0.281	0.653	-0.004
	0.037	0.044	0.058	0.005
Components	0.149	0.271	0.683	-0.005
	0.076	0.039	0.067	0.008
Two-Step	0.136	0.301	0.647	-0.005
	0.471	0.223	0.23	0.036
NPFGLS	0.152	0.292	0.644	-0.007
	0.037	0.031	0.03	0.008
Lin and Carroll	0	0.276	0.766	-0.004
	0.032	0.096	0.029	0.006
Ullah and Roy	-0.036	0.295	0.774	-0.007
	0.061	0.051	0.032	0.006
Wang	0.002	0.14	0.33	-0.006
	0.054	0.038	0.046	0.02
Two-Step $_{\text{RWC}}$	0.153	0.285	0.646	-0.004
	0.036	0.049	0.066	0.004
NPFGLS $_{\text{RWC}}$	0.155	0.281	0.653	-0.004
	0.037	0.044	0.06	0.005
Lin and Carroll $_{\text{RWC}}$	0.155	0.281	0.653	-0.004
	0.038	0.045	0.06	0.005
Ullah and Roy $_{\text{RWC}}$	0.155	0.281	0.653	-0.004
	0.037	0.044	0.058	0.005

known function and very good results in terms of estimation of the varying
coefficient parameter.

The analysis of the Monte Carlo section was complemented by the empirical application. We employed the nonparametric estimators discussed
in this paper to a U.S. state level data set. Specifically, the procedures
were used to estimate the returns to public capital, private capital, and
employment in gross state product from a panel of 48 states over 17 years.
In contrast to previous parametric studies and consistent with past nonparametric studies, we found that, in general, the return to public capital is
positive and significant. Further, for the cases where we found an insignificant result, the estimator generally performed poorly in the Monte Carlo
exercises. This gives us some evidence that the results found in previous
nonparametric studies are robust to alternative nonparametric procedures.

As a last comment we would like to note that throughout the paper
we assume the existence of random individual effects. In practice one way
want to test for the existence of random individual effects.

References

1. Baltagi, B. H. (2001). *Econometric Analysis of Panel Data*. John Wiley and Sons, New York.
2. Baltagi, B. H., Chang, Y.-J. and Li, Q. (1992). Monte Carlo Results on Several New and Existing Tests for the Error Component Model. *Journal of Econometrics* **54**, 95–120.
3. Baltagi, B. H., Hidalgo, J. and Li, Q. (1996). A Nonparametric Test for Poolability using Panel Data. *Journal of Econometrics* **75**, 345–367.
4. Baltagi, B. H. and Pinnoi, N. (1995). Public Capital Stock and State Productivity Growth: Further Evidence from an Error Components Model. *Empirical Economics* **20**, 351–359.
5. Fan, J. and Gijbels, I. (1992). Variable Bandwidth and Local Linear Regression Smoothers. *Ann Statist.* **20**, 2008–2036.
6. Henderson, D. J. and Kumbhakar, S. C. (2006). Public and Private Capital Productivity Puzzle: A Nonparametric Approach. *Southern Economic Journal* **73**, 219–232.
7. Henderson, D. J. and Millimet, D. L. (2005). Environmental Regulation and U.S. State-Level Production. *Economics Letters* **87**, 47–53.
8. Henderson, D. J. and Ullah, A. (2005). A Nonparametric Random Effects Estimator. *Economics Letters* **88**, 403–407.
9. Li, Q. and Racine, J. (2006). *Nonparametric Econometrics: Theory and Practice*. Princeton University Press, Princeton.
10. Li, Q. and Ullah, A. (1998). Estimating Partially Linear Panel Data Models with One-Way Error Components. *Econometric Reviews* **17**, 145–166.
11. Li, Q. and Wang, S. (1998). A Simple Consistent Bootstrap Test for a Parametric Regression Function. *Journal of Econometrics* **87**, 145–165.
12. Lin, X. and Carroll, R. J. (2000). Nonparametric Function Estimation for Clustered Data when the Predictor is Measured Without/With Error. *J. Amer. Statist. Assoc.* **95**, 520–534.
13. Lin, X. and Carroll, R. J. (2001). Semiparametric Regression for Clustered Data using Generalised Estimation Equations. *J. Amer. Statist. Assoc.* **96**, 1045–1056.
14. Lin, X. and Carroll, R. J. (2006). Semiparametric Estimation in General Repeated Measures Problems. *J. Roy. Statist. Soc.* **B 68**, 68–88.
15. Munnell, A. H. (1990). How Does Public Infrastructure Affect Regional Economic Performance? *New England Economic Review*, 11–32.
16. Pagan, A., and Ullah, A. (1999). *Nonparametric Econometrics*. Cambridge University Press, Cambridge.
17. Ruckstuhl, A. F., Welsh, A. H. and Carroll, R. J. (2000). Nonparametric Function Estimation of the Relationship Between Two Repeatedly Measured Variables. *Statistica Sinica* **10**, 51–71.
18. Silverman, B. W. (1986). *Density Estimation for Statistics and Data Analysis*. Chapman and Hall, London.

19. Ullah, A. and Roy, N. (1998). Nonparametric and Semiparametric Econometrics of Panel Data. In *Handbook of Applied Economics Statistics* (A. Ullah and D. E. A. Giles, Eds.), 1, Marcel Dekker, New York, 579–604.
20. Wang, N. (2003). Marginal Nonparametric Kernel Regression Accounting for Within-Subject Correlation. *Biometrika* **90**, 43–52.

Chapter 13

GERT Analysis of Consecutive-k Systems: An Overview

Kanwar Sen[1], Manju Agarwal[2] and Pooja Mohan[3]

[1] *University of Delhi, India*
[2] *Shib Nadar University, India*
[3] *RMS INDIA, India*

This paper presents an overview of Graphical Evaluation and Review Technique (GERT) for the reliability analysis of some consecutive-k systems including m-consecutive-k systems, consecutive-k systems with dependence, multi-state consecutive-k systems. Further, distributions of the waiting times of some patterns are also studied. It is demonstrated that GERT, being straightforward and simple, is generally more efficient than other approaches. This is so since GERT besides providing visual picture of the system makes it possible to analyze the given system in a less inductive manner. GERT is easier to use than the minimal cut set method. In GERT, one has to evaluate a W function, the generating function of the waiting time for the occurrence of the system failure, whereas in minimal cut set method one has to enumerate all possible minimal cut sets leading to system failure. Therefore, once the W function is computed, the reliability of the system consisting of any number of components can then be computed directly. Consecutive-k-out-of-n system models have wide applications in the design of integrated circuits, microwave relay stations in telecommunications, oil pipeline systems, vacuum systems in accelerators, computer ring networks (k-loops), space craft relay stations etc.

Keywords: Graphical Evaluation and Review Technique (GERT), Minimal cut set method, Multi-state consecutive-k system.

Contents

Acronym

GERT	Graphical Evaluation and Review Technique.
$C(k, n : F)$	A consecutive-k-out-of-n : F system.
$C_m(k, n : F)$	An m-consecutive-k-out-of-n : F system.
$C_m^+(k, n : F)$	An m-consecutive-atleast-k-out-of-n : F system.
$C^{k-1}(k, n : F)$	Consecutive-k-out-of-n : F system with $(k-1)$-step Markov dependence.
$C_H(k, n : F)$	Consecutive-k-out-of-n : F system with homogenous Markov dependence.
$C_m^{k-1}(k, n : F)$	An m-consecutive-k-out-of-n : F system with $(k-1)$-step Markov dependence.
$C_m^b(k, n : F)$	An m-consecutive-k-out-of-n : F system with Block-k dependence.
$C_{m_o}(k, n : F)$	An m-consecutive-k-out-of-n : F system for overlapping runs.
$C_{m_0}^{k-1}(k, n : F)$	An m-consecutive-k-out-of-n : F system for overlapping runs with $(k-1)$-step Markov dependence.
$C_{DFM}(k, r, n)$	Consecutive-k, r-out-of-n : F DFM system.

Notations

n : number of components in the system.

m : number of non-overlapping runs of k failed components (number of strings of atleast k consecutive failed components) for $C_m(k, n : F)$ $(C_m^+(k, n : F))$ causing system failure.

Binary Systems

$p(q)$	:	probability that a component is working (failed), $p+q=1$.
$q_1(q_2)$	:	probability of a component in failed-open (failed-short) mode, $p + q_1 + q_2 = 1$.
X_i	:	state of component i; $X_i = 0$ or 1 according as component i is working or failed.
$R_m(k,n)$	:	Reliability of $C_m(k,n\!:\!F)$.
$R_m^+(k,n)$	:	Reliability of $C_m^+(k,n\!:\!F)$.
$R_{k-1}(n)$	:	Reliability of $C^{k-1}(k,n\!:\!F)$.
$R_k^+(n)$	:	Reliability of $C_H(k,n\!:\!F)$.
$R_{k,r}(n)$	:	Reliability of $C_{DFM}(k,r,n)$ system.
$R_{m_0}(n)$	:	Reliability of $C_{m_0}(k,n\!:\!F)$.
$R_{m_0}^{k-1}(n)$	:	Reliability of $C_{m_0}^{k-1}(k,n\!:\!F)$.
$R_m^{k-1}(n)$	:	Reliability of $C_m^{k-1}(k,n\!:\!F)$.
$R_m^b(n)$	:	Reliability of $C_m^b(k,n\!:\!F)$.

Multi-State Systems

$M+1$	:	number of states of the system and its components. State M: perfect functioning. State 0: complete failure.
$\mathbf{S}$	:	$\{0,1,2,\ldots,M\}$, set of all possible states of the system.
x_i	:	state of component i, $x_i \in \mathbf{S}$, $i \in \{1,2,\cdots,n\}$.
$\mathbf{x}$	:	$(x_1,x_2,\ldots,x_n)$, vector of component states.
$\emptyset(\mathbf{x})$	:	system-state structure-function, $\emptyset(\mathbf{x}) \in \mathbf{S}$.
k_j	:	minimum number of components with $x_i \geq j$, $i \in \{1,2,\cdots,n\}$.
P_j	:	probability that a component is in state j or above when all components are i.i.d.
Q_j	:	$1 - P_j$, probability that a component is in state below j.
p_j	:	probability that a component is in state j
$R_{s,j}$	:	$\Pr(\emptyset(\mathbf{x}) \geq j)$.
$r_{s,j}$	:	$\Pr(\emptyset(\mathbf{x}) = j)$.
$F_{s,j}$	:	$1 - R_{s,j}$.
$F_j(n;k_j)$	:	probability that at least k_j consecutive components are in state below j for an n component system.
$R_j(n;k_j)$	:	$1 - F_j(n;k_j)$.

13.1. Introduction

In recent years consecutive-k-out-of-n : F systems $(C(k, n : F))$ have received a great deal of attention by engineering professionals particularly due to the low reliability of series systems and the high cost of the parallel systems (Jenab and Dhillon, 2005). Today, the most challenging problem faced by the researchers is to develop an efficient method for calculating the reliability of k-out-of-n systems. Kontoleon (1980) first studied the problem. However, Chiang and Niu (1981) gave the first solution of linear consecutive-k-out-of-n : F system for binary i.i.d. components. Since then, several other generalized models such as m-consecutive-k-out-of-n:F, k–within-consecutive-m-out-of-n : F, weighted-consecutive-k-out-of-n systems, strict consecutive-k-out-of-n : F systems, etc have been studied by several authors using different approaches. A survey of consecutive-k-out-of-n:F and related systems can also be found in Chao et al. (1995).

The study of dual failure mode (DFM) or three state devices has received continuing research interest since mid-1950s (Dhillon and Rayapati, 1986; Malon, 1989; Page and Perry, 1989; and Satoh et al., 1993). The major areas of substantial advancement are the reliability evaluation and optimal design of various redundant DFM structures. These systems have wide applicability in nuclear industry where the common terminology used is "failure to safety" and "failure to danger"; in fluid flow control networks where a defective valve could be either "stuck open" or "stuck closed"; in electronic/electrical engineering studies, where the modes of failure are usually labeled as "failed-open" and "failed-short", Koutras (1997). Moreover, dependency of the component states has been introduced into the probability models and binary state has been extended to multistate. One can see books: Chang et al. (2000), Kuo and Zuo (2003) and Pham (2003) for detailed discussion about the various approaches, i.e., Recursive Equations, Markov Chain, Combinatorics, to compute the reliabilities and compare their time complexities.

Graphical Evaluation and Review Technique (GERT) has been a well-established technique applied in several areas. However, application of GERT in reliability analysis of consecutive-k systems has not been reported much. Recently, Cheng (1994) and Bao (1989) have proposed $C(k, n : F)$ by GERT. Only recently, reliability evaluation of some important system structures including m-consecutive-k systems, consecutive-k systems with dependence, multistate consecutive-k systems and distributions of the waiting time of patterns have been carried out using GERT (Agarwal et al., 2007 a,b; Agarwal and Mohan, 2008a,b; Mohan et al, 2009; Sen et al., 2013). Reliability analysis of linear strict consecutive-k-out-of-$n : F$ systems, has been introduced and studied by Bollinger (1985). However, GERT is easier to use than the minimal cut set method. In GERT one has to evaluate a W function, the generating function of the waiting time for the occurrence of the system failure, whereas in the minimal cut set method one has to enumerate all possible minimal cut sets leading to system failure. Further, in recursive equation approach one has to set up a recursive equation to compute system reliability. Occasionally, a recursive equation is not self-evident and one has to introduce exogenous variables or equations to form a system of equations whose solution yield the desired recursive equation, Chang et al. (2000).

This paper presents reliability analysis of various consecutive-k systems and further evaluates the generating functions of the waiting time distributions of patterns using GERT. For a quick reference and understanding some models are described in brief. For details one can refer to Mohan (2007). We apologize that some reliability models pertaining to GERT analysis might be omitted in the survey. Section 13.2 provides brief review about GERT and Mason's formula. In sections 13.3-13.5 consecutive-k systems and its various generalizations have been reviewed briefly using GERT.

13.2. Consecutive-k System and its Variants: GERT Approach

Consecutive-k systems are important models in system reliability. They can be used to model many practical systems such as microwave stations of a telecom network, oil pipeline systems, vacuum systems in an electron accelerator etc. The main aim of this paper is to provide concise summary on reliability analysis of various consecutive-k systems using GERT approach.

GERT and Mason's Rule

GERT is a procedure, which combines the disciplines of flow graph theory, MGF (Moment Generating Function), and PERT (Project Evaluation and Review Technique) to obtain a solution to stochastic networks having logical nodes, and directed branches. Each branch has a probability that the activity associated with it will be performed. It, therefore, besides providing a visual picture of the system, makes it possible to analyze the given system in a less inductive manner. The results can be obtained in a straightforward manner based on MGF using Mason's formula, which takes care of all the possible products of transmittances of non-intersecting loops. The transmittance of an arc in a GERT network is the corresponding to W-function. It is used to obtain the information of a relationship, which exists between the nodes. In the GERT network, the variable z is used to multiply the W function associated with a branch; therefore, the power of z specifies the number of times branches were traversed whose values are multiplied by z. When a branch value is multiplied by z, we say the branch is tagged.

If we define $W(s\,|\,r)$, as the conditional W function associated with a network when the branches tagged with a z are taken r times, then the equivalent W generating function can be written as:

$$W(s,z) = W(s\,|\,0) + W(s\,|\,1)z + W(s\,|\,2)z^2 + \ldots + W(s\,|\,r)z^r + \ldots$$

$$= \sum_{r=0}^{\infty} W(s\,|\,r)z^r. \tag{13.1}$$

The relationship between the conditional W function and the conditional MGF is as shown

$$W(s\,|\,r) = \xi(r)M(s\,|\,r)$$

with

$$W(0\,|\,r) = \xi(r) \ (\text{because } M(0\,|\,r) = 1),$$

where $\xi(r)$ is the probability that the network is realized when the branches tagged with a z are traversed r times, and $M(s\,|\,r)$ is the conditional MGF associated with the network given that branches tagged with a z are traversed r times. Thus

$$W(s,z) = \sum_{r=0}^{\infty} \xi(r)M(s\,|\,r)z^r$$

which implies that

$$W(0,z) = \sum_{r=0}^{\infty} W(0\,|\,r)z^r = \sum_{r=0}^{\infty} \xi(r)z^r \ (\text{for } W(0\,|\,r) = \xi(r)). \tag{13.2}$$

The function $W(0,z)$ is the generating function of the waiting time for the network realization.

Mason's Rule (Whitehouse, 1973, pp. 168–172). In an open flow graph, write down the product of transmittances along each path from the independent variable to the dependent variable. Multiply its transmittance by the sum of the non-touching loops to that path. Sum these modified path transmittances, and divide by the sum of all the loops in the open flow graph yielding transmittance T as

$$T = \frac{[\sum(\text{path} * \sum \text{nontouching loops})]}{\sum \text{loops}} \qquad (13.3)$$

where

$$\sum \text{nontouching loops} = 1 - \left(\sum \text{first order nontouching loops}\right)$$
$$+ \left(\sum \text{second order nontouching loops}\right)$$
$$- \left(\sum \text{third order nontouching loops}\right) + \cdots$$

$$\sum \text{loops} = 1 - \left(\sum \text{first order loops}\right)$$
$$+ \left(\sum \text{second order loops}\right) - \cdots$$

13.3. Binary System Models

13.3.1. *m-Consecutive-k-out-of-n : F system* $(C_m(k, n : F))$

Griffith (1986) introduced and studied an m-consecutive-k-out-of-n:F system $(C_m(k, n : F))$ consisting of n components ordered on a line, which fails iff there are at least m non-overlapping runs of k consecutive failed components $(n \geq mk)$. Papastavridis (1990) provides an efficient recursive formula in terms of binomial coefficients for computing the failure probability of such systems having unequal components. Besides that, he also provides an exact formula for the failure probability of the i.i.d. case. Makri and Philippou (1996) have given an exact formula for the reliability of the linear system of i.i.d. components in terms of binomial, and multinomial coefficients. Agarwal, Sen and Mohan (2007) studied $C_m(k, n:F)$ for i.i.d. components using GERT, providing an explicit closed form formula, for its reliability evaluation.

The GERT network for $C_m(k, n : F)$ (Fig. 13.1) consists of m nodes, denoted by $0, k^1, k^2, \cdots, k^{m-1}$, each connected to $k - 1$ nodes denoted by $1, 2, \cdots, (k - 1)$, on the path leading to system failure state represented by the node k^m.

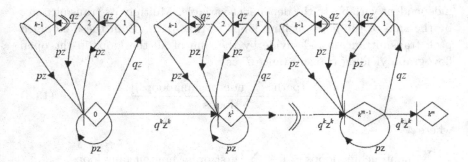

Fig. 13.1.   GERT Network Representing an m-Consecutive-k-out-of-n: F System.

Each node represents a specific state as described below:

0   : no component failure

1   : one component failure

2   : two consecutive components failure

$\vdots$

$k-1$: $k-1$ consecutive components failure

k^i : ith occurrence of k consecutive failures, which may be preceded by isolated strings of consecutive failures of size j $(1 \leq j \leq k-1)$, $i = 1, 2, \cdots, m$.

Then, k^m represents the system failure state.

For positive integers $k, m(mk \leq n)$, the function $W_{0,k^m}(0, z)$ for a $C_m(k, n: F)$, which is the generating function of the waiting time for the occurrence of the system failure, is given by

$$W_{0,k^m}(0, z) = \frac{q^{mk} z^{mk}(1 - qz)^m}{(1 - z + pq^k z^{k+1})^m}. \tag{13.4}$$

From Fig. 13.1, it is clear that the m nodes: $0, k^1, k^2, \cdots, k^{m-1}$ are in series, and identical. For each node,

$$\sum \text{first order loops} = pz + pqz^2 + pq^2 z^3 + \cdots + pq^{k-1} z^k.$$

Second, and higher order loops do not exist.

Therefore, by the procedure of Mason's rule, and (13.1) and (13.3), we obtain the function $W_{0,k^1}(0,z)$ from state 0 to state k^1 as

$$W_{0,k^1}(0,z) = \frac{q^k z^k}{1 - (pz + pqz^2 + \cdots + pq^{k-1}z^k)}$$
$$= \frac{q^k z^k (1 - qz)}{1 - z + pq^k z^{k+1}}. \tag{13.5}$$

As there are m identical nodes in series up to the system failure state k^m, the function $W_{0,k^m}(0,z)$ is given by

$$W_{0,k^m}(0,z) = \left(W_{0,k^1}(0,z)\right)^m$$

which, on using (13.5), yields (13.4).

Theorem 13.1. *For positive integers k, m, and $n (\geq mk)$, the reliability $R_m(k,n)$ of $C_m(k,n\colon F)$ is given by*

$$R_m(k,n) = 1 - \sum_{u=mk}^{n} \sum_{i=0}^{m} \sum_{l=0}^{\left[\frac{u-mk-i}{k+1}\right]} (-1)^l \binom{i-m-l}{i}$$
$$\binom{m+u-mk-i-kl-1}{m-1}\binom{u-mk-i-kl}{l} p^l q^{mk+kl+i} \tag{13.6}$$

where $[x]$ denotes the greatest integer in x.

Proof. The function $W_{0,k^m}(0,z)$ in (1.4) can be expressed as

$$W_{0,k^m}(0,z) = \sum_{u=mk}^{\infty} \xi(u) z^u \tag{13.7}$$

where $\xi(u)$, $u = mk, mk+1, \cdots, n, \ldots$, are the coefficients of series expansion of the RHS of (13.4). The coefficient $\xi(u)$ of z^u is given by

$$\xi(u) = \sum_{i=0}^{m} \sum_{l=0}^{\left[\frac{u-mk-i}{k+1}\right]} (-1)^l \binom{i-m-1}{i}\binom{m+u-mk-i-kl-1}{m-1}$$
$$\binom{u-mk-i-kl}{l} p^l q^{mk+kl+i}. \tag{13.8}$$

Therefore, using (13.3) the failure probability, $\bar{R}_m(k,n)$, of the system is obtained from

$$\bar{R}_m(k,n) = \sum_{u=mk}^{n} \xi(u)$$

which, on using (13.8), leads to (13.6).

Further, to study the time complexity in computing the reliability by the GERT formula, GERT-F, for $C_m(k, n : F)$ in comparison to Papastavridis (1990), and Makri and Philippou (1996) formulae, the reliability values by all the three formulae for several values of n, k, and m, taking $p = 0.75$, are computed on a Pentium 4 computer with a 2.93 GHZ CPU, and 248 MB of RAM under the Windows XP operating system using Mathematica (C Compiler). Some results are given in Table 13.1. A plot of the comparison between computational time (secs) involved in GERT-F and Binomial formulae given by Papastavridis (1990) is presented in Fig. 13.2. In the table, 'count' denotes the number of solutions generated from the equation $\sum_{j=1}^{k} j x_j = n - i - kx$, in the multinomial formula of Makri and Philippou (Theorem 1, 1996).

Thus, it may be concluded that the performance of both the Binomial formula of Papastavridis (1990), and Multinomial formula of Makri and Philippou (1996), in terms of time complexity, is not in general comparable with that of GERT-F.

Table 13.1. Comparison of Computational Time (secs) for $C_m(k, n : F)$ Corresponding to Different Values of n when $p = 0.75$, $k = 2$ and $m = 3$.

n	Reliability	Computational Time (secs.)			Number of 'counts'
		GERT-F	Papastavridis (Binomial)	Makri and Philippou (Multinomial)	
10	0.994821	≈ 0	≈ 0	0.125	27
20	0.940093	≈ 0	0.016	0.187	57
30	0.828963	0.015	0.031	0.401	87
40	0.688856	0.031	0.062	0.531	117
50	0.546094	0.046	0.157	0.845	147
60	0.417133	0.062	0.297	1.267	177
70	0.309325	0.078	0.516	1.470	207
80	0.223937	0.110	0.843	1.923	237
90	0.158947	0.156	1.297	2.375	267

This model has been further generalized to combined m-consecutive-k-out-of-$n : F$ and consecutive-k_c-out-of-$n : F$ systems, consisting of n components ordered in a line, while the system fails iff there exist at least k_c consecutive failed components or at least m non-overlapping runs of k consecutive failed components, where $k_c < mk$. The components are assumed to be i.i.d. An algorithm for system reliability evaluation is suggested based on the analysis of the system using GERT (Mohan et al., 2009).

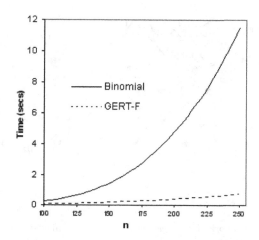

Fig. 13.2.   Graph of Computational Time (secs) involved in Computing $R_m(k,n)$ using GERT-F, and the Binomial Formula for $C_m(k,n:F)$ when $p = 0.75$, $k = 6$ and m = 5 for $n \geq 100$.

13.3.2.   *m-Consecutive-at least-k-out-of-n : F system* $(C_m^+(k, n : F))$

A consecutive-k-out-of-n : F system consisting of an ordered linear sequence of n i.i.d. components fails iff there are at least m non-overlapping runs of at least k consecutive failed components is referred to as m-consecutive-atleast-k-out-of-n:F system (Agarwal, Sen and Mohan, 2007). The GERT network for $C_m^+(k, n : F)$ contains two paths (Fig. 13.3): one leading to the terminal node k^m, and the other leading to the terminal node S_n, both representing system failure states. The first path contains $2m$ nodes, of which the first $(2m - 2)$ nodes form $(m - 1)$ identical pairs $(0, k^1), (0, k^2), \cdots, (0, k^{m-1})$, and the remaining two nodes form a pair $(0, k^m)$. Each pair is connected to $k - 1$ nodes denoted by $1, 2, \cdots, (k - 1)$. Here, nodes 0 to $k - 1$ represent similar specific states as described for $C_m(k, n:F)$. Further,

k^1 : first occurrence of k consecutive failures, which may be preceded by isolated strings of consecutive failures of size $j(1 \leq j \leq k-1)$

k^i : ith occurrence of k consecutive failures, which is preceded by a string of at least one working component that may be preceded by isolated strings of consecutive failures of size $j(1 \leq j \leq k-1)$, $i = 2, \cdots, m$

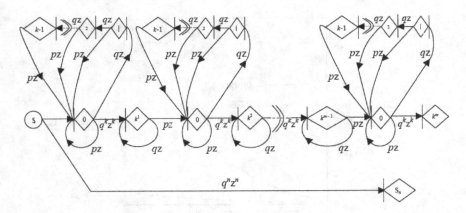

Fig. 13.3. GERT Network Representing an m-Consecutive-atleast-k-out-of-n : F System.

k^m : the system failure state

S_n : all the n components are in the failed state (then, obviously the system is failed).

Theorem 13.2. *For positive integers* $k, m(mk + m - 1 \leq n)$, *the function* $W^+_{0,k^m}(0,z)$ *for* $C^+_m(k, n : F)$, *which is the generating function of the waiting time for the occurrence of the system failure, is given by*

$$W^+_{0,k^m}(0,z) = \frac{q^{mk}p^{m-1}z^{mk+m-1}(1 - qz)}{(1 - z + pq^k z^{k+1})^m} + q^n z^n. \qquad (13.9)$$

It can be proved in the same way as Theorem 13.1.

13.3.3. *Consecutive-k-out-of-n : F system with Markov dependence*

Agarwal and Mohan (2008a) studied reliability analysis of consecutive-k-out-of-n systems with Markov dependence using GERT Approach.

13.3.3.1. *Consecutive-k-out-of-n : F system with (k − 1)-step Markov dependence* $(C^{k-1}(k, n : F))$

A consecutive-k-out-of-n : F system with $(k - 1)$-step Markov dependence consists of n ordered components such that the probability of failure of each component depends on the number of consecutive component failures immediately preceding the component. However, this type of dependence can go back for at the most $(k-1)$ steps. Fu (1986) suggested a recursive equation for reliability evaluation of such systems while Fu and Hu (1987) investigated the reliability bounds for such larger systems. Define

q_m, p_m: failure probability, reliability of a component given that the component is immediately preceded by m consecutive failures, $p_m + q_m = 1$, $m = 0, 1, 2, \cdots, (k-1)$.

The GERT network for $C^{k-1}(k, n : F)$, represented by Fig. 13.4 consists of $(k+1)$ nodes. The probabilities of failure and working are q_m and p_m, respectively, where $m = 0, 1, 2, \cdots, (k-1)$. Here node k represents k consecutive components failure and nodes 0 to $k-1$ represent similar specific states as described for $C_m(k, n : F)$. Obviously k represents the system failure state.

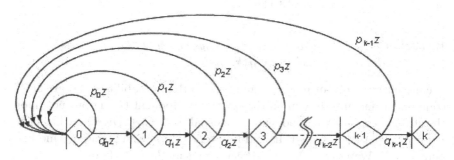

Fig. 13.4. GERT Network Representing Consecutive-k-out-of-n: F System with $(k-1)$-step Markov Dependence.

Theorem 13.3. *For positive integers k and n, the generating function, $W_{0,k}^{(k-1)}(0, z)$, of the waiting time for the occurrence of the system failure, is given by*

$$W_{0,k}^{(k-1)}(0, z) = \frac{\prod_{i=0}^{k-1}(q_i z)}{1 - p_0 z - \sum_{i=1}^{k-1}\left[\left(\prod_{j=0}^{i-1} q_j z\right)(p_i z)\right]}. \quad (13.10)$$

The reliability $R_{k-1}(n)$ of the system is given by

$$R_{k-1}(n) = 1 - \sum_{u=k}^{n} \xi(u), \quad (13.11)$$

where $\xi(u)$ is the coefficient of z^u in the power series expansion of $W_{0,k}^{(k-1)}(0,z)$.

Proof. For $C^{k-1}(k,n:F)$ system there is only one path from node 0 to node k, consisting of $(k+1)$ nodes, $0,1,2,\cdots,k-1,k$, leading to the system failure state, whose value is given by $q_0q_1q_2\cdots q_{k-1}z^k$. Further, only first order loops exist s.t.

$$\sum \text{first order loops} = p_0z + q_0p_1z^2 + q_0q_1p_2z^3 + \cdots$$
$$+ q_0q_1\cdots q_{k-2}p_{k-1}z^k,$$

while second and higher order loops do not exist.

Thus, by the use of Mason's rule and, (13.1) and (13.3), we obtain the function $W_{0,k}^{(k-1)}(0,z)$ as given by (13.10). Hence, the reliability $R_{k-1}(n)$ of the system is given by (13.11).

13.3.3.2. *Consecutive-k-out-of-n : F system with homogenous Markov dependence ($C_H(k,n:F)$)*

A consecutive-k-out-of-n:F system in which the probability of failure of ith component depends only upon the state of component $(i-1)$ but not upon the state of the other components is referred to as $C_H(k,n:F)$. It has been studied earlier by Papstavridis and Lambiris (1987) using recursive method and Ge and Wang (1990) using direct, exact method. Define

$p_{i,1}, q_{i,1}$: $\Pr\{X_i = 0 \,|\, X_{i-1} = 1\}$, probability that component i works given that preceding component fails, for $i = 2,3,\cdots,n; q_{i,1} = 1 - p_{i,1}$.

$p_{i,0}, q_{i,0}$: $\Pr\{X_i = 0 \,|\, X_{i-1} = 0\}$, probability that component i works given that preceding component works, for $i = 2,3,\cdots,n; q_{i,0} = 1-p_{i,0}$.

Assumptions

For the sequence of r.v.'s $X_1, X_2,\cdots,X_n$ to form a homogenous Markov chain, we have:

(i) $\Pr\{X_i = j_i \,|\, X_1 = j_1, X_2 = j_2,\cdots,X_{i-1} = j_{i-1}, X_{i+1} = j_{i+1},$
$\cdots, X_n = j_n\} = \Pr\{X_i = j_i \,|\, X_{i-1} = j_{i-1}\}$, for $j_1, j_2,\cdots,j_n = 0$
or 1.

(ii) $p_{i,0} = p_0, q_{i,0} = q_0, p_{i,1} = p_1, q_{i,1} = q_1$ for $i = 2,3,\cdots,n$.

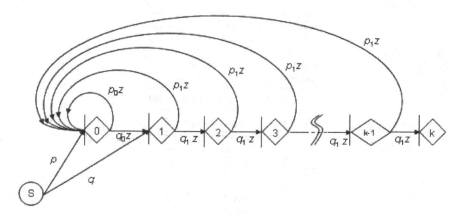

Fig. 13.5. A Consecutive-k-out-of-n:F System with Homogenous Markov Dependence on GERT Network.

GERT network for the corresponding system is shown in Fig. 13.5, where each node represents a specific state as described for $C_m^{k-1}(k, n : F)$ and S denotes the starting node.

Theorem 13.4. *For positive integers k and n, the generating function, $W_{0,k}^+(0, z)$, of the waiting time for the occurrence of the system failure, is given by:*

$$W_{0,k}^+(0, z) = \frac{pq_0 q_1^{k-1} z^{k+1} + qq_1^{k-1} z^k (1 - p_0 z)}{1 - p_0 z - q_0 p_1 z^2 \sum_{i=0}^{k-2} (q_1 z)^i}. \tag{13.12}$$

The reliability $R_k^+(n)$ of the system is given by

$$R_k^+(n) = 1 - \sum_{u=k}^{n} \xi(u), \tag{13.13}$$

where $\xi(u)$ is the coefficient of z^u in the power series expansion of $W_{0,k}^+(0, z)$.

Further, taking $p = p_0$, $q = q_0$ in $C_H(k, n : F)$ and $q_i = q_1$, $p_i = p_1$, $i = 2, 3, \ldots, (k-1)$ in $C^{k-1}(k, n : F)$, (13.10) and (13.12) become identical.

To study computational efficiency of GERT analysis, reliability values corresponding to $q_m = a^m q$, for $m = 0, 1, 2, \cdots, (k-1)$; $q \equiv \lambda/n^{1/k}$ where $0 < q < 1$ and $1 < a(= 1.05)$ and $\lambda = 0.3$, (Proportional increasing model, Fu and Hu, 1987) are computed for different values of n and k, and are given in Table 13.2. Further, in Table 13.3, for $C_H(k, n : F)$, a comparison has

Table 13.2. Computational Time (secs) for $C^{k-1}(k, n : F)$ Corresponding to $q_m = a^m q$ for $m = 0, 1, 2, \cdots, (k-1)$; $q \equiv \lambda/n^{1/k}$ where $0 < q < 1$, $1 < a(= 1.05)$ and $\lambda = 0.3$.

n	k	Reliability	Time (secs)
25	3	0.974322	0.015
	5	0.997181	0.015
50	3	0.972722	0.016
	5	0.996855	0.016
100	5	0.996657	0.032
	7	0.999516	0.032
150	5	0.996577	0.078
	7	0.999501	0.078

Table 13.3. Computational Time (secs) for $C_H(k, n : F)$ Corresponding to Different Values of n and k when $p = 0.90, p_0 = 0.80$ and $p_1 = 0.75$.

n	k	Reliability	GERT Time (secs)	Ge and Wang Time (secs)
25	3	0.792571	0.015	0.047
	7	0.999276	0.015	0.047
75	3	0.475132	0.031	0.797
	7	0.997351	0.031	0.531
100	5	0.942450	0.047	1.438
	7	0.996390	0.047	1.203
150	5	0.913732	0.078	5.329
	7	0.994471	0.078	4.218
175	5	0.899703	0.109	8.859
	7	0.993513	0.109	6.937

been made between GERT and, Ge and Wang (1990) formulae, illustrating the efficiency of GERT.

13.3.4. *m-Consecutive-k-out-of-n : F system with dependence*

Agarwal, Mohan and Sen (2007) extended m-consecutive-k-out-of-n : F system to the case of $(k - 1)$-step dependence and Block-k dependence. In addition to this, Agarwal and Mohan (2008b) also extended m-consecutive-k-out-of-n : F consisting of non-overlapping runs of k consecutive failed components to the case of overlapping runs of k consecutive failures.

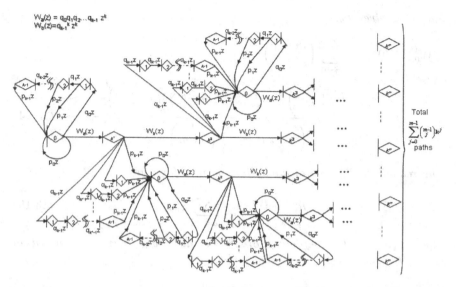

Fig. 13.6. Network Representing m-Consecutive-k-out-of-n:F System with $(k-1)$-step Markov Dependence.

13.3.4.1. m-Consecutive-k-out-of-n : F system with $(k-1)$-step Markov dependence $(C_m^{k-1}(k, n : F))$

Define

q_t, p_t: failure probability, reliability of a component given that the component is immediately preceded by t consecutive failures, $q_t + p_t = 1, t = 0, 1, 2, \ldots, (k-1)$.

An m-consecutive-k-out-of-n : F system with $(k-1)$-step Markov dependence consists of n components linearly ordered which fails iff there are at least m non-overlapping runs of k consecutive failures having $(k-1)$-step Markov dependence. In the GERT network (Fig. 13.6) nodes 0 to $k-1$ represent similar specific states as described for $C_m(k, n:F)$. Further,

k^i : ith occurrence of k consecutive failures having $(k-1)$-step Markov dependence which may be preceded by isolated strings of consecutive failures of size j $(1 \le j \le k-1), i = 1, 2, \ldots, m$.

Obviously, k^m represents system failure state.

Theorem 13.5. *For positive integers k, m and $n(\ge mk)$, the generating function $W_{o,k^m}^{(k-1)}(0, z)$, of the waiting time for the occurrence of the system*

failure, is given by:

$$W_{0,k^m}^{(k-1)}(0,z) = \sum_{j=0}^{m-1} \binom{m-1}{j}$$

$$\frac{\left(\prod_{i=0}^{k-1} q_i z^k\right)^{j+1} (q_{k-1}^k z^k)^{m-j-1}\left(\sum_{i=0}^{k-1}(q_{k-1}z)^i\right)^j p_{k-1}^j z^j}{\left(1 - p_0 z - \sum_{l=1}^{k-1}\left(\prod_{i=0}^{l-1} q_i\right) p_l z^{l+1}\right)^{j+1}} \qquad (13.14)$$

The reliability $R_m^{k-1}(n)$ of the system is, therefore, given by:

$$R_m^{k-1}(n) = 1 - \sum_{u=mk}^{n} \xi(u), \qquad (13.15)$$

where $\xi(u)$ is the coefficient of z^u in the power series expansion of $W_{o,k^m}^{(k-1)}(0,z)$.

The whole process can be summarized as:

We have $(m+k)$ nodes: $0, 1, 2, \ldots, k-1, k^1, k^2, \ldots, k^m$ on the paths leading to system failure. The total number of paths is counted keeping in view the following cases:

(i) Each of the nodes, node 0 and the nodes $k^j (j = 1, 2, \ldots, m-1)$ is followed by k contiguous failures, giving the path

$$0 \to k^1 \to k^2 \to \ldots k^j \to \ldots k^{m-1} \to k^m.$$

(ii) Only one of the nodes k^j $(j = 1, 2, \ldots, m-1)$ is followed by isolated failure strings of less than k contiguous failures, i.e., strings of size $0, 1, 2, \ldots, k-1$ yielding $\binom{m-1}{1}k$ paths.

(iii) Only two nodes, say k^{j_1} and $k^{j_2} (j_1 \neq j_2 = 1, 2, \ldots, m-1)$ are followed by isolated strings of less than k contiguous failures, leading to $\binom{m-1}{2}k^2$ paths.

⋮

(m) All nodes $k^j (j = 1, 2, \ldots, m-1)$ are followed by isolated strings of less than k contiguous failures, thus leading to $\binom{m-1}{m-1}k^{m-1}$ paths.

Therefore, from cases (i) to (m), the total number of paths leading to system failure is $1 + \sum_{j=1}^{m-1} \binom{m-1}{j}k^j = \sum_{j=0}^{m-1} \binom{m-1}{j}k^j = (k+1)^{m-1}$. It can otherwise also be argued like this: there is only one path from node 0 to k^1 and number of paths to reach node k^{i+1} from k^i $(i = 1, 2, \ldots, m-1)$

is $(k+1)$. Thus, total number of paths from node 0 to k^m is $(k+1)^{m-1}$. Further, corresponding to each node '0',

$$\sum \text{first order loops} = p_0 z + q_0 p_1 z^2 + q_0 q_1 p_2 z^2 + \ldots + q_0 q_1 q_2 \ldots q_{k-2} p_{k-1} z^k,$$

while second and higher orders do not exist.

Applying Mason's formula, $W_{0,k^m}^{k-1}(0,z)$, is obtained as given in (13.14).

Now, $W_{0,k^m}^{k-1}(0,z)$ can be expressed as:

$$W_{o,k^m}^{(k-1)}(0,z) = \xi(mk)z^{mk} + \xi(mk+1)z^{mk+1} + \ldots + \xi(n)z^n$$
$$+ \xi(n+1)z^{n+1} + \xi(n+2)z^{n+2} + \ldots \quad (13.16)$$

and so system unreliability is obtained as $\sum_{u=mk}^{n} \xi(u)$. Hence the reliability $R_m^{k-1}(n)$ is given by (13.15).

13.3.4.2. *m-Consecutive-k-out-of-n : F system with Block-k dependence* $(C_m^b(k,n:F))$

Define

q_i, p_i: failure probability, reliability of a component given that the component is preceded by i blocks each of k consecutive failures, $q_i + p_i = 1$, $i = 0, 1, 2, \ldots, (m-1)$.

An m-consecutive-k-out-of-n:F system in which after the occurrence of ith block of k consecutive failures system deteriorates and component failure probability becomes $q_i, i = 1, 2, \ldots, (m-1)$, is referred to as m-consecutive-k-out-of-n : F system with Block-k dependence. The network (Fig. 13.7) consists of $(m+k)$ nodes such that the nodes 0 to $k-1$ represent similar specific states as described for $C_m(k,n:F)$. Further,

k^i : ith occurrence of k consecutive failures having Block-k dependence which may be preceded by isolated strings of consecutive failures of size j $(1 \le j \le k-1)$, $i = 1, 2, 3, \ldots, m$.

Obviously, k^m represents system failure state.

Theorem 13.6. *For positive integers k, m and $n (\ge mk)$, the generating function $W_{o,k^m}^b(0,z)$, of the waiting time for the occurrence of the system failure, is given by:*

$$W_{0,k^m}^b(0,z) = \prod_{i=0}^{m-1} \frac{(q_i z)^k}{(1 - p_i z - q_i p_i z^2 - q_i^2 p_i z^3 - \ldots - q_i^{k-1} p_i z^k)}. \quad (13.17)$$

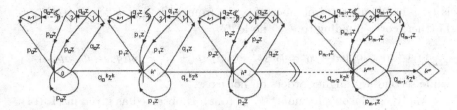

Fig. 13.7. Network Representing m-Consecutive-k-out-of-n : F System with Block-k Dependence.

The reliability $R_m^b(n)$ of the system is, therefore, given by:

$$R_m^b(n) = 1 - \sum_{u=mk}^{n} \xi(u), \qquad (13.18)$$

where $\xi(u)$ is the coefficient of z^u in the power series expansion of $W_{o,k^m}^b(0, z)$.

Proof. For $C_m^b(k, n : F)$, we can summarize the steps as follows: If one component is normal, then there is conditional probability p_0 from state $v-1$ to state 0 ($v = 1, 2, 3, \ldots, k$). If one component fails, then there is conditional failure probability q_0 from state $v-1$ to state v ($v = 1, 2, 3, \ldots, k$). System moves to state k^1 on occurrence of k contiguous failures for the first time and, conditional working and failure probability of a component now becomes p_1 and q_1, respectively. In this way after each subsequent ith occurrence of block of k consecutive failures system deteriorates and component failure probability becomes q_i, $i = 1, 2, \ldots, (m-1)$. System fails when node k^m is entered.

There exists only one path from 0 to k^1 to k^2 to $\ldots k^m$ and in this path the terminal node k^m is preceded by m similar nodes $0, k^1, k^2, \ldots, k^{m-1}$, which differ only with respect to the probability of failure of a component, such that correspondingly

$$\sum \text{first order loops} = p_i z + q_i p_i z^2 + q_i^2 p_i z^3 + \ldots + q_i^{k-1} p_i z^k,$$
$$i = 0, 1, 2, \ldots, (m-1),$$

and second and higher order loops do not exist.

Applying Mason's formula $W_{o,k^m}^b(0, z)$ is obtained as given in (13.17). Now, the $W_{o,k^m}^b(0, z)$ function in (13.17) can be expressed as:

$$W_{o,k^m}^b(0, z) = \xi(mk)z^{mk} + \xi(mk+1)z^{mk+1} + \ldots$$
$$+ \xi(n)z^n + \xi(n+1)z^{n+1} + \xi(n+2)z^{n+2} + \ldots$$

and so, the system unreliability is obtained as $\sum_{u=mk}^{n} \xi(u)$. Hence, the reliability $R_m^b(n)$ is given by (13.18).

To study computational efficiency of GERT analysis, reliability values of $C_m^{k-1}(k, n : F)$ and $C_m^b(k, n : F)$ systems for several pairs of values of n and m taking $k = 4$ are computed and their computing time (secs) is given in Tables 13.4 and 13.5, respectively.

Table 13.4. Computational Time (secs) for $C_m^3(4, n : F)$ System Corresponding to Different Values of n and m when $p_0 = 0.90$, $p_1 = 0.80$, $p_2 = 0.70$, $p_3 = 0.65$.

n	m	Reliability	Time
30	2	0.9985248	0.032
	3	0.9999695	0.032
	5	≈ 1	0.032
50	2	0.9958910	0.046
	3	0.9998581	0.047
	5	≈ 1	0.047
100	2	0.9844201	0.125
	3	0.9989695	0.125
	5	0.9999977	0.132

Table 13.5. Computational Time (secs) for $C_m^b(4, n : F)$ System Corresponding to Different Values of n and m when $p_0 = 0.90$, $p_1 = 0.80$, $p_2 = 0.75$, $p_3 = 0.70$, $p_4 = 0.65$.

n	m	Reliability	Time
30	2	0.9999674	0.015
	3	0.9999995	0.016
	5	≈ 1	0.017
50	2	0.9998908	0.031
	3	0.9999963	0.031
	5	≈ 1	0.032
100	2	0.9995109	0.156
	3	0.9999612	0.157
	5	0.9999994	0.158

13.3.4.3. *m-Consecutive-k-out-of-n : F system with overlapping*
        runs $(C_{m_0}(k, n : F))$

An m-consecutive-k-out-of-$n : F$ system, consisting of an ordered linear
sequence of n i.i.d. components that fails if and only if there are at least
m overlapping runs of k consecutive failed components is referred to as
$C_{m_0}(k, n : F)$ (Agarwal and Mohan, 2008b). The GERT network is given
in Fig. 13.8, where nodes 0 to $k - 1$ represent specific states as described
for $C_m(k, n : F)$. Further,

k	:	occurrence of first run of k consecutive failures, which may be preceded by isolated strings of consecutive failures of size $j (1 \le j \le k - 1)$
$k + 1$	:	occurrence of second run of k consecutive failures which may or may not be preceded by isolated failure strings of j consecutive failures $(1 \le j \le k - 1)$
$\vdots$		
$m + k - 1$	:	occurrence of mth run of k consecutive failures, may or may not be preceded by isolated failure strings of j consecutive failures $(1 \le j \le k - 1)$ (Obviously the system is in failed state).

It can be summarized as follows. If one component is normal, then there
is conditional probability p from state $v - 1$ to state 0 $(v = 1, 2, 3, \ldots, k)$.
If one component fails, then there is a failure conditional probability q
from state $v - 1$ to state $v(v = 1, 2, 3, \ldots, k)$. Once when k contiguous
components are failed system moves to state k and if the next contiguous
component is in failed state then system enters state $k + 1$, otherwise, in
case of normal component system enters state 0; same procedure is followed
for the remaining set of components until next set of k contiguous failures
occurs. In this way same procedure is followed until state $m + k - 1$ is
reached. The system fails as soon as it enters this state.

Theorem 13.7. *For positive integers $k, m, n (\ge mk)$, the $W_{0,m+k-1}(0, z)$
function, generating function of the waiting time for the occurrence of system
failure, is given by:*

$$W_{0,m+k-1}(0, z)$$
$$= \sum_{j=0}^{m-1} \binom{m-1}{j} \frac{q^{m+k+(k-1)j-1}p^j z^{m+k+kj-1}}{(1 - pz - qpz^2 - q^2pz^3 - \ldots - q^{k-1}pz^k)^{j+1}}. \quad (13.19)$$

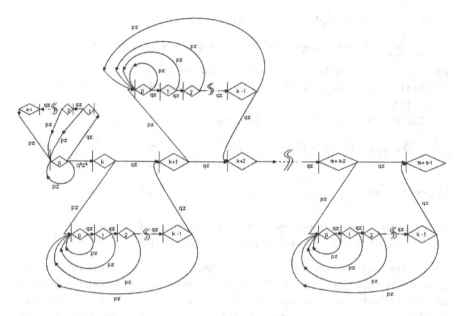

Fig. 13.8. GERT Network Representing m-Consecutive-k-out-of-n : F System with Overlapping Runs.

The reliability $R_{m_0}(n)$ of the system is given by:

$$R_{m_0}(n) = 1 - \sum_{u=m+k-1}^{n} \xi(u), \qquad (13.20)$$

where $\xi(u)$ is the coefficient of z^u in the power series expansion of $W_{0,m+k-1}(0,z)$.

Proof. In $C_{m_0}(k,n:F)$ system there are $(m+k)$ nodes: $0,1,2,\ldots,k-1,k,k+1,\ldots,m+k-1$ on the paths leading to system failure. There is only one path from node 0 to node k, the first occurrence of k consecutive failures, which may or may not be preceded by isolated strings of less than k contiguous failures. Further, from each of the nodes $k,k+1,k+2,\ldots,k+m-2$, there are 2 paths to reach the subsequent node, i.e., from node $k+i$ to $k+i+1(i=0,1,2,\ldots,m-2)$ either directly or via nodes $0,1,2,\ldots,k-1$ depending upon whether the next contiguous component is failed or working (which may or may not be succeeded by isolated strings of less than k contiguous failures). Thus, there are 2^{m-1} paths leading to system failure.

Further, corresponding to each node '0':

$$\sum \text{first order loops} = pz + qpz^2 + q^2pz^3 + \ldots + q^{k-1}pz^k,$$

while second and higher order loops do not exist.

By the use of Mason's rule $W_{0,m+k-1}(0, z)$ function is obtained as given in (13.19).

Now the $W_{0,m+k-1}(0, z)$ function can be expressed as:

$$W_{0,m+k-1}(0, z) = \xi(m + k - 1)z^{m+k-1} + \xi(m + k)z^{m+k} + \ldots$$
$$+ \xi(n)z^n + \xi(n + 1)z^{n+1} + \ldots$$

Hence, the reliability $R_{m_0}(n)$ of the system is given by (13.20).

13.3.4.4. *m-Consecutive-k-out-of-n : F system with overlapping runs for $(k-1)$-step Markov dependence ($C_{m_0}^{k-1}(k, n : F)$)*

An m-consecutive-k-out-of-n:F system with overlapping runs for $(k-1)$-step Markov dependence $C_{m_0}^{k-1}(k, n : F)$ consists of n components linearly arranged which fails iff there are at least m overlapping runs of k consecutive failures having $(k-1)$-step Markov dependence (Agarwal and Mohan, 2008b).

The GERT network (Fig. 13.9) can be summarized as follows: If one component is normal, then there is conditional probability p_{v-1} from state $v-1$ to state 0 ($v = 1, 2, 3, \ldots, k$). If one component fails, then there is conditional failure probability q_{v-1} from state v-1 to state v ($v = 1, 2, 3, \ldots, k$). Once when k contiguous components are failed system moves to state k and if the next contiguous component is in failed state then system enters state $k + 1$ otherwise in case of normal component system enters state 0; same procedure is followed for the remaining set of components until next set of k contiguous failures occurs. In this way same procedure is followed until state $m + k - 1$ is reached. The system fails as soon as it enters this state.

Theorem 13.8. *For positive integers $k, m, n(\geq mk)$, the $W_{0,m+k-1}^{k-1}(0, z)$ function, i.e., generating function of the waiting time for the occurrence of system failure is given by:*

$$W_{0,m+k-1}^{(k-1)}(0, z)$$
$$= \sum_{j=0}^{m-1} \binom{m-1}{j} \frac{(q_0 q_1 q_2 \cdots q_{k-2} z^{k-1})^{j+1}(q_{k-1}z)^m p_{k-1}^j z^j}{(1 - p_0 z - q_0 p_1 z^2 - q_0 q_1 p_2 z^3 - \ldots - q_0 q_1 \cdots q_{k-2} p_{k-1} z^k)^{j+1}}.$$

$$(13.21)$$

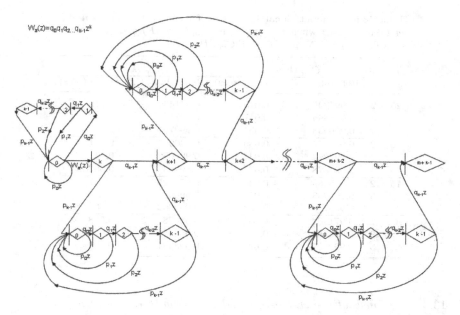

Fig. 13.9. GERT Network Representing m-Consecutive-k-out-of-n : F System with Overlapping Runs for $(k-1)$-step Markov Dependence.

The reliability $R_{m_o}^{k-1}(n)$ of the system is given by:

$$R_{m_o}^{k-1}(n) = 1 - \sum_{u=m+k-1}^{n} \xi(u), \tag{13.22}$$

where $\xi(u)$ is the coefficient of z^u in the power series expansion of $W_{0,m+k-1}^{k-1}(0,z)$.

Proof. Follows as of Theorem 7.

To study the computational efficiency of GERT analysis, reliability values of both the systems, $C_{m_0}(k,n:F)$, and $C_{m_0}^{k-1}(k,n:F)$ for several values of n for different pairs of values of m and k ($n \geq mk$) were computed using Mathematica. It may be observed from Table 13.6 that CPU time taken in both the systems is very small. Even for $n = 100$ it is less than 0.11 secs, illustrating the efficiency of GERT in reliability analysis of such systems.

Table 13.6. Computational Time (secs) for Different Values of n and m taking $k = 4$ when $p = 0.\ 80$ in $C_{m_0}(k, n : F)$ and $p_0 = 0.90$, $p_1 = 0.80$, $p_2 = 0.75$, $p_3 = 0.70$ in $C_{m_0}^{k-1}(k, n : F)$.

n	m	$C_{m_0}(k, n : F)$		$C_{m_0}^{k-1}(k, n : F)$	
		Reliability	CPU Time (secs)	Reliability	CPU Time (secs)
25	2	0.994395	0.015	0.991434	0.016
	3	0.998896	0.017	0.997501	0.019
	5	0.999958	0.031	0.999791	0.032
50	2	0.987262	0.030	0.980921	0.031
	3	0.997274	0.046	0.994078	0.047
	5	0.999878	0.051	0.999437	0.052
75	2	0.979583	0.035	0.970025	0.037
	3	0.995363	0.056	0.990302	0.057
	5	0.999768	0.064	0.999002	0.068
100	2	0.971401	0.057	0.958783	0.058
	3	0.993165	0.073	0.986186	0.076
	5	0.999625	0.094	0.998481	0.104

13.3.4.5. *Consecutive-k, r-out-of-n : DFM system $(C_{DFM}(k, r, n))$*

A consecutive-k, r-out-of-n : DFM system consists of n components linearly arranged which fails if and only if at least k consecutive components are failed-open or at least r consecutive components are failed-short. The GERT network (Agarwal and Mohan, 2007) for $C_{DFM}(k, r, n)$ is given in Fig. 13.10, where each node represents a specific state as described below:

S	:	starting node
0^1	:	no component in failed-open mode
0^2	:	no component in failed-short mode
1	:	one component in failed-open or failed-short mode
2	:	two consecutive components in failed-open or failed-short mode
$\vdots$		
$k - 1 (\text{or } (r - 1))$	:	$k - 1$ (or $(r - 1)$) consecutive components in failed-open mode (or in failed-short mode)
k (or r)	:	(k or r) consecutive components in failed-open mode (or in failed-short mode).

For $C_{DFM}(k, r, n)$, GERT network (Fig. 13.10) can be explained as follows: There are two paths from the starting node S one leading to terminal

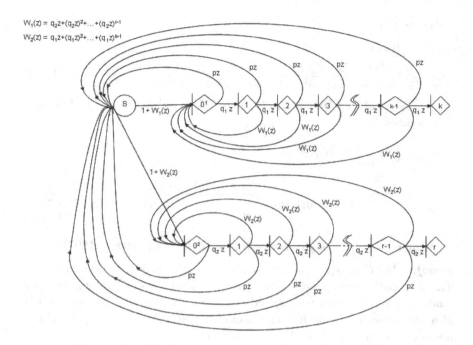

Fig. 13.10. GERT Network Representing $C_{DFM}(k,r,n)$.

node k in which system fails due to open mode and second leading to terminal node r in which system fails due to short mode. The first path can be summarized as: System on occurrence of first open mode failure (which may be preceded by sub strings consisting of t consecutive short-mode failures with probability q_2^t, $t = 1, 2, \cdots, r-1$ and/or normal component with probability p) moves to state 1 via 0^1 from S. Further, if the next contiguous component also fails due to open mode then system moves to state 2 from state 1 with conditional probability q_1. Otherwise, system either moves to state S with conditional probability p (in case next contiguous component is normal) or to state 0^1 with conditional probability q_2^t, $t = 1, 2, \cdots, r-1$ (in case there occur t contiguous short-mode failures). In this way same procedure is followed until state k is reached i.e., k contiguous open mode failures occur. Similar procedure is followed for the second path in which system fails due to short mode failure.

Theorem 13.9. *For positive integers k, r, n ($n \geq \max(k,r)$), $W_{DFM}^{k,r}(0, z)$, i.e., generating function of the waiting time for the occurrence of system* *failure is given by:*

$$W_{DFM}^{k,r}(0,z)$$

$$= \frac{(q_1z)^k(1-(q_2z)^r)(1-q_1z)+(q_2z)^r(1-(q_1z)^k)(1-q_2z)}{1-z+(1-q_1)z(q_1z)^k+(1-q_2)z(q_2z)^r-(1+pz)(q_1z)^k(q_2z)^r} \quad (13.23)$$

and the generating function $R_{k,r}(0,z)$ for the reliability $R_{k,r}(n)$ of the system is given by :

$$R_{k,r}(0,z) = \sum_{n=0}^{\infty} R_{k,r}(n)z^n$$

$$= \frac{(1-(q_1z)^k)(1-(q_2z)^r)}{1-z+(1-q_1)z(q_1z)^k+(1-q_2)z(q_2z)^r-(1+pz)(q_1z)^k(q_2z)^r}. \quad (13.24)$$

It matches with the generating function $G(z)$ of Koutras (1997).

Proof.   For $C_{DFM}(k,r,n)$ there exist two paths leading to system failure. The first path corresponds to the case in which system fails due to open mode and second path corresponds to system failure due to short mode. Moreover, each of the paths consists of only first order loops, as second and higher orders do not exist. For Path No. 1:

$$\sum \text{first order loops} = pz\left(1+\sum_{i=1}^{k-1}(q_1z)^i\right)\left(1+\sum_{j=1}^{r-1}(q_2z)^j\right)$$

$$+(q_1z)(q_2z)\left(\sum_{i=0}^{k-2}(q_1z)^i\right)\left(\sum_{j=0}^{r-2}(q_2z)^j\right)$$

$$= \frac{\begin{pmatrix} pz(1-(q_1z)^k)(1-(q_2z)^r)+(q_1z) \\ \times(q_2z)(1-(q_1z)^{k-1})(1-(q_2z)^{r-1}) \end{pmatrix}}{(1-q_1z)(1-q_2z)}. \quad (13.25)$$

Similarly, we obtain $\sum$ first order loops for Path No.2, which is same as (13.25). As $\sum$ nontouching loops do not exist for both the paths, therefore, by the use of Mason's rule, $W_{DFM}^{k,r}(0,z)$ function is:

$$W_{DFM}^{k,r}(0,z) = \frac{(q_1z)^k(1-(q_2z)^r)(1-q_1z)+(q_2z)^r(1-(q_1z)^k)(1-q_2z)}{\begin{pmatrix} (1-q_1z)(1-q_2z)-pz(1-(q_1z)^k)(1-(q_2z)^r)) \\ -q_1q_2z^2(1-(q_1z)^{k-1})(1-(q_2z)^{r-1}) \end{pmatrix}}$$

$$= \frac{(q_1z)^k(1-(q_2z)^r)(1-q_1z)+(q_2z)^r(1-(q_1z)^k)(1-q_2z)}{\begin{pmatrix} 1-z+(1-q_1)z(q_1z)^k \\ +(1-q_2)z(q_2z)^r-(1+pz)(q_1z)^k(q_2z)^r \end{pmatrix}}.$$

Hence (13.23) is proved.

Thus, the generating function $R_{k,r}(0,z)$ for the reliability $R_{k,r}(n)$ of the system (Feller, 1968) is:

$$R_{k,r}(0,z) = \frac{1 - W_{DFM}^{k,r}(0,z)}{1 - z}, \qquad (13.26)$$

which on solving yields (13.24).

Further, for $q_2 = 0$, the generating function $R_{k,r}(0, z)$ reduces to reliability generating function of ordinary consecutive-k-out-of-$n : F$ system. It can be observed that using GERT, the generating function for the reliability of the $C_{DFM}(k, r, n)$ can be obtained in a much easier way than Koutras (1997).

13.4. Multi-state Models

Many practical systems can perform their intended function at more than two different levels ranging from partially working to completely failed and referred to as multi-state (MS) systems, for example, see Zuo and Liang (1994), Malinowski and Preuss (1995, 1996), Huang et al. (2000), Yamamoto et al. (2006). Mohan and Agarwal (2007a) studied reliability analysis of increasing MS k-out-of-n:G system and decreasing MS consecutive-k-out-of-$n : F$ system using GERT.

13.4.1. *Increasing MS k-out-of-n : G system*

As defined in Huang et al. (2000), $\emptyset(x){\geq}j(j = 1, 2, \cdots, M)$ if there exists an integer value $l(j \leq l \leq M)$ such that at least k_l components are in states at least as good as l. An n-component system with such a property is a MS k-out-of-n:G system.

When $k_1 \leq k_2 \leq \cdots \leq k_M$, the system is called increasing MS k-out-of-n:G system. However, when k_j is a constant, i.e., $k_1 = k_2 = \cdots = k_M = k$, the system reduces to MS k-out-of-n:G system.

Theorem 13.10. *For positive integers k_j and n, the function $W_{s,j}(0, z)$ for increasing MS k-out-of-n:G system, which is the generating function of the waiting time for the occurrence of system state level j, i.e., if $\emptyset(x){\geq}j$ is given by:*

$$W_{s,j}(0, z) = \left(\frac{P_j z}{1 - Q_j z} \right)^{k_j}. \qquad (13.27)$$

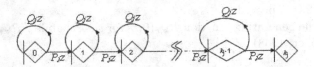

Fig. 13.11.   GERT Network Representing MS Increasing k-out-of-n: G System at System State Level j.

The reliability $R_{s,j}$ for increasing MS k-out-of-n: G system is, therefore, given by:

$$R_{s,j} = P_j^{k_j} \sum_{u=0}^{n-k_j} \binom{u+k_j-1}{k_j-1} Q_j^u. \qquad (13.28)$$

Proof.   GERT network for MS increasing k-out-of-n: G at system state level j is represented by Fig. 13.11. It consists of $k_j + 1$ nodes denoted by $0, 1, 2, \ldots, k_j - 1, k_j$ on the path leading to the required system state. Each node represents a specific state as described below:

0   :   a component in state $< j$
1   :   one component in state $\geq j$
2   :   second component in state $\geq j$

$\vdots$

k_j : k_jth component in state $\geq j$

It can be summarized as: If one component is in state $\geq j$, then there is conditional probability P_j from state $v - 1$ to state $v(v = 1, 2, \ldots, k_j)$. If one component is below state j then there is conditional probability Q_j from state $v - 1$ to state $v(v = 1, 2, \ldots, k_j)$. System state is reached as soon as node k_j is reached.

From Fig. 13.11, we see that there are k_j identical nodes. Thus, by the procedure of Mason's rule, we obtain $W_{s,j}(0, z)$ function as given in (13.28).

Now, the function $W_{s,j}(0, z)$ in (1.28) can be expressed as:

$$W_{s,j}(0, z) = \sum_{u=k_j}^{\infty} \xi(u) z^u, \qquad (13.29)$$

where $\xi(u), u = k_j, k_j + 1, \cdots, n, \ldots$, is the coefficient of z^u in the power series expansion of $W_{s,j}(0, z)$.

Thus, the reliability $R_{s,j}$ of the system is given by

$$R_{s,j} = \sum_{u=k_j}^{n} \xi(u),$$

which on solving yields (13.28).This proves Theorem 13.10.
Now, the probability that the system is in state j is given by:

$$r_{s,j} = R_{s,j} - R_{s,j+1}, \quad R_{s,0} = 1. \tag{13.30}$$

13.4.2. *Decreasing MS consecutive-k-out-of-n : F system*

As defined in Huang et al. (2003), $\emptyset(x) < j$ $(j = 1, 2, \ldots, M)$ if at least k_l consecutive components are in states below l for all l such that $j \leq l \leq M$. An n-component system with such a property is called a multi-state consecutive-k-out-of-n: F system.

When $k_1 \geq k_2 \geq \cdots \geq k_M$, the system is called a decreasing multi-state consecutive-k-out-of-n: F system. However, when k_j is constant, i.e., $k_1 = k_2 = \cdots = k_M = k$, the structure of the system is the same for all the system state levels. This reduces to the definition of the MS consecutive-k-out-of-n system provided by Haim and Porat (1991) and the system is called constant MS consecutive-k-out-of-n: F system.

Theorem 13.11. *For positive integers k_j and n, the function $W_{s,k_j}(0,z)$, the generating function of the waiting time for the occurrence of system state to be below j in case of decreasing MS consecutive-k-out-of-n : F system is given by:*

$$W_{s,k_j}(0,z) = \frac{(Q_j z)^{k_j}}{1 - P_j z \sum_{i=0}^{k_j-1}(Q_j z)^i}. \tag{13.31}$$

Thus, the probability that system state level is below j is given by:

$$F_j(n; k_j) = \sum_{u=k_j}^{n} Q_j^{k_j} \sum_{l=0}^{\left[\frac{u-k_j}{k_j+1}\right]} (-1)^l \binom{u - k_j - lk_j}{l} (P_j Q_j^{k_j})^l$$

$$- \sum_{u=k_j+1}^{n} Q_j^{k_j+1} \sum_{l=0}^{\left[\frac{u-k_j-1}{k_j+1}\right]} (-1)^l \binom{u - k_j - lk_j - 1}{l} (P_j Q_j^{k_j})^l$$

$$j = 1, 2, \ldots, M. \tag{13.32}$$

Proof. GERT network for MS decreasing k-out-of-$n : F$ system to be below level j is represented by Fig. 13.12. It consists of $k_j + 1$ nodes denoted by $0, 1, 2, \ldots, k_j - 1, k_j$ on the path leading to the required system state. Each node represents a specific state as described below:

0 : state of component $\geq j; j = 1, 2, \ldots, M$

1 : one component in state $< j$

2 : second consecutive component in state $< j$

$\vdots$

k_j : k_j consecutive components in state $< j$.

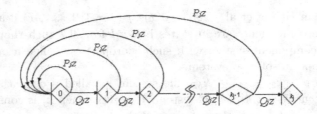

Fig. 13.12. GERT Network Representing Decreasing MS Consecutive-k-out-of-$n : F$ System below Level j.

It can be summarized as: If one component is in state below j, then there is conditional probability Q_j from state $v - 1$ to state $v(v = 1, 2, \ldots, k_j)$. If one component is in state $\geq j$, then there is conditional probability P_j from state $v-1$ to state $0(v = 1, 2, \ldots, k_j)$. System state is reached as soon as node k_j is reached.

By the procedure of Mason's rule we obtain $W_{s,k_j}(0, z)$ function as given in (13.31).

Now, the function $W_{s,k_j}(0, z)$ in (1.31) can be expressed as:

$$W_{s,k_j}(0, z) = \sum_{u=k_j}^{\infty} \xi(u) z^u, \qquad (13.33)$$

where $\xi(u), u = k_j, k_j + 1, \cdots, n, \ldots$, are the coefficients of series expansion of RHS of (13.31). Thus, the probability that the system is below level j is given by:

$$F_j(n; k_j) = \sum_{u=k_j}^{n} \xi(u), \quad j = 1, 2, \ldots, M, \qquad (13.34)$$

which on solving yields (13.32).This proves Theorem 13.11.

The probability that the system is in state j for $j = 0, 1, 2, \ldots, M$ is given by:

$\Pr(\emptyset(x){<}j) = F_j(n; k_j), \ \text{for} \ j = 0, 1, 2, \ldots, M,$

$\Pr(\emptyset(x) = 0) = \Pr(\emptyset(x){<}1),$

$\Pr(\emptyset(x) = M) = 1 - \Pr(\emptyset(x) < M),$

$\Pr(\emptyset(x) = j) = \Pr(\emptyset(x) < j + 1) - \Pr(\emptyset(x) < j), \ \text{for} \ j = 1, 2, \ldots, M - 1.$

To study computational efficiency, system state distribution for both, the increasing MS k-out-of-$n : G$ systems and, decreasing MS consecutive-k-out-of-$n : F$ systems are obtained for different values of n, using Software Mathematica. Computing Time (secs) is given in Tables 13.7 and 13.8, respectively. The average computation times given are from 5 trials. "N/A" in the results of Huang et al. (2003) means that computation time is greater than one hour. It can be observed that CPU time involved in both the systems using our formulae, to be referred as IGERT and DGERT, is much less as compared to Huang et al. (2000) and Huang et al. (2003), respectively, thereby illustrating the efficiency of GERT in reliability analysis of such systems.

Table 13.7. System State Distribution, and Computation Time for Increasing MS k-out-of-$n : G$ System for Different Values of n when $p_0 = 0.1$, $p_1 = 0.12$, $p_2 = 0.13$, $p_3 = 0.14$, $p_4 = 0.15$, $p_5 = 0.16$, $p_6 = 0.2$ and $k_1 = 1$, $k_2 = 2$, $k_3 = 3$, $k_4 = 4$, $k_5 = 5$ and $k_6 = 6$.

n	$R_{s,1}$	$R_{s,2}$	$R_{s,3}$	$R_{s,4}$	$R_{s,5}$	$R_{s,6}$	CPU Time (secs) Huang et al.	CPU Time (secs) IGERT
6	0.999999	0.997475	0.882576	0.362682	0.0253958	0.000064	≈ 0	≈ 0
10	≈ 1	0.999990	0.995179	0.844039	0.270842	0.006369	≈ 0	≈ 0
20	$\approx$	≈ 1	0.999999	0.999037	0.898937	0.195792	≈ 0	≈ 0
40	≈ 1	≈ 1	≈ 1	≈ 1	0.999803	0.838671	0.006	0.003
50	≈ 1	≈ 1	≈ 1	≈ 1	0.999995	0.951973	0.009	0.003
100	≈ 1	≈ 1	≈ 1	≈ 1	≈ 1	0.999981	0.019	0.006

13.5. Waiting Time Distribution of the Patterns

In recent years distribution theory related to runs has received a great deal of attention (see, Aki et al., 1996; Antzoulakos and Philippou, 1999; Han and Aki, 2000; Han and Hirano, 2003; Fu and Chang, 2002; Fu and Lou,

Table 13.8. System State Distribution, and Computation Time for Decreasing MS Consecutive-k-out-of-$n:F$ System for Different Values of n when $p_0 = 0.1$, $p_1 = 0.12$, $p_2 = 0.14$, $p_3 = 0.16$, $p_4 = 0.22$, $p_5 = 0.26$, and $k_1 = 5$, $k_2 = 4$, $k_3 = 3$, $k_4 = 2$, $k_5 = 1$.

n	$F_1(n, k_1)$	$F_2(n, k_2)$	$F_3(n, k_3)$	$F_4(n, k_4)$	$F_5(n, k_5)$	CPU Time (secs) Huang et al.	CPU Time (secs) DGERT
5	0.0000100	0.00416976	0.1063757	0.6246802	0.998812	≈ 0	≈ 0
10	0.0000550	0.0132938	0.2447527	0.881549	0.99999	0.006	0.003
15	0.0001000	0.0223417	0.3614931	0.962559	≈ 1	0.078	0.015
25	0.0001900	0.0401892	0.5436293	0.996259	≈ 1	7.086	0.031
40	0.0003250	0.0663515	0.7242295	0.999882	≈ 1	N/A	0.047
50	0.0004149	0.0833956	0.8028936	0.999988	≈ 1	N/A	0.078

2006). This is because of their widespread applications, especially in DNA sequence analysis, quality management, educational statistics, health care sector to name a few. The books by Godbole and Papastavridis (1994), Balakrishnan and Koutras (2002), Fu and Lou (2003) provide excellent information on past and current developments in this area. Such patterns have applications in reliability also. Mohan et al. (2009) and Sen et al. (2013) studied various patterns using GERT. One of the patterns is mentioned below.

The pattern $\Lambda^{k_1, k_2, k} = S \underbrace{FF \ldots F}_{k_1 \leq k_A \leq k_2} \underbrace{SS \ldots S}_{k}$ is a compound pattern consisting of at least k_1 but not more than k_2 failures preceded by a working component and followed by a run of at least k working components, $k_1, k > 0$.

Theorem 13.12. *The probability generating function for the waiting time distribution for the 1st occurrence of the pattern*

$$\Lambda^{k_1, k_2, k} = S \underbrace{FF \ldots F}_{k_1 \leq k_A \leq k_2} \underbrace{SS \ldots S}_{k}, \quad k_1, \ k > 0$$

involving homogenous Markov dependence is given by

$$W_{S \underbrace{FF \ldots F}_{k_1 \leq k_A \leq k_2} \underbrace{SS \ldots S}_{k}}(0, z)$$

$$= \frac{q_1 q_2^{k_1-1} p_2 p_1^{k-1} z^{k_1+k} (q_0 p_2 z^2 + p_0 z(1 - q_2 z))(1 - (q_2 z)^{k_2 - k_1 + 1})}{\begin{pmatrix} (1 - q_2 z)(1 - q_2 z - p_1 z - q_1 p_2 z^2 + q_2 p_1 z^2) \\ + q_1 q_2^{k_1-1} p_2 p_1^{k-1} z^{k_1+k} (1 - (q_2 z)^{k_2 - k_1 + 1}) \end{pmatrix}}$$

Proof.   GERT network for this pattern is represented by Fig. 13.13, where each node represents a specific state as described below:

S : initial state

1^A : a component is in failed state

0 : a component is in working state

1 : a component failure is preceded by at least one working component

2 : second consecutive failure occurs preceded by a working component

⋮

k_1 : k_1 consecutive failures occur preceded by a working component

$k_1 + 1$: $k_1 + 1$ consecutive failures occur preceded by a working component

⋮

k_2 : k_2 consecutive failures occur preceded by a working component

$k_2 + 1$: $k_2 + 1$ consecutive failures occur preceded by a working component

0^1 : first working component immediately preceded by a substring consisting of at least one working and at least k_1 but not more than k_2 consecutive failures

0^2 : second consecutive working component preceded by a 0^1 state

⋮

0^k : kth consecutive working component preceded by a 0^{k-1} state.

It can be summarized as: If first component is working then there is conditional probability p_0 from state S to 0 otherwise q_0 from state S to 1^A. Now, if the system is in state 1^A and next contiguous component(s) are also in failed state then it continues to move in that state with conditional probability q_2 until a working component occurs at which system moves to state 0 with conditional probability p_2. Now, if the system is in state 0 and the next contiguous component is in failed state then system moves to state 1 with conditional probability q_1 otherwise it continues to move in state 0 with conditional probability p_1. Further, if the system is in state 1 and next contiguous component is also in failed state then the system moves to state 2 with conditional probability q_2 otherwise to state 0 with conditional probability p_2. In short, if one component is working then there is a conditional probability p_2 from state $v - 1$ to state 0 ($v = 2, 3, \cdots, k_1$) and if one component fails then there is a conditional probability q_2 from state $v - 1$ to state $v (v = 2, 3, \cdots, k_1, k_1 + 1, k_1 + 2, \cdots k_2 - 1, k_2)$. Now, in case there occur k_1 contiguous failed components and the next contiguous component is working then system moves to state 0^1 with conditional probability p_2 otherwise it moves to state $k_1 + 1$ with conditional probability

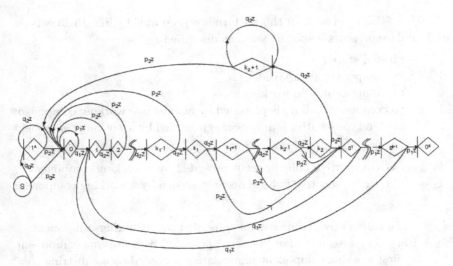

Fig. 13.13.   GERT Network Representing the Occurrence of the Pattern GERT Network Representing the Occurrence of the Pattern $\Lambda^{k_1,k_2,k} = S \underbrace{FF\ldots F}_{k_1 \leq k_A \leq k_2} \underbrace{SS\ldots S}_{k}$ involving

Homogenous Markov Dependence.

q_2. Further, if the system is in state $k_1 + 1$ and next contiguous component is in working state then system moves to state 0^1 with conditional probability p_2 otherwise to state $k_1 + 2$ with conditional probability q_2. Similar procedure is followed for the remaining contiguous components till state k_2 is reached. If after the occurrence of k_2 contiguous failures the next contiguous component is again in the failed state then system moves to $k_2 + 1$ with conditional probability q_2 and continues to remain in that state until a working component occurs in which case system moves to state 0 with conditional probability p_2. However, if next contiguous component after k_2 contiguous failures is normal then system moves to state 0^1. Now, if the next contiguous component is normal system moves to 0^2 with conditional probability p_1 otherwise moves back to state 1 with conditional probability q_1. Similar procedure is followed till state 0^k is reached, resulting in 1st occurrence of the pattern $\Lambda^{k_1,k_2,k}$.

There are in all $k_2 + k + 4$ nodes designated as $S, 1^A, 0, 1, 2, \ldots, k_1, k_1 + 1, \ldots, k_2 - 1, k_2, k_2 + 1, 0^1, 0^2, \ldots, 0^{k-1}, 0^k$ representing specific states of the system as described above. Moreover, there are $2(k_2 - k_1 + 1)$ paths leading to the required system state. They are as given:

No.	Paths	Value
1	S to 1^A to 0 to 1 to...to k_1 to 0^1 to 0^2... to 0^k	$(q_0 z)(q_1 z)(q_2 z)^{k_1-1}(p_1 z)^{k-1}(p_2 z)^2$
2	S to 1^A to 0 to 1 to...to k_1 to k_1+1 to 0^1 to 0^2... to 0^k	$(q_0 z)(q_1 z)(q_2 z)^{k_1}(p_1 z)^{k-1}(p_2 z)^2$
3	S to 1^A to 0 to 1 to...to k_1+1 to k_1+2 to 0^1 to 0^2... to 0^k	$(q_0 z)(q_1 z)(q_2 z)^{k_1+1}(p_1 z)^{k-1}(p_2 z)^2$
$\vdots$	$\vdots$	$\vdots$
k_2-k_1+1	S to 1^A to 0 to 1 to...to k_2-1 to k_2 to 0^1 to 0^2... to 0^k	$(q_0 z)(q_1 z)(q_2 z)^{k_2-1}(p_1 z)^{k-1}(p_2 z)^2$
k_2-k_1+2	S to 0 to 1 to...to k_1 to 0^1 to 0^2... to 0^k	$(p_0 z)(q_1 z)(q_2 z)^{k_1-1}(p_1 z)^{k-1}(p_2 z)$
k_2-k_1+3	S to 0 to 1 to...to k_1 to k_1+1 to 0^1 to 0^2... to 0^k	$(p_0 z)(q_1 z)(q_2 z)^{k_1}(p_1 z)^{k-1}(p_2 z)$
k_2-k_1+4	S to 0 to 1 to...to k_1+1 to k_1+2 to 0^1 to 0^2... to 0^k	$(p_0 z)(q_1 z)(q_2 z)^{k_1+1}(p_1 z)^{k-1}(p_2 z)$
$\vdots$	$\vdots$	$\vdots$
$2(k_2-k_1+1)$	S to 0 to 1 to...to k_2-1 to k_2 to 0^1 to 0^2... to 0^k	$(p_0 z)(q_1 z)(q_2 z)^{k_2-1}(p_1 z)^{k-1}(p_2 z)$

Further, only first, second and third order loops exist and:

$$\sum \text{first order loops} = q_2 z + p_1 z + (q_1 z)(p_2 z) \sum_{i=0}^{k_1-2} (q_2 z)^i + (q_1 z)(p_2 z)$$

$$(q_2 z)^{k_1-1} \sum_{i=0}^{k-2}(p_1 z)^i \sum_{j=0}^{k_2-k_1} (q_2 z)^j + (q_1 z)(q_2 z)^{k_2}(p_2 z)$$

$$\sum \text{second order loops} = (q_2 z)(p_1 z)$$

$$+(q_1 z)(p_2 z)(q_2 z)^{k_1-1}(p_1 z) \sum_{i=0}^{k-2} (p_1 z)^i \sum_{j=0}^{k_2-k_1} (q_2 z)^j$$

$$+(q_1 z)(p_2 z)(q_2 z) \sum_{i=0}^{k_1-2} (q_2 z)^i$$

$$+(q_1 z)(p_2 z)(q_2 z)^{k_1}(p_1 z) \sum_{i=0}^{k-2}(p_1 z)^i \sum_{j=0}^{k_2-k_1} (q_2 z)^j$$

and

$$\sum \text{third order loops} = (q_1 z)(p_2 z)(q_2 z)^{k_1}(p_1 z) \sum_{i=0}^{k-2}(p_1 z)^i \sum_{j=0}^{k_2-k_1} (q_2 z)^j.$$

Further, paths numbered $k_2 - k_1 + 2, k_2 - k_1 + 3, \ldots, 2(k_2 - k_1 + 1)$ also contain first order nontouching loop (1^A to 1^A), whose value is given as $q_2 z$.

Therefore, by applying Mason's rule we obtain the generating function $W_{S\underbrace{FF\ldots F}_{k_1 \le k_A \le k_2}\underbrace{SS\ldots S}_{k}}(0, z)$ of the waiting time for the occurrence of the required system state:

$$
W_{S\underbrace{FF\ldots F}_{k_1 \le k_A \le k_2}\underbrace{SS\ldots S}_{k}}(0, z)
$$

$$
= \frac{(q_1 z)(q_2 z)^{k_1-1}(p_2 z)(p_1 z)^{k-1}((q_0 z)(p_2 z) + p_0 z(1 - q_2 z))(1 - (q_2 z)^{k_2 - k_1 + 1})}{(1 - q_2 z)(1 - q_2 z - p_1 z - (q_1 z)(p_2 z) + (q_2 z)(p_1 z))}
$$
$$
+ (q_1 z)(q_2 z)^{k_1-1}(p_2 z)(p_1 z)^{k-1}(1 - (q_2 z)^{k_2 - k_1 + 1}))
$$

which is (13.34).

$$
E[W(\Lambda^{k_1, k_2, k})] = E \text{ (Minimum number of trials required to obtain}
$$
$$
\text{the pattern } \Lambda^{k_1, k_2, k})
$$

$$
= \frac{p_1^2 q_2 + p_1(p_2^2 - 1) + q_1 p_1^k q_2^{k_2+1}(q_0 + p_2)(1 - q_2^{k_1 - k_2 - 1})}{q_1 p_2 p_1^k q_2^{k_2+1}(1 - q_2^{k_1 - k_2 - 1})}.
$$
$$
(13.35)
$$

and

$$
Var[W(\Lambda^{k_1, k_2, k})] = \frac{d^2 W_{S\underbrace{FF\ldots F}_{k_1 \le k_A \le k_2}\underbrace{SS\ldots S}_{k}}(0, z)}{dz^2} \Bigg|_{z=1} + E[W(\Lambda^{k_1, k_2, k})]
$$
$$
- \left(E[W(\Lambda^{k_1, k_2, k})] \right)^2.
$$
$$
(13.36)
$$

Particular cases

Take $p_0 = p_1 = p_2 = p, q_0 = q_1 = q_2 = q$, i.e., components are i.i.d.

(i) For $k_1 = k_2 = k_A, k = 1$, i.e., for a run of exactly k_A failures bounded by successes, (13.34) becomes

$$
W_{S\underbrace{FF\ldots F}_{k_A}S}(0, z) = \frac{(qz)^{k_A}(pz)^2}{(1 - z + (qz)^{k_A}(pz)(1 - qz))}
$$
$$
(13.37)
$$

and interchanging p and q, verifies the results of Sen and Goyal (2004).

(ii) For $k_1 = 0, k_2 = k_A - 2, k = 1$ i.e., for a run of at most $k_A - 2$ failures bounded by successes, (13.34) becomes

$$W_{S\underbrace{FF\ldots F}_{k_{A-2}}S}(0, z) = \frac{(pz)^2(1 - (qz)^{k_A-1})}{(1 - qz)[1 - z + (pz(1 - (qz)^{k_A-1}))]}. \quad (13.38)$$

Thus, (13.34), (13.35) and value of the variance obtained verify the results of Koutras (1996, Theorem 3.2).

13.6. Conclusion

In this paper an attempt has been made to present a brief review of GERT and its efficiency in reliability analysis of various consecutive-k systems and patterns. From the models discussed in the paper it can be observed that GERT:

a is generally more transparent and easy to understand.
b provides reasonable attraction to other models of consecutive systems.
c provides visual picture of the system which helps to analyze the system in a better way.

Besides, numerical computations reveal the efficiency of GERT in reliability analysis of these systems, in comparison with those existing in the literature.

References

1. Agarwal, M. and Mohan, P. (2007). Reliability analysis of consecutive k, r-out-of-n:DFM system using GERT. *Int. J. Operations Research.* **4**, 110–117.
2. Agarwal, M. and Mohan, P. (2008a). GERT analysis of consecutive-k-out-of-$n:F$ system with $(k-1)$-step and homogenous Markov dependence. *Int. J. App. Math. and Stat.* **13**, 58–68.
3. Agarwal, M. and Mohan, P. (2008b). GERT analysis of m-consecutive-k-out-of-$n:F$ system with overlapping runs and $(k-1)$-step Markov dependence. *Int. J. Operational Research.* **3**, 36–51.
4. Agarwal, M., Mohan, P. and Sen, K. (2007). GERT analysis of m-consecutive-k-out-of-$n:F$ systems with dependence, *EQC (Int. J. Qual. and Reliab).* **22**, 141–157.
5. Agarwal, M., Sen, K. and Mohan, P. (2007). GERT analysis of m-consecutive-k-out-of-n systems, *IEEE Trans. Reliability.* **56**, 26–34.
6. Aki, S., Balakrishnan, N. and Mohanty, S. G. (1996). Sooner and later waiting time problems for success and failure runs in higher order Markov dependent trials. *Ann. Inst. Stat. Math.* **48**, 773–787.

7. Antzoulakos, D. L. and Philippou, A. N. (1999). On waiting time problems associated with runs in Markov dependent trials. *Ann. Inst. Stat. Math.* **51**, 323–330.

8. Balakrishnan, N. and Koutras, M. V. (2002). *Runs and Scans with Applications.* John Wiley and Sons, New York.

9. Bao, C. P. (1989). A new estimate method for consecutive-k-out-of-n : F system reliability. *J. Chinese Stat. Assoc.*

10. Bollinger, R. C. (1985). Strict consecutive-k-out-of-n : F systems, *IEEE Trans. Reliability.* **34**, 50-52.

11. Chang, G. K., Cui, L. and Hwang, F. K. (2000). *Reliabilities of Consecutive-k-Systems.* Dordrecht, Netherlands, Kluwer.

12. Chao, M. T., Fu, J. C. and Koutras, M. V. (1995). Survey of reliability studies of consecutive-k-out-of-n : F and related systems, *IEEE Trans. Reliability.* **44**, 120–127.

13. Cheng, C. H. (1994). Fuzzy consecutive-k-out-of-n : F system reliability, *Microelectron. Reliability.* **34**, 1909–1922.

14. Chiang, D. T. and Niu, S. C. (1981). Reliability of a consecutive-k-out-of-n : F system, *IEEE Trans. Reliability.* **30**, 87-89.

15. Dhillon, B. S. and Rayapati, S. N. (1986). A method to evaluate reliability of three-state device networks, *Microelectron. Reliability.* **26**, 535–554.

16. Feller, W. (1968). *An Introduction to Probability Theory and Its Applications,* Vol. 1, 3rd edition. John Wiley & Sons, New York.

17. Fu, J. C. (1986). Reliability of consecutive-k-out-of-n : F systems with $(k-1)$-step Markov dependence, *IEEE Trans. Reliability.* **35**, 602–606.

18. Fu, J. C. and Chang, Y. M. (2002). On probability generating functions for waiting time distributions of compound patterns in a sequence of multi-state trials, *J. Appl. Prob.* **39**, 70–80.

19. Fu, J. C. and Hu, B. (1987). On reliability of a large consecutive-k-out-of-n : F system with $(k - 1)$-step Markov dependence. *IEEE Trans. Reliability.* **36**, 75–77.

20. Fu, J. C. and Lou, W. Y. W. (2003). *Distribution Theory of Runs and Patterns and Its Applications: A Finite Markov Chain Imbedding Approach,* World Scientific Publishing Co.

21. Fu, J. C. and Lou, W. Y. W. (2006). Waiting time distributions of simple and compound patterns in a sequence of r-th order Markov dependent multi-state trials, *Ann. Inst. Stat. Math.* **58**, 291–310.

22. Ge, G. and Wang, L. (1990). Exact reliability formula for consecutive-k-out-of-n : F systems with homogenous Markov dependence, *IEEE Trans. Reliability.* **39**, 600–602.

23. Godbole, A. P. and Papastavridis, S. G. (1994). *Runs and Patterns in Probability: Selected papers.* Kluwer Academic Publishers, Netherlands.

24. Griffith, W. S. (1986). On consecutive-k-out-of-n failure systems and their generalizations, *Reliability and Quality Control,* 157–165.

25. Haim, M. and Porat, Z. (1991). Bayes reliability modeling of a multi-state consecutive-k-out-of-n : F system, *Proceedings of the Annual Reliability and Maintainability Symposium,* 582–586.

26. Han, Q. and Aki, S. (2000). Sooner and later waiting time problems based on a dependent sequence, *Ann. Inst. Stat. Math.* **52**, 407–414.

27. Han, Q. and Hirano, K. (2003). Sooner and later waiting time problems for patterns in Markov dependent trials, *J. App. Prob.* **40**, 73-86.

28. Huang, J., Zuo, M. J. and Fang, Z. (2003). Multi-state consecutive-k-out-of-n systems. *IIE Trans.* **35**, 527–534.

29. Huang, J., Zuo, M. J. and Wu, Y. (2000). Generalized multi-state k-out-of-$n:G$ systems, *IEEE Trans. Reliability.* **49**, 105–111.

30. Jenab, K. and Dhillon, B. S. (2005). k-out-of-n system with self-loop units, *Int. J. Reliab., Qual. & Safety Eng.* **12**(1), 61–73.

31. Kontoleon, J. M. (1980). Reliability determination of r-successive-out-of-$n:F$ system. *IEEE Trans. Reliability* **29**, 437.

32. Koutras, M. V. (1996). On a waiting time distribution in a sequence of Bernoulli trials, *Ann. Inst. Stat. Math.* **48**, 789–806.

33. Koutras, M. V. (1997). Consecutive-k, r-out-of-n :DFM systems, *Microelectron. Reliability.* **37**, 597-603.

34. Kuo, W. and Zuo, M. (2003). *Optimal Reliability Modelling – Principles and Applications.* John Wiley and Sons, New Jersey.

35. Makri, F. S. and Philippou, A. N. (1996). Exact reliability formulas for linear and circular m-consecutive-k-out-of-$n:F$ systems, *Microelectron. Reliability.* **36**, 657–660.

36. Malinowski, J. and Preuss, W. (1995). Reliability of circular consecutively-connected systems with multi-state components, *IEEE Trans. Reliability.* **44**, 532–534.

37. Malinowski, J. and Preuss, W. (1996). Reliability of reverse-tree-structured systems with multi-state components, *Microelecrton. Reliability* **36**(1), 1–7.

38. Malon, D. M. (1989). On a common error in open and short-circuit reliability computation, *IEEE Trans. Reliability* **38**, 275–276.

39. Mohan, P. (2007). *Reliability Analysis of Consecutive-k Systems: GERT Approach*, Ph.D. Thesis, University of Delhi, Delhi.

40. Mohan, P. and Agarwal, M. (2007a). Strict consecutive k-out-of-$n:F$ system: GERT analysis, *OPSEARCH* **44**, 338-346.

41. Mohan, P. and Agarwal, M. (2007b). Reliability analysis of multi-state systems: GERT approach, *Communicated.*

42. Mohan, P., Agarwal, M. and Sen, K. (2009). Combined m-consecutive-k-out-of-n systems, and consecutive k_c-out-of-$n:F$ systems, *IEEE Trans. Reliablility* **58**, 328–337.

43. Page, L. B. and Perry, J. E. (1989). A note on three-state systems, *IEEE Trans. Reliability* **38**, 277.

44. Papastavridis, S. (1990). m-consecutive-k-out-of-$n:F$ systems, *IEEE Trans. Reliability* **39**, 386–387.

45. Papastavridis, S. and Lambiris, M. (1987). Reliability of consecutive-k-out-of-$n:F$ system for Markov dependent components, *IEEE Trans. Reliability* **36**, 78–79.

46. Pham, H. (2003). *Handbook of Reliability Engineering.* Springer-Verlag, London.

47. Satoh, N., Sasaki, M., Yuge, T. and Yanagi, S. (1993). Reliability of three-state device systems with simultaneous failures. *IEEE Trans. Reliability* **42**, 470–477.

48. Sen, K., Agarwal, M. and Mohan, P. (2008). Waiting time distributions for mth occurrence of patterns involving Markov-dependent trials using GERT, *to be communicated*.

49. Sen, K. and Goyal, B. (2004). Distribution of patterns of two failures separated by success runs of length k, *J. Korean Stat. Soc.* **33**(1), 35–58.

50. Sen, K., Mohan, P. and Agarwal, M. (2013). Distributions of patterns of pair of Successes Separated by Failure Runs of Length at Least k_1 and at Most k_2 Involving Markov Dependent Trials: GERT Approach. *J. Quality And Reliability Engineering*, Article ID 494976, 9 pages.

51. Whitehouse, G. E. (1973). *Systems Analysis And Design Using Network Techniques*, Prentice-Hall, Englewood Cliffs, New Jersey.

52. Yamamoto, H., Zuo, M. J., Akiba, T. and Tian, Z. (2006). Recursive formulas for the reliability of multi-state consecutive-k-out-of-n : G systems, *IEEE Trans. Reliability* **55**, 98–104.

53. Zuo, M. J. and Liang, M. (1994). Reliability of multi-state consecutively-connected systems, *Reliab. Eng. & System Safety.* **44**, 173–176.

Chapter 14

Moment Bounds for Strong-Mixing Processes with Applications

Ratan Dasgupta

Indian Statistical Institute, India

A sharp moment bound for sample sum of truncated non-stationary strong-mixing random variables with polynomial decay is obtained when some finite order (> 2) moment of individual random variables exist. It is also shown that this technique of estimation originally due to Ibragimov (1962), later extended by Ghosh and Babu (1977), Babu, Ghosh and Singh (1978), Babu and Singh (1978), Babu (1980), and Dasgupta (1997) breaks down for strong-mixing processes with decay lower than polynomial order. The moment bounds have applications to obtain the rates of convergence in Marcinkiewicz-Zygmund strong law of large numbers, central limit theorem, probabilities of deviations and in quantile processes of the random variables.

Keywords: Davydov's inequality, Growth curve model, Linear process, Marcinkiewicz-Zygmund strong law of large numbers.

Contents

14.1. Introduction

Let $\{X_i, i \geq 1\}$ be a non-stationary strong mixing process with mixing coefficients $\{\alpha(i)\}$. Under moderate assumptions the standardized sum of X_i's converges to the normal law, see e.g., Merlevde et al. (2000). The moment bounds on the sample sum of truncated random variables $Y_i = X_i I(\mid X_i \mid \leq d)$, with a suitable truncation point d, are of interest in convergence rates to the limiting normal distribution, Marcinkiewicz-

Zygmund strong law of large numbers (MZSLLN), deviation probabilities and also in computing deviation between quantile and empirical processes. See e.g., [2]-[11]. With polynomial decay of mixing coefficients for strong-mixing random variables, the above mentioned results may be refined via these moment bounds. This technique of estimating moments breaks down for strong-mixing processes with decay lower than polynomial order. In this paper, we prove a moment bound for strong-mixing processes with polynomial decay and point out where the technique breaks down for lower order decay. Since Davydov's (1970) inequality, which is repeatedly used in the proof is quite sharp, such results may not be possible with decay lower than polynomial order.

For stationary strong-mixing process with exponentially decaying mixing coefficients, Tikhomirov (1980) obtained the Berry-Essen bound $O(n^{-1/2}(\log n)^2)$ for the standardized sample sum, under the existence of the third moment of the random variables. The question of attaining the optimal bound $O(n^{-1/2})$ is still unresolved for strong-mixing random variables.

In Section 14.2, we state and prove a sharp moment bound. In Section 14.3 these moment bounds are used to obtain MZSLLN for strong-mixing random variables with polynomial decay of the mixing coefficients, extending the results of ϕ-mixing random variables given in Babu (1980). In this context, we mention that the results of Rio (1995) are applicable to variables with *same marginal distribution* rather than general nonstationary sequences. An example of strong mixing random variables with polynomially decaying mixing coefficients is cited in the last section.

14.2. The Moment Bound

The following moment bound improves a bound stated in Dasgupta (1988). A detailed proof is also provided.

Theorem 1. Let $\{X_i, i \geq 1\}$ be a non-stationary strong mixing process with mixing coefficients $\alpha(t) \leq q \, t^{-\lambda}$, $q \geq 1, \lambda > 0$ and $\mathrm{var}(\sum_{i=1}^{u} X_{i+h}) = O(u)$, $\forall \, h \geq 0$. Let $EX_i = 0$, $\sup_{i \geq 1} E \mid X_i \mid^{\delta + \varepsilon^*} < \infty$, where $\delta > 2$, $0 < \varepsilon^* < 1$.

For fixed constants $v_0 (\geq 2)$ and $\beta(> 1)$, let λ be so large that

$$v_0 = 1 + \frac{\log(1 + D_0)}{\log 2} < \frac{\delta}{2} \tag{14.1}$$

where $D_0 = 5q(2^{v_0} - 2)\beta v_0(v_0 + \varepsilon^*)/(\lambda \varepsilon^* \log 2)$ $(\to 0 \text{ as } \lambda \to \infty)$. Let

$$Y_i = Y_{i,d} = X_i I(| X_i | \leq d), 1 < d < u^{(v_0 - \eta)/\varepsilon^*}, \ \eta > 0. \tag{14.2}$$

Then for any $v \leq v_0$, there exists a constant $D(v) > 0$, not dependent on d such that for all $u \leq d^2$, the relation

$$E \mid \sum_{i=1}^{u} Y_{i+h} \mid^v \leq D(v)(u^{v/2} + u^v R(v)) \tag{14.3}$$

holds, where

$$R(v) = d^{v-\delta}, \ 1 \leq \nu \leq \nu_0.$$

Further, for $u \leq d^2 \leq u(\log u)^s$, where $s > 0$ is an arbitrary but fixed constant, the moment bound (14.3) holds with $R(v) = d^{v-(\delta+\varepsilon^*)}$.

Remark 1. We explain the result. Here $\{X_i, i \geq 1\}$ is a sequence of nonstationary strong-mixing random variables with polynomially decaying mixing coefficients, $\alpha(t) \leq q \, t^{-\lambda}$. The variables are centered at origin and the variance of sum of u number of random variables at a stretch is of order u. Further, let a moment of order higher than 2, viz., $(\delta + \varepsilon^*)$th absolute moment of the individual random variables be uniformly bounded. Theorem 1 then states that for a fixed v_0 and for a selected constant $\beta > 1$, if the power of the polynomial decay λ is large enough: $\lambda \gg 2^{v_0}$, then moments of the sum of the truncated random variables can be bounded above by (14.3) for all $v \leq v_0$.

For v-th moment, the leading term of u in the r.h.s. of the bound (14.3) is of order $u^{v/2}$, which is also the case for sum of u i.i.d random variables. Thus the order of u is sharp in the bound.

Definition of ν_0 in (14.1) states that $\nu_0 \to 1$, as $\lambda \to \infty$. Hence condition (14.1) is satisfied for large λ and consequently the moment bound (14.3) holds for the sum of truncated random variables. Truncation of random variables with $d^2 = O(u \log u)$ is often used to compute the probabilities of moderate deviations. In such a situation taking $s = 1$, one may get an improved moment bound with $R(v) = d^{v-(\delta+\varepsilon^*)}$ in (14.3), rather than $R(v) = d^{v-\delta}$. With $R(v) = d^{v-(\delta+\varepsilon^*)}$, contribution from the moment assumption $\sup_{i \geq 1} E \mid X_i \mid^{\delta+\varepsilon^*} < \infty$, is fully utilized in the order of $R(v)$, for that includes the ε^* part. The result is comparable with Babu et al. (1978a) for ϕ-mixing sequence, where $\nu = 1$.

Proof of Theorem 1. First consider (14.3) with $R(v) = d^{v-\delta}$. Fix an integer $h \geq 0$. Denote $Z_u = \sum_{i=1}^{u} Y_{i+h}$, $Z_{u,t} = Z_{2u+t} - Z_{u+t}$, $S_{u,t} =$

$\sum_{i=1}^{t} Y_{u+i+h}$, $c(u, v, h) = E|\sum_{i=1}^{u} Y_{i+h}|^{v}$, $c(u, v) = \sup_{h \geq 0} c(u, v, h)$. By Davydov's (1970) inequality,

$$c(u, 2, h) \ll u, \quad \text{if} \quad \sum_{j=1}^{\infty} [\alpha(j)]^{1-2/(\delta+\varepsilon^{*})} < \infty.$$

The lemma is proved by induction from $c(u, m)$ to $c(u, m + \varepsilon)$, $0 < \varepsilon \leq 1$, $m + \varepsilon \leq v_{0}$. Write,

$$E|Z_{u} + Z_{u,t}|^{m+\varepsilon} \leq E[(|Z_{u}| + |Z_{u,t}|)^{m}(|Z_{u}|^{\varepsilon} + |Z_{u,t}|^{\varepsilon})]$$

$$\leq 2c(u, m + \varepsilon) + \sum_{j=1}^{m-1} \binom{m}{j} E|Z_{u}|^{m-j+\varepsilon}|Z_{u,t}|^{j}.$$

By Davydov's inequality, one gets

$$|E|Z_{u}|^{m-j+\varepsilon}|Z_{u,t}|^{j} - E|Z_{u}|^{m-j+\varepsilon}E|Z_{u,t}|^{j}|$$
$$\leq 10(\alpha(t))^{\varepsilon_{1}/(m+\varepsilon+\varepsilon_{1})}[c(u, m + \varepsilon + \varepsilon_{1})]^{(m+\varepsilon)/(m+\varepsilon+\varepsilon_{1})},$$

where $\varepsilon_{1} > 0$ will be chosen later; see equations (7)-(9) of Babu et al. (1978a) for ϕ-mixing sequence. Denote $N = \sup_{i \geq 1} E|X_{i}|^{\delta+\varepsilon^{*}}$. Then, $\sup_{h \geq 0} E|Y_{h}|^{v} \leq H(v)$, where $H(v) = 2N$, if $v \leq \delta$ and $H(v) = NR(v)$, if $v > \delta$, i.e., $H(v)$ is of same order as $R(v)$. Note that

$$[u^{v}R(m)]^{(m+\varepsilon)/m} \leq u^{v}R(m + \varepsilon), \quad \text{if} \quad \nu \leq \frac{\delta}{2} \quad \text{as} \quad u \leq d^{2}. \tag{14.4}$$

Let $k > 0$ be a generic constant. Proceeding as in Babu et al. (1978a), write

$$c(2u, m + \varepsilon, h)$$

$$= E|\sum_{i=1}^{2u} Y_{i+h}|^{m+\varepsilon} = E|Z_{u} + Z_{u,t} + S_{u,t} - S_{2u,t}|^{m+\varepsilon}$$

$$\leq [\{E|Z_{u} + Z_{u,t}|^{m+\varepsilon}\}^{\frac{1}{m+\varepsilon}} + 2t\{E|Y_{h}|^{m+\varepsilon}\}^{\frac{1}{m+\varepsilon}}]^{m+\varepsilon}$$

$$\leq 2[\{(1 + 5(2^{m} - 2)(\alpha(t))^{\varepsilon_{1}/(m+\varepsilon+\varepsilon_{1})})c(u, m + \varepsilon + \varepsilon_{1})$$

$$+k(u^{(m+\varepsilon)/2} + u^{v}H(m + \varepsilon))\}^{1/(m+\varepsilon)} + 2tH^{1/(m+\varepsilon)}(m + \varepsilon)]^{m+\varepsilon}.$$

$$\tag{14.5}$$

In the right hand side of the second line of the above set of equations, the first two terms involving Z are considered as main parts with time lag t. These parts become asymptotically independent as $t \to \infty$. The two terms involving S are treated as the remainder. Minkowski's inequality is used to separate the main parts from the remainder.

With $2u = 2^r \leq d^2$, $t = [2^{\frac{r\nu}{\beta v_0}}]$ and $\varepsilon_1 = \varepsilon^*/r$, we obtain

$$c(2^r, m + \varepsilon, h)$$

$$\leq [2\{1 + 5b(2^m - 2)2^{-\frac{\lambda\varepsilon^*\nu}{\beta v_0(m+\varepsilon+\varepsilon^*/r)}}\}c(2^{r-1}, m + \varepsilon + \varepsilon^*/r)$$

$$+ k\{2^{(r-1)(m+\varepsilon)/2} + 2^{(r-1)\nu}H(m+\varepsilon)\}](1+\beta_r)^{m+\varepsilon} \qquad (14.6)$$

as $\alpha(t) \leq qt^{-\lambda}$, $1 \leq b \simeq q^{\frac{\varepsilon^*/r}{(m+\varepsilon+\varepsilon^*/r)}} \simeq q^{\frac{\varepsilon^*}{(r v_0)}} < q$, and

$$\beta_r = 2t/u^{\frac{\nu}{v_0}} \simeq 2^{1-(r-1)\nu(1-\beta^{-1})/v_0}.$$

Repeating the above steps with ε replaced by $\varepsilon + \varepsilon^*/r$ and $\varepsilon_1 = \varepsilon^*/r$, one gets

$$c(2^r, m + \varepsilon, h)$$

$$\leq [2^2 c(2^{r-2}, m + \varepsilon + 2\varepsilon^*/r)\{1 + 5q(2^m - 2)2^{-\frac{\lambda\varepsilon^*\nu}{\beta v_0(m+\varepsilon+\varepsilon^*/r)}}\}$$

$$\times \{1 + 5q(2^m - 2)2^{\frac{-\lambda(r-1)\nu}{\beta v_0}\frac{\varepsilon^*/r}{(m+\varepsilon+2\varepsilon^*/r)}}\}$$

$$+ k\{2^{(r-1)(m+\varepsilon+\varepsilon^*/r)/2} + 2^{(r-1)/\nu}H(m+\varepsilon)(2^{(r-1)\nu}R(m))^{\frac{\varepsilon^*}{mr}}\}$$

$$+ k\{2^{r(m+\varepsilon)/2} + 2^{r\nu}H(m+\varepsilon)\}]\{(1+\beta_r)(1+\beta_{r-1})\}^{m+\varepsilon} \qquad (14.7)$$

Note that at the i-th stage of iteration, $i = 0, 1, 2, \ldots, r$, the term $c^{(m+\varepsilon+i\varepsilon^*/r)/m}(u, m)$ gives rise to the following term,

$$[u^\nu R(m)]^{(m+\varepsilon+i\varepsilon^*/r)/m} \leq [u^\nu R(m+\varepsilon)][u^\nu R(m)]^{\frac{i\varepsilon^*}{rm}}. \qquad (14.8)$$

The above inequality follows from (14.4). The above term is seen in (14.7) as $2^{(r-1)/\nu}H(m+\varepsilon)(2^{(r-1)\nu}R(m))^{\varepsilon^*/(mr)}$. Note further that, at i-th iteration 2^{r-i} appears in place of u. Thus, the term in the left hand side of (14.8) is less than or equal to $2^{r\nu}R(m+\varepsilon)[(2^{r\nu}R(m))^{\varepsilon^*/(rm)}/2^\nu]^i$.

Now, the ratio

$$p = (2^{r\nu}R(m))^{\varepsilon^*/(rm)}/2^\nu < (1 - \varepsilon') \qquad (14.9)$$

if $d < u^{\nu/(\varepsilon^*(1-\varepsilon')r/\varepsilon^*)}$ holds, since $u \leq d^2$, $\nu \leq \nu_0 < \delta/2$ with $u = 2^r$. Now, $1 < d < u^{(\nu_0-\eta)/\varepsilon}$, $\eta > 0$, from (14.2), therefore $p < (1 - \varepsilon')$, for small $\varepsilon' > 0$ holds from (14.9). Hence,

$$c(2^r, m + \varepsilon, h)$$

$$\leq [2^r \prod_{i=1}^{r} \{1 + 5q(2^m - 2)(2^{\lambda\varepsilon^*\nu/\beta(m+\varepsilon+\varepsilon^*)})^{-i/r}\}c(1, m + \varepsilon + \varepsilon^*)$$

$$+ k2^{r\nu}H(m+\varepsilon)(1 + p + p^2 + \cdots)$$

$$+ k2^{r(m+\varepsilon)/2}(1 + p^* + p^{*2} + \cdots)]\prod_{i=1}^{r}(1+\beta_i)^{m+\varepsilon} \qquad (14.10)$$

where $p < (1 - \varepsilon^*)$ and $p^* = 2^{-(m+\varepsilon+\varepsilon^*/r)/2+\varepsilon^*/2} < 1$ if $\varepsilon^* < m + \varepsilon \leq v_0$.
Now $\prod_{i=1}^{\infty}(1 + \beta_i) \simeq \prod_{i=1}^{\infty}(1 + 2^{1-(i-1)\nu(1-\beta^{-1})/v_0}) < \infty$ and

$$\prod_{i=1}^{r}(1 + AB^i)]^{\frac{1}{r}} \leq \frac{1}{r}\sum_{i=1}^{r}(1 + AB^i) \simeq 1 + A(1 - C^{-1})\log C$$
$$\leq (1 + D). \tag{14.11}$$

Approximation by integration is valid for small grid size $1/r$, i.e. when r is large. However this is valid as an upper bound. Here,

$$A = 5(2^m - 2)q, \ B = C^{-1/r}, \ C = 2^{\lambda\varepsilon^*\nu/\beta v_0(m+\varepsilon+\varepsilon^*)},$$

and

$$D = A(1 - C^{-1})/\log C.$$

Hence the first term in the r.h.s. of (14.10) $\leq 2^{\nu^* r}$, $\nu^* = \log(1 + D)/\log 2$. Now $C \geq 2^{\lambda\varepsilon^*\nu/\beta v_0(v_0+\varepsilon^*)}$, since $m + \varepsilon \leq v_0$. Then $D \leq D_0$, where D_0 is defined in (14.1). Therefore,

$$c(2^r, m + \varepsilon, h) \leq k2^{\nu r}c(1, m + \varepsilon + \varepsilon^*) + k2^{\nu r}H(m + \varepsilon) + k2^{r(m+\varepsilon)/2}$$

$$= k(2^{r(m+\varepsilon)/2} + 2^{\nu r}R(m + \varepsilon)) \tag{14.12}$$

where $\nu = 1 + \nu^* \leq v_0 \to 1$, as $\lambda \to \infty$. Hence the theorem holds for any integer of the form 2^r, r integer. For general u, one may use the usual binary decomposition of u.

To prove a stronger form of the theorem for $d^2 \leq u(\log u)^s$, note that in equation (14.9) we now have a different ν^* replacing ν, and we require

$$[u^{\nu^*}R(m)]^{(m+\varepsilon)/m} = (u^{\nu^*}d^{m-\delta-\varepsilon^*})^{1+\varepsilon/m}.$$

This is less than $u^{\nu^*}R(m + \varepsilon) = u^{\nu^*}d^{m+\varepsilon-\delta-\varepsilon^*}$, if $u^{\nu^*}d^{m-\delta-\varepsilon^*} \leq d^m$, i.e., if $\nu^* \leq (\delta + \varepsilon^*)/2$, as $u \leq d^2$.

Now consider $\nu^* = \nu + \varepsilon^*/2 + \varepsilon''$, where $\varepsilon'' > 0$. This choice serves the purpose $\nu + \varepsilon^*/2 + \varepsilon'' \leq (\delta + \varepsilon^*)/2$ in view of the fact that $\nu < \delta/2$. Similarly (14.9) holds with $R(v) = d^{v-\delta-\varepsilon^*}$, $\nu^*(= \nu + \varepsilon^*/2 + \varepsilon'')$ replacing ν, since we have

$$d^m < u^{\nu^* m/\varepsilon^*}u^{-\nu^*}d^{\delta+\varepsilon^*}(1 - \varepsilon')^{rm/\varepsilon^*},$$

if $d < u^{\nu^*/\varepsilon^*}(1 - \varepsilon')^{r/\varepsilon^*}$, and if $\nu^* < (\delta + \varepsilon^*)/2$. These hold, since $\nu < \delta/2$ and $d^2 \leq u(\log u)^s$.

Finally, for the step in (14.12), we have

$$c(2^r, m + \varepsilon, h) \leq k(2^{r(m+\varepsilon)/2} + 2^{\nu r} d^{\varepsilon^*} R(m + \varepsilon))$$
$$= k(u^{(m+\varepsilon)/2} + u^{\nu^*} R(m + \varepsilon))$$

with $\nu^* = \nu + \varepsilon^*/2 + \varepsilon''$, $d^2 \leq u(\log u)^s$. Observe that, one may select $\varepsilon^* > 0$, $\varepsilon'' > 0$; small enough such that $\nu^*(= \nu + \varepsilon^*/2 + \varepsilon'') < \delta/2$, as $\nu < \delta/2$.

Unfortunately, the technique of proof breaks down for lower order decay. In (14.10) we need $p^* < 1$, so that the sum converges and the order of the last term is $u^{(m+\varepsilon)/2}$. Now

$$p^* = 2^{-(m+\varepsilon+\varepsilon_1)/2 \ + \ r\varepsilon_1/2} < 1, \quad \text{if } \varepsilon_1 < (m + \varepsilon)/(r - 1).$$

Since the lag between Z_u and $Z_{u,t}$ is necessarily less than 2^r, as $t \leq u = 2^r$; taking the upper bound of ε_1 we have the order of the product term of (14.10) as

$$\prod_{i=1}^{r} [1 + 5(2^m - 2)\{\alpha(2^{r-i})\}^{(m+\varepsilon)/\{(r-1)(m+\varepsilon+\varepsilon_1)\}}].$$

Now $[\alpha(t)]^{1/\log u} \to 1$, $u \to \infty$ when $\alpha(n) \gg n^{-\lambda}$, $\forall \lambda > 0$. Hence the product term is of order $2^{mr} = u^m$, which makes the bound crude.

14.3. Application: Rates in MZSLLN

In this section we obtain convergence rates in Marcinkiewicz-Zygmund strong law of large numbers for strong-mixing random variables with polynomial decay, using the moment bounds given in Theorem 1. Similar convergence results are obtained in Babu (1980) for ϕ-mixing process. We prove the following result.

Theorem 2. Let $\{X_i, i \geq 1\}$ be a non-stationary strong-mixing process with polynomially decaying mixing coefficients. Under the conditions of Theorem 1, the following holds for each fixed $\alpha > 1/2$.

$$\sum_{n=1}^{\infty} n^{\delta\alpha-(\nu_o+1)} P(\sup_{1\leq j\leq n} |\sum_{i=1}^{j} X_i| > n^{\alpha}) < \infty. \tag{14.13}$$

Remark 2. The rates specified in (14.13) get sharper as λ, the power of polynomial-decay increases; note that $\nu_o \downarrow 1$ as $\lambda \uparrow \infty$.

We first prove the following lemmas, modifying the necessary steps in Section 14.3 of Babu (1980).

Lemma 1. Let $A > 0$, $\alpha > 1/2$ and $\beta > 0$. Under the assumptions of Theorem 1, there exists a constant $D(A, \beta) > 0$ such that for all $n \geq 3$, $h \geq 0$ and $1 \leq j \leq n - h$, the following holds.

$$P(|\sum_{i=1}^{j} X_{i+h,n}| > An^\alpha) \leq D(A, \beta) j^{\nu_o} n^{-\alpha\delta} (\log n)^{-2} \qquad (14.14)$$

where $X_{i,n} = X_i I(|X_i| \leq n^\alpha (\log n)^{-\beta})$.

Proof. Let $S_o = 0$ and $S_j = \sum_{i \leq j} X_{i,n}$. By Theorem 1, for each $\theta \geq 2$, there exist a constant $D(\theta)$ such that

$$E|S_{j+h} - S_h|^\theta \leq D(\theta)(j^{\theta/2} + j^\nu n^{\alpha(\theta-\delta)}(\log n)^{-\beta(\theta-\delta)}) \qquad (14.15)$$

Since $2\alpha > 1$, take $\theta = \max(2(\alpha\delta - \nu)/(2\alpha - 1), \delta + 2\beta^{-1})$ and use Markov inequality, to obtain

$$P(|S_{j+h} - S_h| > An^\alpha) \leq A^{-\theta} n^{-\theta\alpha} E|S_{j+h} - S_h|^\theta$$
$$\leq 2A^{-\theta} D(\theta) j^{\nu_o} n^{-\alpha\delta} (\log n)^{-2}, \qquad (14.16)$$

$1 \leq \nu \leq \nu_o$. This completes the proof.

Lemma 2. Let $A > 0$ and $\beta > 0$. Let $X_{i,n}$ and $S_j = \sum_{i \leq j} X_{i,n}$ be defined as in Lemma 1. Then

$$P(\sup_{j \leq n} |S_j| > 3A \, n^\alpha) = O(n^{\nu_o - \alpha\delta}(\log n)^{-2}) \qquad (14.17)$$

Proof. For $1 \leq i \leq n$, write $E_i = (\sup_{j < i} |S_j| \leq 3A \, n^\alpha, |S_i| > 3A \, n^\alpha)$, as the event that at i th stage the partial sums cross the specified boundary for the first time. The events E_i s are disjoint and union of these sets provides

$$E = \cup_{i=1}^n E_i = (\sup_{j \leq n} |S_j| > 3A \, n^\alpha).$$

Next write,

$$P(E) \leq \sum_{i=1}^{n} P(E_i \cap (|S_n| \leq A \, n^\alpha)) + P(|S_n| > A \, n^\alpha)$$
$$\leq \sum_{i=1}^{n} P(E_i \cap (|S_n - S_i| > 2A \, n^\alpha)) + P(|S_n| > A \, n^\alpha). \quad (14.18)$$

For $1 \leq i \leq n - q$, where $q = q(n)$ is a positive integer to be chosen later, write

$$P(E_i \cap (|S_n - S_i| > 2A\, n^\alpha))$$
$$\leq P(E_i \cap (|S_n - S_{i+q}| > A\, n^\alpha)) + P(|S_{i+q} - S_i| > A\, n^\alpha)$$
$$\leq P(E_i)P(|S_n - S_{i+q}| > A\, n^\alpha) + \alpha(q) + P(|S_{i+q} - S_i| > A\, n^\alpha).$$
$$(14.19)$$

Let the mixing-coefficients $\alpha(.)$ and the time-lag $q = q(n)$ satisfy $n\alpha(q(n)) = O(n^{\nu_o - \delta\alpha}(\log n)^{-2})$. This is satisfied for $q = q(n) = [n^\mu]$, $0 < \mu < 1$, and for polynomially decaying mixing-coefficients with power $\lambda > (\alpha\delta - \nu_o + 1)/\mu$. From Lemma 1, (14.18) and (14.19), one may then write

$$P(E) \leq \sum_{i=n-q+1}^{n} P(|S_n - S_i| > 2A\, n^\alpha) + \sum_{i=1}^{n-q} P(|S_{i+q} - S_i| > A\, n^\alpha)$$
$$+ P(|S_{n^*}| > A\, n^\alpha) + O(n^{\nu_o - \delta\alpha}(\log n)^{-2}) \qquad (14.20)$$

where $n^* = n - [n^\mu] = O_e(n)$ and O_e represents exact order. Hence

$$P(E) = O(n^{\nu_o - \delta\alpha}(\log n)^{-2}).$$

Remark 3. The condition $\lambda > (\alpha\delta - \nu_o + 1)/\mu$, used in the proof of Lemma 2 is not a restrictive assumption. In view of the requirements $\nu_o < \delta/2$ from Theorem 1; $\alpha > 1/2$ from Lemma 1; and $\mu < 1$ from the choice of $q(n)$ made above; the said condition on λ is seen to be satisfied by taking $\lambda > 1$.

Proof of Theorem 2. By Lemma 2, we have for any $\beta > 0$,

$$\sum_{n=1}^{\infty} n^{\delta\alpha - (\nu_o + 1)} P(\sup_{j \leq n} |S_j| > n^\alpha) < \infty.$$

Since,

$$P(\sup_{1 \leq j \leq n} |\sum_{i=1}^{j} X_i| > n^\alpha)$$
$$\leq P(\sup_{1 \leq j \leq n} |S_j| > n^\alpha) + \sum_{i=1}^{n} P(|X_i| > n^\alpha(\log n)^{-\beta}), \qquad (14.21)$$

and since $\nu_o \geq 1$, it is sufficient to show the more stringent condition

$$\sum_{n=1}^{\infty} n^{\delta\alpha - 2} \sum_{i=1}^{n} P(|X_i| > n^\alpha(\log n)^{-\beta}) < \infty.$$

Let $W_i = |X_i|^\delta(\log(1 + |X_i|))^\epsilon$, $\epsilon > 1$. There exists $t > 0$, $n_1 > 0$ such that for all $n \geq n_1$,

$$\sum_{i=1}^{n} P(|X_i| > n^\alpha(\log n)^{-\beta})$$

$$\leq \sum_{i=1}^{n} P(|W_i| > tn^{\alpha\delta}(\log n)^{\epsilon-\beta\delta}) \leq t^{-1}(\log n)^{\beta\delta-\epsilon}n^{1-\alpha\delta}(\sup_{i\geq 1} E|W_i|).$$

Now take $\beta = \frac{(\epsilon-1)}{2\delta}$, so that $\beta\delta - \epsilon < -1$. Theorem 2 then follows from the assumption: $\sup_{i\geq 1} E \mid X_i \mid^{\delta+\epsilon^*} = N < \infty$, made in Theorem 1.

14.4. Example: Linear Process

Consider $X_n = \Sigma_{i=1}^{\infty}a_i\xi_{n-i+1}$ or, $X_n = \Sigma_{i=1}^{\infty}a_i\xi_{n+i-1}$, where a_i is a sequence of constants with $\Sigma_{i=1}^{\infty}a_i^2 < \infty$ and $\xi_i s$ are pure white noise. Without loss of generality, let $E\,\xi = 0$ and $E\,\xi^2 = 1$. Let a finite segment of the distribution of the sum in X_n has a range on which it is absolutely continuous. Then from Withers (1981), see also Athreya and Pantula (1986), Theorem 2, it follows that under the condition $\Sigma_{n=1}^{\infty}\max(A_n^{1/3}, \sqrt{A_n|\log A_n|}) < \infty$, where $A_n = \Sigma_{i=n}^{\infty}a_i^2$, the process X_n is strong mixing.

In Dasgupta (1992, 2006) rates of convergence in CLT for linear process is computed under the condition $\Sigma_{i=1}^{\infty}i|a_i| < \infty$ which appears in the estimate of remainder term. If $a_i = O(i^{-\delta})$, $\delta > 2$, then the above two conditions are satisfied. The order of the mixing coefficients is then

$$\alpha(t) = O(\Sigma_{n=t}^{\infty}\max(A_n^{1/3}, \sqrt{A_n|\log A_n|})) = O(t^{-(2\delta-4)/3}),$$

which decays at a polynomial rate.

Note that under the restriction of absolute continuity, i.e., existence of density, the innovations ξ s cannot be purely discrete; there is an absolutely continuous component in it.

Note: Linear process has applications in different branches of science. The model $X_n = \Sigma_{i=1}^{\infty}a_i\xi_{n+i-1}$, may serve as a growth model for a plant at nth time segment. One may interpret $\{\xi_j, j \geq i, (i \geq 1)\}$ to be net effect of ith time segment, accumulated with dormant near future effects, on the plant growth when incorporated by scaling these with decreasing coefficients $a_1, a_2, a_3, \ldots$ Then X_n represents modeled growth of a plant on nth time segment and $\sum_n X_n$ may model the total growth of a plant in its lifetime. Similarly the process $X_n = \Sigma_{i=1}^{\infty}a_i\xi_{n-i+1}$ may model a growth curve with accumulated past effects.

References

1. Athreya, K. B. and Pantula, S. G. (1986). Mixing properties of Harris chains and autoregressive processes. *J. Appl. Probab.* **23**, 880–892.

2. Babu, G. J., Ghosh, M. and Singh, K. (1978a). On rates of convergence to normality for ϕ mixing processes. *Sankhyā* **A 40**, 278–293.

3. Babu, G. J., and Singh, K. (1978b). Probabilities of moderate deviation for some stationary strong mixing processes. *Sankhyā* **A 40**, 38–43.

4. Babu, G. J., and Singh, K. (1978c). On deviations between empirical and quantile processes for mixing random variables. *J. Mult. Anal.* **8**, 532–549.

5. Babu, G. J. (1980). An inequality for moments of sums of truncated ϕ-mixing random variables and its applications. *Sankhyā* **A 42**, 1–8.

6. Dasgupta, R. (1988). Non-uniform rates of convergence to normality for strong mixing processes. *Sankhyā* **A 51**, 436–451.

7. Dasgupta, R. (1997). Non-uniform rates of convergence to normality for some non-stationary ϕ-mixing processes. In *Selected Papers in Probability and its Applications* (M. C. Bhattacharjee and S. K. Basu, Eds.) Oxford Univ. Press, Delhi, 47–79.

8. Dasgupta, R. (1992). Rates of convergence to normality for some variables with entire characteristic function. *Sankhyā* **A 54**, 198–214.

9. Dasgupta, R. (2006). Nonuniform rates of convergence to normality. *Sankhyā* **68**, 620–635.

10. Davydov, Yu, A. (1970). The invariance principle for stationary processes. *Theory Probab. Appl.* **15**, 487–498.

11. Ghosh, M. and Babu, G. J. (1977). Probabilities of moderate deviation for some stationary ϕ-mixing processes. *Ann. Probab.* **5**, 222–234.

12. Ibragimov, I. A. (1962). Some limit theorems for stationary processes. *Theory Probab. Appl.* **7**, 349–382.

13. Merlevde, F. and Peligrad, M. (2000). The functional central limit theorem under the strong mixing condition. *Ann. Probab.* **28**, 1336–1352.

14. Rio, E. (1995). A maximal inequality and dependent Marcinkiewicz-Zygmund strong laws. *Ann. Probab.* **23**, 918–937.

15. Tikhomirov, A. N. (1980). On the rate of convergence in the central limit theorem for weakly dependent random variables. *Theory Probab. Appl.* **25**, 800–818.

16. Withers, C. S. (1981). Conditions for linear processes to be strong mixing. *Z. Wahrsch. Verw. Gebiete* **57**, 479–480.